Inorganic Chemistry Concepts
Volume 9

W0234387

Editors

Christian K. Jørgensen, Geneva · Michael F. Lappert, Brighton
Stephen J. Lippard, Cambridge, MA · John L. Margrave, Houston
Kurt Niedenzu, Lexington · Heinrich Nöth, Munich
Robert W. Parry, Salt Lake City · Hideo Yamatera, Nagoya

Boon K. Teo

EXAFS:
Basic Principles and Data Analysis

With 88 Figures and 17 Tables

Springer-Verlag
Berlin Heidelberg New York Tokyo

Dr. Boon K. Teo

AT & T
Bell Laboratories
600 Mountain Avenue
Murray Hill, New Jersey 0 79 74
USA

Library of Congress Cataloging-in-Publication Data

Teo, B. K. – EXAFS: basic principles and data analysis.
(Inorganic chemistry concepts ; v. 9)
Bibliography: p. Includes index.
1. Extended X-ray absorption fine structure.
I. Title. II. Series.
QC482.S6T45 1986   539.7'222   86-1811

ISBN 978-3-642-50033-6   ISBN 978-3-642-50031-2 (eBook)
DOI 10.1007/978-3-642-50031-2

The use of registered names, trademarks, etc. in this publication does not imply, even in the absence of a specific statement, that such names are exempt from the relevant protective laws and regulations and therefore free for general use.

2152/3020-543210

To

My Parents

and

My Family

PREFACE

The phenomenon of Extended X-Ray Absorption Fine Structure (EXAFS) has been known for some time and was first treated theoretically by Kronig in the 1930s. Recent developments, initiated by Sayers, Stern, and Lytle in the early 1970s, have led to the recognition of the structural content of this technique. At the same time, the availability of synchrotron radiation has greatly improved both the acquisition and the quality of the EXAFS data over those obtainable from conventional X-ray sources. Such developments have established EXAFS as a powerful tool for structure studies.

EXAFS has been successfully applied to a wide range of significant scientific and technological systems in many diverse fields such as inorganic chemistry, biochemistry, catalysis, material sciences, etc. It is extremely useful for systems where single-crystal diffraction techniques are not readily applicable (e.g., gas, liquid, solution, amorphous and polycrystalline solids, surfaces, polymer, etc.). Despite the fact that the EXAFS technique and applications have matured tremendously over the past decade or so, no introductory textbook exists. *EXAFS: Basic Principles and Data Analysis* represents my modest attempt to fill such a gap. In this book, I aim to introduce the subject matter to the novice and to help alleviate the confusion in EXAFS data analysis, which, although becoming more and more routine, is still a rather tricky endeavor and may, at times, discourage the beginners.

My involvement with EXAFS started almost a decade ago when Bell Labs helped to build the first EXAFS spectrometer at the then Stanford Synchrotron Radiation Project (SSRP). As a chemist, it took me very little time to appreciate the importance of the technique as a structural tool. In 1975, I initiated a series of concurrent collaborations with many people of vastly different backgrounds, including theoretical physicist Dr. P. A. Lee, experimental physicists Drs. P. Eisenberger and B. M. Kincaid, and biophysicist Dr. R. G. Shulman. These collaborations proved to be extremely fruitful and I learned a great deal from these people.

In 1978, when the field of EXAFS was still at its infancy, I started writing this book (at the suggestion of Prof. S. J. Lippard). In the subsequent year, I was invited, along with my colleague Dr. David Joy, by the Materials Research Society to organize a symposium on EXAFS which was held on Nov. 26-30, 1979 in Boston, MA. A book entitled *EXAFS Spectroscopy: Techniques and Applications* was published in 1981 based upon the proceedings of the conference.

Thanks to many contributions from first rate scientists of many seemingly unrelated disciplines of science, EXAFS has now grown and matured. Nonetheless, it is still a rapidly expanding field as manifested by the thousands of publications covering all aspects of EXAFS theory, analysis, techniques and applications. It is in the first two general areas of EXAFS theory and data analysis that this book is built upon.

In writing a book such as this, one cannot avoid a certain amount of oversimplification and, at the same time, neglecting many important contributions in the history of EXAFS. Furthermore, despite the fact that the manuscript has been reviewed and proofread by many experts (see Acknowledgements), I am sure there are still misprints, errors, and oversights. I shall bear full responsibility and, at the same time, would like to apologize for whatever inadequacies the reader may find in this book. I also welcome receiving suggestions and comments concerning ways in which to improve the book; they will be given serious consideration in future editions.

If *EXAFS: Basic Principles and Data Analysis* succeeds in serving the scientific community as an introductory book for scientists preparing to enter the field, as a reference book for researchers already in it, and as a textbook for students wanting to learn a new structural technique, then the countless nights and weekends spent in its preparation will not have been wasted in vain.

Boon Keng Teo
AT&T Bell Labs
September 27, 1985   Murray Hill, New Jersey

ABOUT THE BOOK

In the writing of *EXAFS: Basic Principles and Data Analysis*, I have constantly kept in mind the different needs of the novice, the specialist, and the student. The result is a book with the following strategy: the reader is first led from basic principles to advanced theories and then from simple analyses to sophisticated manipulations. Chapter 1 describes the properties of X-rays and electrons as well as their various interactions with matter. In Chapter 2 the phenomenon of EXAFS is introduced in qualitative terms. Chapter 3 presents an account of the variables and functions of EXAFS along with graphical illustrations of their effects and correlations. Chapter 4 describes the various theories of the EXAFS phenomenon. In Chapter 5, real world problems are addressed, including a discussion of the physical significance of phase and amplitude transferabilities, disorder effects and multiple scattering. The relevance of the generalized EXAFS formalism to data analysis is emphasized. Chapter 6 provides a discussion of the advantages and disadvantages of various data analysis practices. In Chapter 7, theoretical calculations of amplitude and phase functions are discussed. The book concludes with a description of multiple scattering and bond angle determination in Chapter 8. Seven appendices provide a great deal of useful information for the interpretation of EXAFS data.

There are several important features of the book that are worth mentioning. First, each chapter is designed to be self-contained; thereby the reader is spared the trouble of cross referencing. As such, the book can serve both as an introduction to the field of EXAFS as well as a reference book. For example, a novice, who does not wish to be bothered with details, may find that Chapters 1, 2, 3, 5 and 6 are just enough to understand the technique and to analyze the data. An experienced EXAFS practitioner may find that this book is a useful source for pertinent equations, figures, tables, appendices, as well as for more detailed treatments, such as those found in Chapters 4, 7, and 8. Second, with very few exceptions, each equation is derived in a step-by-step manner. Hence, the book can be read with ease and the reader will not be surprised by phrases such as "it is obvious" and "it can be shown." To those who are neither interested in, nor have time for all of the details, the intermediate steps can be skipped. Third, the physical significance of the variables and functions are explained in detail. This is usually done following the introduction or the discussion of the terms in an equation.

The layout of the book is as follows: The sections are arranged in the numberical form J.K.L.M. Here J is the chapter number. The section numbers **J.K** are in bold, whereas the subsection numbers *J.K.L* are in italics. Further subdivision, if needed, are in Roman. The figures (abbreviated as Fig. or Figs.) and the equations (abbreviated as Eq. or Eqs.) are numbered consecutively in the form J.N, where J is the chapter number. The references are designated in the text by the last names of the authors followed by the year of publication.

References are provided in the Bibliography section at the end of the book. They are arranged in alphabetical order in terms of the authors (first, second, third, and so on) and, for the same group of authors, in numerical order in terms of year of publication (in bold). Further designation, if necessary, is provided by appending **a**, **b**, **c**,... to the **year** of publication. The *journal* names and the *volume* numbers are in italic.

The Bibliography contains more than just the corresponding number of references cited throughout the book. In fact, it represents an extensive literature search up to the end of 1984. Some 1985 publications are also included. It covers not only EXAFS theory and data analysis, but also includes citations for radiation sources, techniques, applications, and X-ray absorption near edge structure (XANES).

ACKNOWLEDGEMENTS

I wish to acknowledge the authors and copyright holders for the use of their figures, tables, and other materials, which are clearly identified in the book.

In writing this book, I relied heavily upon numerous review articles and books, all of which are listed in the Bibliography section.

I am indebted to my collaborators and colleagues, both past and present, for their invaluable participation in my endeavor in the field of EXAFS. Among them are Drs. P. A. Lee, P. Eisenberger, B. M. Kincaid, R. A. Shulman, G. S. Brown, A. L. Simons, J. Reed, S. J. Lippard, J. K. Barton, R. Bau, K. Kijima, B. A. Averill, M. R. Antonio, H. S. Chen, D. Coucouvanis, S. M. Kauzlarich, W. H. Orme-Johnson, P. A. Lindahl, M. J. Nelson, S. E. Groh, M. A. Bruck, J. L. Dye, O. Fussa, D. Shriver, P. Blonsky (roughly in chronological order), and the list goes on. Over the years, I also enjoyed interactions with pioneers like Drs. E. A. Stern, F. W. Lytle, J. Sinfelt, and D. E. Sayers, who are always eager to help. I also treasure my friendship with Drs. J. Wong and T. K. Sham. I must admit that I benefited a lot through these collaborations and/or interactions.

Many of my colleagues have been kind enough to review, at various stages, this work and make many valuable suggestions and improvements. In particular, I would like to express my most sincere gratitude for the time and effort Drs. P. A. Lee, B. M. Kincaid, Y. Saito, and L. F. Dahl spent in reading and commenting on the final manuscript. I am also grateful to Drs. P. A. Lee and Y. Saito for expressing their impression about the book in the Forewords. Dr. A. P. Ginsberg, V. Bakirtzis, and K. Keating read earlier versions of some of the chapters.

Without the superb text processing ability of Joan Alder this book would have taken a much longer time to publish. Finally, I would like to thank the publisher and the editors of the series for their patience and encouragement. Last, but not least, I would like to thank my parents and my family for their understanding, patience, and support in the pursuit of my scientific career.

FOREWORD

This book begins with a thorough review of the fundamental properties of X-rays and electrons, followed by the introduction of the EXAFS phenomenon and the development of the theory of EXAFS, as well as various aspects of data analysis. As one of those who have long standing interests in physical measurement using X-rays and electrons, I enjoyed reading the book from the beginning to the end. This book presents lots of new and useful information. In this book, EXAFS is first explained in simple terms by emphasizing the fundamental physical phenomena, with the help of examples and illustrations, avoiding the complicated details. Then the reader is presented the more advanced and sophisticated theories. Since the equations are derived in a step-by-step manner designed to guide the reader, one can read through the book with little difficulty. For people who are interested in, but not specialists of, the EXAFS technique, this book, the first of its kind, is the best place to start. Since *EXAFS: Basic Principles and Data Analysis* also contains much up-to-date information, it is also very useful for scientists who are already working in the field of EXAFS spectroscopy.

Yahachi Saito
Toyohashi University of Technology
Toyohashi, Japan

FOREWORD

The past decade has seen the establishment of extended X-ray absorption fine structure (EXAFS) as an important structural tool and Dr. Boon K. Teo has been one of the major contributors to this field. Dr. Teo's involvement is particularly unique in that his contributions cover all aspects of EXAFS, including the selection of significant problems, acquisition of data at the synchrotron radiation centers, data analysis, and the development of the theory. It is in the last aspect of the problem that I had the good fortune of collaborating with him. Furthermore, one of the most rewarding aspects of doing research in EXAFS is the opportunity to know and work with first rate people from other disciplines and, among them, Dr. Teo certainly distinguishes himself as a bridge builder connecting such diverse disciplines as inorganic chemistry, synchrotron radiation research and theoretical physics.

Dr. Boon Teo has pulled together his extraordinarily wide experience and broad perspectives in the writing of this book. The result is a book that will have much appeal to chemists and physicists alike. It covers the basic principles in a thorough way, paving way for the development of more sophisticated theories of EXAFS. Many important and potentially useful equations, pertinent to EXAFS theory and analysis, are derived in a detailed manner easily understandable by a novice. This book also contains a great number of tables and figures of theoretically calculated amplitude and phase functions which are useful in EXAFS data analysis.

I highly recommend *EXAFS: Basic Principles and Data Analysis* to researchers planning on (or already) using EXAFS in their research. It is also an invaluable reference book for researchers working in the fields of physics, chemistry, biology, and materials science.

P. A. Lee
Professor of Physics
Massachusetts Institute of Technology
Cambridge, Massachusetts

CONTENTS

1.   **X-Rays and Electrons**..1

1.1   Introduction 1
1.2   Generation of X-Rays...2
1.3   Properties of X-Rays and Electrons..4
1.3.1   Wave-Particle Duality of Photons...4
1.3.2   Photoelectric Effect 4
1.3.3   Wave-Particle Duality of Electrons 5
1.4   Electronic Structure of Atoms 5
1.4.1   Models of the Atom...5
1.4.2   Electronic Transitions 8
1.5   Absorption Coefficients and Absorption Edges 10
1.5.1   Absorption Coefficients 10
1.5.2   Absorption Edges 11
1.5.3   True Absorption and Scattering..12
1.6   Interactions of Photons and Electrons with Matter 13
1.6.1   Excitations and Relaxations 14
1.6.2   Scattering 18
1.6.3   Electrons 19

2.   **Extended X-Ray Absorption Fine Structure**
      (EXAFS) Spectroscopy..21

2.1   EXAFS Spectroscopy 22
2.2   Theory..24
2.3   Data Analysis 31

3.   **EXAFS Parameters** 34

3.1   Variables and Functions 34

3.2   Effects of Important Parameters 45

3.3   Convention of Changing E_0 52

4.   **Theory of EXAFS** 53

4.1   Introduction 53

4.2   Derivations of EXAFS Theory 54

4.2.1   Lee and Pendry (1975) 55

4.2.2   Lee (1976) 58

4.2.3   Boland, Crane, and Baldeschwieler (1982) 64

4.2.4   Curve-Wave Theory 71

4.3   EXAFS of L Edges 72

4.4   The Photoelectron and the Excited Atom 75

4.4.1   Lifetime of the Core Hole 75

4.4.2   Core Hole Relaxation 77

4.4.3   Potential Experienced by Photoelectron 78

4.4.4   Multi-electron Excitations 78

5.   **Improvement of EXAFS Theory** 79

5.1   Energy Threshold — The Phase Problem........................80

5.1.1   Choosing E_0 80

5.1.2   Phase Transferability 80

5.1.3   Varying E_0 81

5.2   Inelastic Scatterings — The Amplitude Problem.............84

5.2.1   Central Atom: Shake Up/Off Processes 85

5.2.2   Scatterer: Electron Inelastic Mean Free Path 89

5.3   Static and Thermal Disorder Effects 91

5.3.1   Small Disorders...94

5.3.1.1 Symmetric Pair Distribution 94

5.3.1.2 Discrete Bonds...96

5.3.1.3 Harmonic Vibration 99

5.3.2   Large Disorders..101

5.3.2.1 Derivation of the Generalized EXAFS Formalism 102

5.3.2.2 Moments of $g(r)$ 103

5.3.2.3 Symmetric Pair Distributions..105

5.3.2.4   Asymmetric Pair Distributions...106
5.3.2.5   Anharmonic Vibration Potentials...111
5.3.2.6   Comparison of EXAFS and Diffraction 112
5.4   Multiple Scattering EXAFS Formalism...................................113

6.   **Data Analysis in Practice** 114
6.1   Data Reduction 114
6.1.1   Conversion of Experimental Variables....................................114
6.1.2   Background Removal...115
6.1.3   Normalization and μ_0 Correction...119
6.1.4   Conversion of E to k 120
6.1.5   Weighting Scheme 121
6.1.6   Deglitching and Truncation...124
6.2   Fourier Transform (FT) 124
6.3   Fourier Filtering (FF) 127
6.4   Curve Fitting (CF) 128
6.4.1   Parameterization 128
6.4.2   Phenomenological EXAFS Models...130
6.4.3   Least-squares Refinements 130
6.4.4   Correlations 131
6.4.5   Errors 132
6.5   Parameter Correlation and the FABM Method........................133
6.5.1   Fine Adjustment Based on Models 134
6.5.2   Criteria for the Selection of Good Models 136
6.6   The "Difference" Technique 139
6.7   The Min-Max Method..142
6.8   Decomposition into Amplitude and Phase...............................146
6.8.1   Phase Information 149
6.8.2   Amplitude Information (The Ratio Method)............................149
6.9   The Beat-node Method 151
6.9.1   Derivations of Eq. 6.33-40..152
6.10   The Lee and Beni Method..153
6.11   The r Space Method..156
6.12   The Phase Linearization Method 157
6.13   The Regularization Algorithm 157
6.14   Other More Specialized Methods 157

7.   **Theoretical Amplitude and Phase Functions** 158

7.1   Introduction 158

7.2   Theoretical Methods 160

7.3   Theoretical Amplitude and Phase Functions.............................163

7.4   Properties of Amplitude and Phase Functions 164

7.4.1   Amplitude 165

7.4.2   Scatterer Phase 167

7.4.3   Central Atom Phase Shift 171

7.4.4   Effect of Electronic Configuration 171

7.4.5   Charge Effect 175

7.4.6   Comparison of ϕ_a^l ($l = 0, 1, 2$) Functions.............................175

7.4.7   Relativistic Effect...175

7.5   Comparison of Theory and Experiment....................................181

8.   **Multiple Scattering and Bond Angle Determination** 183

8.1   Scattering Amplitude and Phase...184

8.1.1 $F(\beta, k)$ and $\theta(\beta, k)$ 187

8.2   Multiple Scattering 192

8.2.1 ABC Systems..194

8.2.1.1   Approximations 196

8.2.1.2   Anisotropic ABC Systems...207

8.2.2   Multiple Scattering: AB_1B_2C Systems....................................208

8.2.3   Multiple Scattering: Linear or Nearly Linear
         $AB_1B_2 \cdots B_nC$ Systems 210

8.3   Comparison of Theory and Experiment....................................212

8.4   Angle Determination..213

8.4.1 ABC Systems 213

8.4.1.1   Empirical Approach..219

8.4.2 $AB_1 \cdots B_nC$ Systems 220

8.5   Conclusion 220

Bibliography 223

Appendix I.   The Periodic Table...285

Appendix II.   X-Ray Absorption Edges and Characteristic
                        X-Ray Emission Lines...287

Appendix III.   Victoreen's C and D Values for True Absorption.................291

XVIII

Appendix IV.   Fluorescence Yields 297

Appendix V.   Backscattering Amplitude, Backscattering Phase,
                and Central Atom Phase 301

Appendix VI.   Tables of Scattering Amplitude, $F(\beta,k)$ in Å,
                and Phase, $\theta(\beta,k)$ in Radians, Functions..............................315

Appendix VII.   Graphs of Scattering Amplitude, $F(\beta,k)$ in Å,
                and Phase, $\theta(\beta,k)$ in Radians, vs. Photoelectron
                Wavevector k in Å^{-1} for Some Elements at
                Different Scattering Angles, β in Degrees................................337

Index 347

Chapter 1

X-Rays and Electrons

1.1 Introduction

X-rays are electromagnetic radiations which lie between ultraviolet light and gamma rays in the electromagnetic spectrum. X-rays are characterized by the relatively short wavelengths of 0.01Å to 100Å, with hard X-rays on one end and soft X-rays on the other. They are conventionally produced by either the conversion of the kinetic energy of charged particles into radiation or the excitation of atoms in a target upon which fast moving electrons impinge. The former produces a *continuous spectrum* of X-rays whereas the latter gives rises to *characteristic lines* of nearly monochromatic X-rays.

Throughout the history of modern physical science, X-rays have been used as powerful tools in analytical, physical, chemical, biological, and structural characterization of matter. For example, X-ray fluorescence is widely used in qualitative and quantitative elemental analysis; X-ray photoelectron spectroscopy (XPS) can be used to study the electronic structure of materials; X-ray crystallography provides three-dimensional structures of crystalline materials at atomic resolution; and various X-ray scattering techniques can yield structural information for amorphous materials at varying degrees of resolution (dependent upon the technique and the wavelength used). Since the production and the properties of X-rays are intimately related to electrons, in this chapter we shall discuss both in an attempt to give a broader conceptual picture of the interactions of these fundamental particle/waves with matter.

1.2 Generation of X-Rays

Traditionally, X-rays are produced by sudden deceleration of fast moving electrons at a target material. In an X-ray tube, electrons generated by a heated tungsten filament are accelerated through a voltage V at the cathode and impinged upon a target anode material in vacuum to produce the X-ray spectrum shown in Fig. 1.1. If all the energy acquired by the electron, eV, were

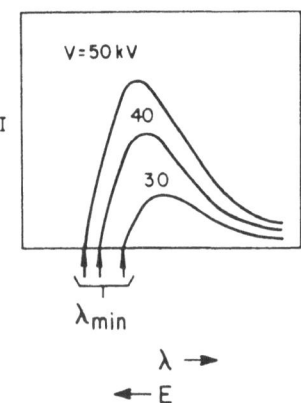

Fig. 1.1. Intensity (I) vs wavelength (λ) or energy (E) for continuous X-ray spectra (white radiation or bremsstrahlung) as a function of accelerating voltage (V).

converted to the X-ray photon energy, $h\nu$, the wavelength of the resulting X-rays will be given by the Duane-Hunt law

$$\lambda_{min} = \frac{hc}{eV} = \frac{12398}{V} \qquad (1.1)$$

where h is the Planck's constant, c is the velocity of light, e is the charge of an electron, and λ_{min} and V are in Å and volts, respectively. λ_{min} represents the shortest wavelength (highest energy) obtainable for a given applied accelerating voltage V since not all the electrons lose all their energy in this manner. In fact, the majority of electrons undergo multiple collisions with, and transfer part of their energy to, the target material. This results in a continuous *bremsstrahlung* spectrum, generally referred to as *white radiation*, shown in Fig. 1.1. The intensity peaks at a wavelength somewhat greater than λ_{min}. As the voltage is increased, λ_{min} decreases and the total intensity increases.

When the accelerating voltage reaches a certain *critical potential*, dependent upon the target material, the electrons are capable of knocking electrons out of the inner atomic orbitals of the target material, thereby producing a vacancy in the core shell. As the electrons from the higher shells descend to fill the core hole, they produce a number of sharp spikes known as the *characteristic lines,* superimposed on the continuous bremsstrahlung spectrum and some 10^3 greater in intensity, as shown in Fig. 1.2. The energies of these characteristic lines are well-defined and they depend upon the energies of the two atomic shells involved (e.g. K_{α_1}, K_{α_2} lines for $L \rightarrow K$ and K_{β_1}, K_{β_2} lines for $M \rightarrow K$). As such, they can be considered as monochromatic radiation. Furthermore, the wavelength of the characteristic lines decreases with increasing atomic number Z of the target material. For tables relating to the production, wavelengths, and intensities of X-rays, see "International Tables for X-Ray Crystallography," Vol. III (Ref. (a)).

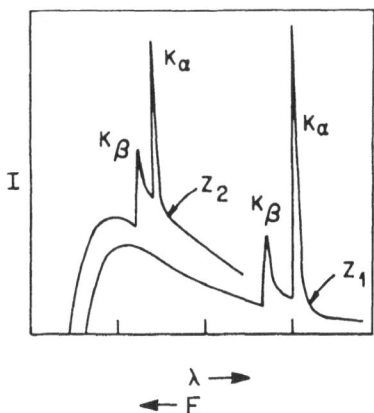

Fig. 1.2. X-ray spectra with characteristic peaks K_α and K_β for two different elements with atomic numbers $Z_1 < Z_2$.

It is thus apparent that conventional X-ray tubes are capable of producing both a continuous X-ray spectrum and discrete characteristic lines. The latter are widely used in X-ray diffraction studies, while the former may be utilized in EXAFS measurements, among other applications. However the continuous white radiation suffers from being a thousand times weaker in intensity than the characteristic lines.

Three new sources of continuous X-radiation have recently been developed. The first is the rotating anode X-ray tube. Using bent crystal optics, as much as 10^2-10^3 mrad^2 of the bremsstrahlung radiation can be collected which amounts to a 10^2-10^3 enhancement in intensity (see, for example, Knapp, *et al,*

1978). The second is the high-intensity, nanosecond-pulsed *soft* X-rays (≤ 3 keV) generated from a laser-produced plasma (Mallozzi, *et al*, 1979). And finally the recently available synchrotron radiations, ranging from soft to hard X-rays, from storage rings have revolutionized research involving X-rays. Synchrotron radiation (see, for example, Winick and Doniach, 1980) is emitted when charged particles such as electrons or positrons travel with a speed approaching that of light in curved paths in a magnetic field. The advantages of synchrotron radiation are: (1) high intensity (10^3 times over characteristic lines and 10^6 times over white lines of conventional X-ray tubes); (2) tunability over a wide energy range with a continuous spectrum; (3) high collimation; (4) plane-polarization; and (5) precisely pulsed time structure (nanosecond pulses separated by micro-seconds). These three new or improved X-ray sources and their relevance to EXAFS spectroscopy will be discussed elsewhere.

1.3 Properties of X-Rays and Electrons

1.3.1 Wave-Particle Duality of Photons

X-rays, like any other electromagnetic radiation, can be treated as either a *wave* or a *particle*. This wave-particle duality of light is defined by the Einstein relation

$$E = h\nu = \frac{hc}{\lambda} \qquad (1.2)$$

where $h\nu$ is the energy of the photon with h being Planck's constant and ν and λ are the frequency and wavelength of the wave. For E in eV and λ in Å, $E = 12398/\lambda$.

1.3.2 Photoelectric Effect

The concept of the wave-particle duality of light enabled Einstein, in 1905, to explain the photoelectric effect, the ejection of an electron from an atom by the absorption of a photon. Einstein suggested that the maximum kinetic energy (K.E.) of the emitted photoelectron from a material is given by

$$K.E. = \frac{1}{2}mv^2 = E - E_o \qquad (1.3)$$

where $E = h\nu$ is the quantized energy of the incident photons and $E_o = e\phi$ is the threshold energy with ϕ being the work function. That is, E_o is the minimum energy required to eject an electron from a particular atomic or molecular orbital of the material. This relation explains why the kinetic energies of the emitted electrons are independent of the intensity of light but the number of electrons is proportional to the light intensity. Furthermore, the time delay in the emission of photoelectrons, which normally does not exceed 3×10^{-9} sec, is thus consistent with the time scale of electronic excitation. The Einstein photoelectric equation applies to both ultraviolet and X-ray radiations, with the former effecting the emission of electrons from valence (outer) orbitals while the latter, generally speaking, causes emission from core (inner) shells.

1.3.3 Wave-Particle Duality of Electrons

Like photons, electrons possess wave-particle duality. The wavelength λ and the momentum p of an electron are related by the de Broglie equation

$$\lambda = \frac{h}{p} \tag{1.4}$$

where h is Planck's constant. Defining the wave vector

$$k = \frac{2\pi}{\lambda} \tag{1.5}$$

we can write the kinetic energy of an electron as

$$K.E. = \frac{p^2}{2m} = \frac{\hbar^2 k^2}{2m} \tag{1.6}$$

where $\hbar = h/2\pi$ and m is the mass of an electron.

For the photoelectron with a kinetic energy given by Eq. (1.3), we have

$$k = \sqrt{\frac{2m}{\hbar^2}(E-E_o)} \tag{1.7}$$

which relates the photoelectron wave vector k to the incident photon energy E and to the threshold energy E_o of a particular electronic shell of the atom.

1.4 Electronic Structure of Atoms

1.4.1 Models of the Atom

The Bohr model of an atom consists of a positively charged nucleus and electrons occupying successive *electronic shells* (or *levels*) based on their relative energies (Aufbau principle). This is shown schematically in **Fig. 1.3**. For a hydrogen-like atom with a nuclear charge Z, the one-electron energy of

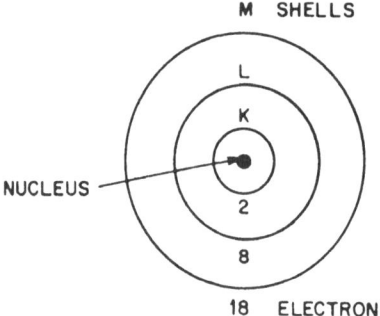

Fig. 1.3. A simplified Bohr atom showing the first three inner shells (core levels) and their maximum number of electrons.

each shell is

$$E_n = -\frac{1}{2}\frac{Z^2 me^4}{n^2\hbar^2} = -\frac{Z^2}{n^2} \times 13.6 \text{ eV} \qquad (1.8)$$

and the radius of the electron orbit is

$$r = \frac{n^2\hbar^2}{Zme^2} = \frac{n^2}{Z} \times 0.529\text{Å} \qquad (1.9)$$

Here 13.6eV and 0.529Å are the atomic units of energy and distance, respectively. n is the principal quantum number and the shells (as well as the electrons in them) are classified as K, L, M, N, etc. according to the n values of 1, 2, 3, 4, etc. Each shell has a degeneracy of n^2 and is capable of housing $2n^2$ electrons.

Modern theory of many-electron atoms differs substantially from the Bohr model. In the Schrödinger theory, the electrons in a many-electron atom occupy a series of *atomic orbitals*, whose wavefunctions and eigenvalues are solutions to the Schrödinger equation, according to the Aufbau principle which specifies that electrons occupy the lowest available atomic orbitals first. Each of these orbitals is characterized by four quantum numbers n, l, m, s which are the principal, the orbital (azimuthal), the magnetic, and the spin quantum number, respectively. The energetic difference between atomic orbitals with different quantum numbers is shown schematically in Fig. 1.4. Pauli's *exclusion principle* forbids any two electrons in an atom having the same set of quantum numbers, n, l, m, s. In a many-electron atom, each electron feels not only the Coulomb field of the nucleus but also the field of the other electrons; one must therefore take into account the interelectron repulsions. The net result is that

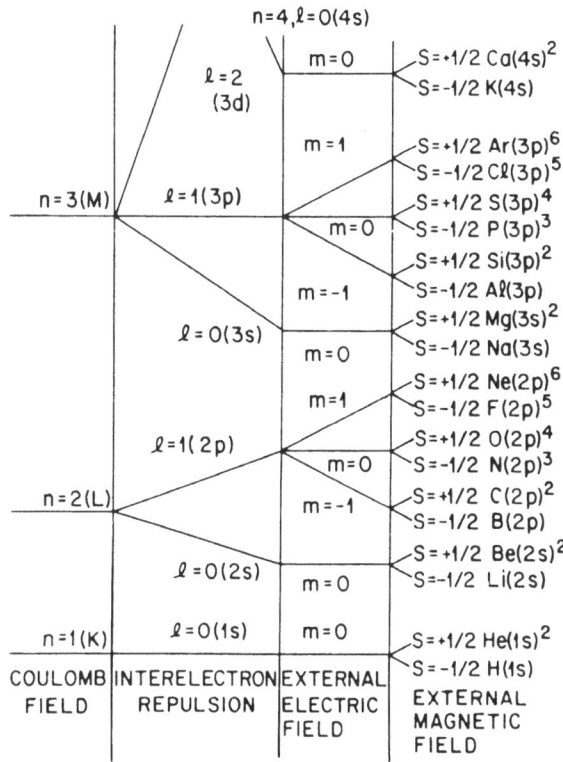

Fig. 1.4. Schematic diagram of orbital energetics in a many-electron atom. For electrons moving in a pure coulomb field, each shell (n) has $2n^2$-fold degeneracy. Interelectron repulsions give rise to subshells (l) with $2(2l+1)$-fold degeneracy. Finally, in the presence of external electric and magnetic fields, the degeneracies due to the quantum numbers m and s, respectively, are lifted (see text). (Adapted from Avery, 1972).

the energy of the orbital is dependent not only upon the quantum number n, but also on the quantum number l. The energy levels corresponding to the same n but different l values are called subshells designated as s, p, d, f, $l = 0, 1, 2, 3, \cdots$. Each subshell maintains a $2(2l+1)$ - fold degeneracy. The energy differences between the subshells within each shell however, are much smaller than that between the different shells. With externally applied non-central electric field, the energy also depends on m where $m = -l, \cdots 0$, $\cdots, l - 1, l$. And finally in the presence of a magnetic field, the spin degeneracy $s = \pm 1/2$ is also lifted (spin orbital included). The electronic configuration of the elements in the Periodic Table (Appendix I) can be found in standard textbooks.

1.4.2 Electronic Transitions

Electrons in an atom can jump from one energy level to another by gaining (absorbing) or losing (emitting) one quantum, $h\nu$, of electromagnetic radiation where

$$h\nu = E_f - E_i \qquad (1.10)$$

Here E_i and E_f are the energies of the initial ($nlms$) and final ($n'l'm's'$) states, respectively. The dipole transition probabilities of these radiative transitions are governed by the matrix elements of the rectangular components of the electric dipole moment $e\vec{r}$ of the atomic system:

$$X = e <n'l'm's'| \; x \; |nlms> \qquad (1.11a)$$

$$Y = e <n'l'm's'| \; y \; |nlms> \qquad (1.11b)$$

$$Z = e <n'l'm's'| \; z \; |nlms> \qquad (1.11c)$$

In the electric-dipole-induced transition, the charge distributions in the initial and final states differ in a manner corresponding to an electric dipole. Such a transition can therefore couple with electromagnetic radiation by interaction with the oscillating electric vector, thereby transferring energy to (emission) or from (absorption) the electromagnetic field. Nonzero values of X, Y, or Z mean that the transition is allowed by interacting with an electric vector oscillating in the x, the y, or the z direction, respectively; *i.e.*, the transition is x-, y-, or z-polarized, respectively. Ignoring the spin angular momentum, the electric dipole selection rules are: $\Delta l = \pm 1$, $\Delta m = \pm 1$, 0, and $\Delta n =$ unrestricted, or, individually,

$$X \text{ or } Y = 0 \text{ unless } \Delta l = \pm 1 \text{ and } \Delta m = \pm 1$$

$$Z = 0 \text{ unless } \Delta l = \pm 1 \text{ and } \Delta m = 0$$

The relative intensities of various radiative optical transitions can in principle be calculated using Eq. 1.11a-c. If both the initial and the final states of a transition are bound, it is called a *resonance transition*.

Since the characteristic emission line spectra of X-rays from an X-ray tube originate from the electronic transitions between two atomic levels (or states) of an atom, we can relate these radiative transitions to the X-ray term diagram shown in Fig. 1.5. Adopting Siegbahn's notation, the series of lines which result from higher shell electrons dropping into a K hole (unoccupied K level) is

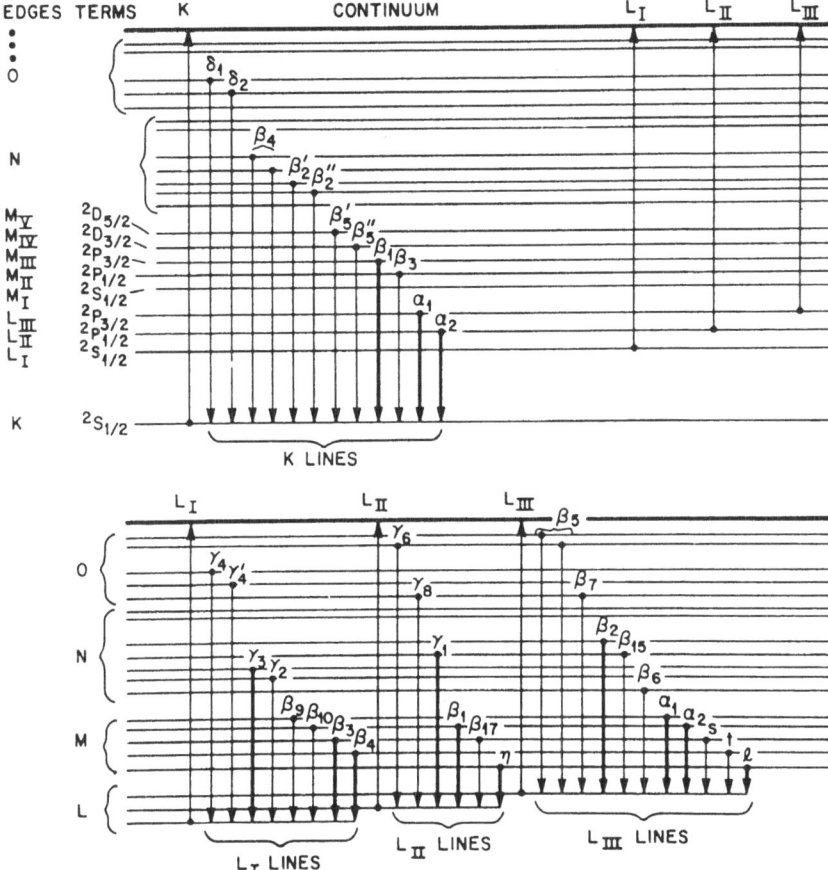

Fig. 1.5. Some of the X-ray absorption edges and the corresponding characteristic emission lines. The notation for the shells is that of Sommerfeld (K, L_I, L_{II}, L_{III}, etc.) and that for the transitions is that of Siegbahn (K_{α_1}, K_{α_2}, K_{β_1}, etc.).

called the K series and the various lines of the series are called K_α, K_β, K_γ, etc., in the order of increasing energy. Similarly, the series which results from transitions into an L hole (unoccupied L level) is called the L series, and so on. This is illustrated in Fig. 1.5. The relative intensities within each series of a given element are, approximately, K_{α_1} : K_{α_2} : K_{β_1} : K_{β_2} : K_{β_3} $\approx 1 : 0.50 : (0.15\text{-}0.30) : (0.01\text{-}0.1) : (0.06\text{-}0.15)$; L_{α_1} : L_{α_2} : L_{β_1} : L_{β_2} : L_{β_3} : L_{β_4} : L_{γ_1} : $\approx 1 : 0.10 : 0.50 : 0.20 : (0.01\text{-}0.06) : (0.03\text{-}0.05) : (0.01\text{-}0.1)$ and so on. Finally, the variation in the frequencies of each series of characteristic X-ray emission lines with atomic number Z follows Moseley's law

$$\nu = K^2(Z - \sigma)^2 \qquad (1.12)$$

where K and σ are constants characteristic of the series, *viz.*, different K and σ values for different edge (*vide infra*). The energies of the characteristic X-rays lines are tabulated in Appendix II.

If, on the other hand, the electron in an occupied level is excited to a virtual (unoccupied) bound state by an absorption of a photon, we obtain an *absorption spectrum*. For such radiative transitions involving core state(s), we have an X-ray *absorption spectrum*. Furthermore, if the electron is excited from a core state to continuum, an X-ray *absorption edge* results. The minimum photon energy required to eject an electron out of a particular atomic state is called the *threshold* energy $E_o = h\nu_o$ which is equal to the absolute value of the binding energy of the electron. In the notation of Sommerfeld, the absorption edges are labeled, in the order of increasing energy, by K, L_I, L_{II}, L_{III}, etc., corresponding to the excitation of an electron from the $1s(^2S_{1/2})$, $2s(^2S_{1/2})$, $2p(^2P_{1/2})$, $2p(^2P_{3/2})$, etc., orbitals (states), respectively. For example, L_{III} refers to a state of excitation in which an electron is missing from the state $2p\ ^2P_{3/2}$. This is also schematically depicted in Fig. 1.5. The edge energies are tabulated in Appendix II. It is evident that the characteristic lines and the edge series are related. For example, in the case of tungsten, $K_{\alpha_1} = K - L_{III}$, $K_{\alpha_2} = K - L_{II}$, $K_{\beta_1} = K - M_{III}$, $K_{\beta_2} = K - N_{II,III}$, $K_{\beta_3} = K - M_{II}$ for the K series; $L_{\alpha_1} = L_{III} - M_V$, $L_{\alpha_2} = L_{III} - M_{IV}$, $L_{\beta_1} = L_{II} - M_{IV}$ for the L series, and so on. Note also that for the K series, $K_\alpha \ll K_\beta < K$ in increasing order of energy. Similar relationship holds for the higher series.

1.5 Absorption Coefficients and Absorption Edges

1.5.1 Absorption Coefficients

When a collimated beam of monochromatic X-rays travels through matter, it inevitably loses its intensity via interaction with the material. The loss in intensity I is proportional to the original intensity and the thickness x:

$$dI = -\mu I dx \qquad (1.13)$$

with the proportionality constant μ being the *linear absorption coefficient*. Integrating Eq (1.13) gives rise to

$$\frac{I}{I_0} = e^{-\mu x} \qquad (1.14)$$

where I_0 and I are the incident and the transmitted X-ray intensities upon passage through a homogeneous sample of thickness X.

Since the attenuation (or absorption) of intensity is determined by the quantity of matter traversed by the X-ray beam, the absorption coefficient is sometimes expressed as *mass absorption coefficient*, μ/ρ, obtained by dividing the linear absorption coefficient μ by the density ρ of the material. For μ in cm^{-1} and ρ in gm/cm^3, μ/ρ has a unit of cm^2/gm. μ/ρ is roughly independent of the physical state of the material and is approximately additive:

$$\mu/\rho = \sum_j g_j \ (\mu/\rho)_j \qquad (1.15)$$

where g_j is the mass fraction of the element j with mass absorption coefficient $(\mu/\rho)_j$. Another commonly used unit is the *atomic absorption coefficient* μ_a which is defined as

$$(\mu_a)_j = (\mu/\rho)_j \ (A_j/N) \qquad (1.16)$$

where A_j is the atomic weight of the element j and N is Avogadro's number. μ_a has a unit of cm^2. The absorption cross-section of $10^{-24} cm^2$ is sometimes referred to as 1 barn.

1.5.2 Absorption Edges

As the wavelength λ of the X-rays is gradually decreased (or equivalently as the energy of the photons is increased), the absorption coefficient μ generally decreases until a certain critical wavelength is reached where the absorption coefficient increases abruptly by several-fold. This discontinuity in the absorption coefficient corresponds to the ejection of a core electron from an atom and is called the *absorption edge*. Further decrease in λ causes a similar decrease in μ, but at a somewhat different rate, until another absorption edge is reached. As is evident from the previous section, there are one, three, five, ... absorption edges for the K, L, M, ... shells, respectively.

Fig. 1.6 depicts schematically the K and L-absorption edges of an element. Note that both the abscissa and the ordinate are in logarithmic scales. Several observations can be made. First, it is seen that the probability for ejecting an electron is largest for a photon with just sufficient energy for the process. Within each edge, the absorption cross section decreases with increasing energy. Second, the difference' in energy between the successive absorption edges K, L, M, etc., decreases dramatically with increasing principal quantum

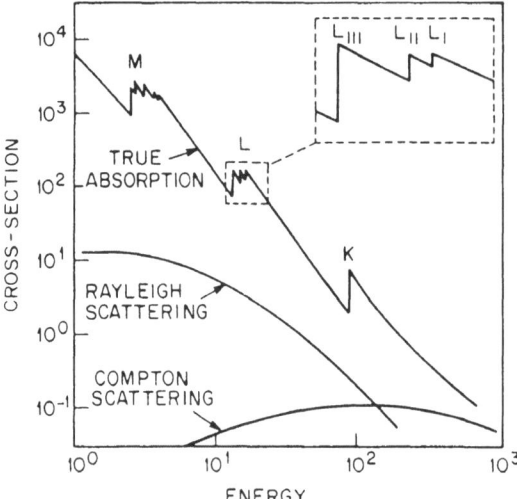

Fig. 1.6. Schematic representations of the relative cross-sections for true photoelectric absorption, elastic (Rayleigh), and inelastic (Compton) scatterings as a function of photon energy. Note that both axes are in logarithmic scales. (Adapted from McMaster, *et al.*, 1969)

number n. In other words, the edges are farther apart in energy in going from low to high-energy edges (*e.g.*, $M \rightarrow L \rightarrow K$). Third, the "jump ratios" between the two values of absorption coefficients at the edge, the so-called "edge jumps," generally decrease in going from low- to high-energy edges, e.g., $L_{III} > L_{II} > L_I > K$. Finally, with increasing atomic number, the absorption edges shift to higher energies and the edge jumps show a decrease.

1.5.3 *True Absorption and Scattering*

There are two components in the X-ray absorption coefficient which reflect two distinctly different modes of interactions between the X-rays and the matter through which it passes. The total absorption coefficient μ is the sum of a true absorption coefficient τ and a scattering coefficient σ:

$$\mu = \tau + \sigma \tag{1.17}$$

The true absorption coefficient τ relates to the photoelectric effect when a core electron is ejected and carries with it the excess energy in terms of kinetic energy. Beyond each absorption edge,

$$\tau = C' Z^m \lambda^n \tag{1.18}$$

where $m \approx 4$, $n \approx 3$, and C' is a constant. These constants vary with the absorption edge. According to Walter (1927), $m = 3.94$, $n = 3$, and $C' = 2.64 \times 10^{-26}$ for $\lambda < \lambda_K$ and $m = 4.30$, $n = 3$, and $C' = 8.52 \times 10^{-28}$ for $\lambda_K < \lambda < \lambda_L$ (λ's are in Å). The approximate power law $Z^4 \lambda^3$ can qualitatively be understood in the following manner. The interaction between the photoelectron and the nucleus is essential to the photoelectric process in order to conserve the linear momentum. The larger the momentum that must be transferred, the less likely would be the ejection of the photoelectron. With increasing Z, such interaction is enhanced and thus the absorption coefficient increases. The Z^4 dependence arises because the matrix element for the electric-dipole transition (of an electron from a bound state to an unbound state) contains a Z^2 factor and the transition probability is the square of the matrix element. On the other hand, for a given Z, the momentum that must be transferred decreases with increasing λ which leads to an increase in τ.

The scattering coefficient σ relates to the fact that X-ray photons can be deflected from their original direction of propagation, either with or without loss of energy, by collision with an electron or an atom. σ is a weak function of Z and λ. It is normally negligible except for the lightest elements and for short wavelengths.

A useful empirical relation for calculating the absorption coefficients of the elements with atomic number $Z = 1$ to 83 is given by Victoreen (1948),

$$\mu/\rho = C\lambda^3 - D\lambda^4 + \sigma_{K-N} NZ/A \qquad (1.19)$$

where the constants C and D are functions of Z, changing abruptly at the absorption edges. These constants are tabulated in "International Tables for X-Ray Crystallography," Vol. III (c) and in Appendix III. σ_{K-N} is the Klein-Nishina (1929) coefficient for the scattering of photons with $\lambda \lesssim 0.04$ Å by free electrons which is a good approximation to the total scattering coefficient (including elastic and inelastic scatterings). Values of σ_{K-N} are tabulated in "International Tables for X-Ray Crystallography," Vol. III (d). Other details and tables concerning absorption coefficients and absorption edges can be found in the "International Tables for X-Ray Crystallography," Vol. III (Ref. (b)).

1.6 Interactions of Photons and Electrons with Matter

Electromagnetic radiations such as X-rays, and charged particle beams such as electrons, interact with matter in a number of distinctly different ways. The interactions inevitably involve excitation of, or scattering with, the medium. The various modes of interactions of X-rays with matter are categorized in Table 1.1.

1.6.1 Excitations and Relaxations

The excitation of an atom requires the *absorption* of a photon. In the case of X-ray *photoionization*, a photoelectron is ejected from a core level after absorbing a photon. The excited atom can then relax through several mechanisms, giving rise to fluorescence X-rays, *Auger* electrons, or *secondary* electrons as shown schematically in Fig. 1.7.

The filling of an inner shell vacancy by an outer shell electron produces X-ray *fluorescence* characteristic of the absorbing material. The phenomenon of fluorescent radiation is therefore a *radiative* process and is important for core levels with energies $\geq$ 10 keV and elements with high atomic number Z. The energy of the fluorescent radiation is the difference in energy between the two shells and is just the characteristic radiation of the absorber discussed in a previous section (1.4.2) and tabulated in Appendix II. For example, a K-shell

Table 1.1.

Interaction of X-Rays with Matter and the Relaxation Mechanisms

True Absorption	Photoionization	radiative,	one-step
	X-ray Fluorescence	radiative	two-step
	Auger (autoionization)	nonradiative	two-step
	Secondary Electrons	nonradiative	multi-step
Scattering	Coherent	elastic (Rayleigh)	unmodified
		Diffraction (Bragg)	unmodified
	Incoherent	inelastic (Compton)	modified

ONE-STEP

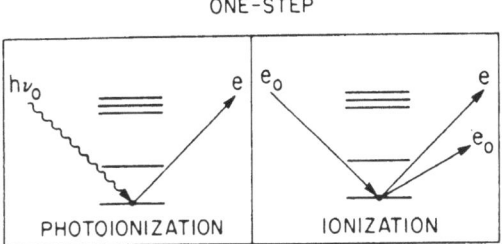

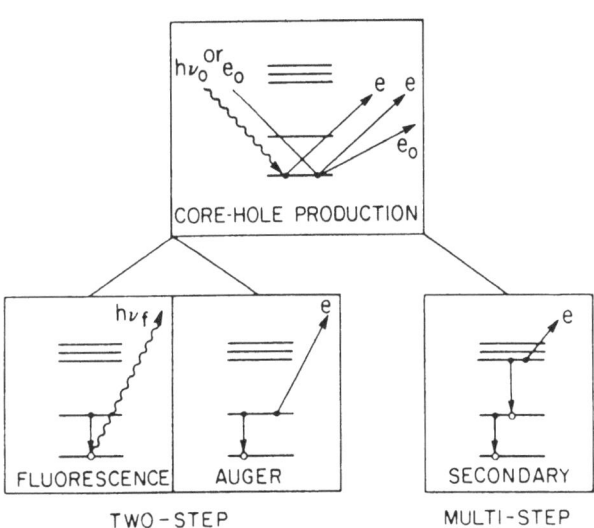

Fig. 1.7 Interactions of photons $(h\nu)$ or electrons (e_o) with the core levels of an atom, emitting fluorescent photons $(h\nu_f)$ or electrons (e). (Adapted from Carlson, 1972.)

vacancy in an iron atom produced by photoionization with an X-ray photon of 7.112 keV can return to a ground state by emitting X-ray photons K_{α_1}, K_{α_2}, K_{β_1}, $\cdots$ with energies of 6.403, 6.390, 7.057, ... keV and intensity ratios of 1 : 0.5 : 0.167 ... . The K_α, K_β, and K_γ lines, in descending order of intensity, correspond to electronic transitions from the L, M, and N shells to the K-shell, respectively.

The fluorescent yield, or the radiative probability, ϵ_f can be defined as the ratio of emitted X-rays to the number of primary vacancies created. It is a monotonically increasing function of atomic number Z, and is larger for K line emissions than L or subsequent line emissions. For example, the K-shell fluorescent yields for O, Cl, Fe, Mo, and Pt are 0.0058, 0.094, 0.347, 0.764, and 0.963, respectively. For comparison, the L-shell fluorescent yields of Mo and Pt

16

are 0.067 and 0.33, respectively. The K fluorescence yield as a function of atomic number Z is shown in Fig. 1.8 (Dyson, 1973). The K and L fluorescence yields for elements with $Z - 4-92$ are tabulated in Appendix IV. Basically, for $Z \geq 50$, ϵ_f for K fluorescence approaches unity.

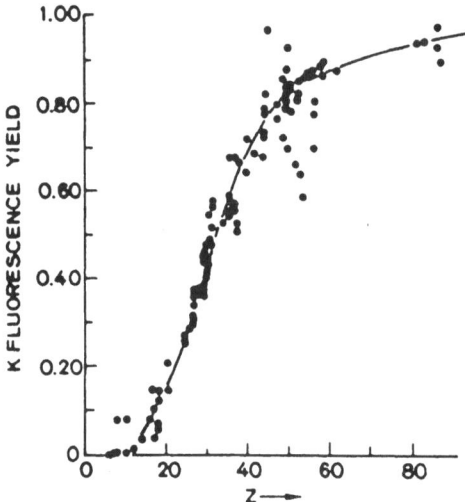

Fig. 1.8. K fluorescence yield as a function of atomic number Z (Dyson, 1973).

The other major phenomenon following photoelectric absorption is the Auger effect. In this process, the core vacancy is filled by an electron dropping from a higher shell while the atom simultaneously ejects another electron, usually from the same higher shell. The Auger effect is therefore a nonradiative or *radiationless* process. For example, a K-$L_I L_{II}$ Auger transition has an initial state of a K hole and the final state of two vacancies in the L shell. The kinetic energy of the Auger electron is just the difference in energy between the initial and the final state. For a K-LM transition, it amounts to

$$E_{\text{Auger}} - E_K - E_L - E_M' \tag{1.20}$$

where E_K and E_L are the binding energies of the K and L shells of a neutral atom and E_M' is the binding energy of an electron in the M shell of an ion having a single vacancy in the L shell. Obviously E_{Auger} must be greater than zero for the Auger transition(s) to occur. Qualitatively one can think of the Auger phenomenon as an *autoionization* process in which the "fluorescent photon" with energy $E_K - E_L$ is energetic enough to ionize another electron from the M shell of the same atom. In reality, the Auger process is a "one-step two-electron" Coulombic readjustment to the initial core hole rather than a two-step transition.

Auger transitions occur with quite high probability, especially for light elements and/or at low energies. Furthermore, an atom may emit two or more Auger electrons simultaneously. If the Auger process involves only core electrons, there remain inner shell vacancies that can further be filled by additional Auger processes, creating more ionization. A series of such processes is called a vacancy cascade.

The transition probability for an individual Auger process can be calculated from

$$P_{i \to f} = \frac{2\pi}{\hbar} | < \phi_1^f \phi_2^f | \frac{e^2}{\vec{r}_1 - \vec{r}_2} | \phi_1^i \phi_2^i > |^2 \qquad (1.21)$$

where ϕ_1^i and ϕ_2^i represent the initial bound states for the electrons and, following the nonradiative Coulombic readjustment, the final states of the electrons are represented by a bound orbital ϕ_1^f (the filled vacancy) and the continuum wave function ϕ_2^f. The selection rules for an Auger transition are that the initial and final states must have the same *symmetry* and *parity*. A particular type of Auger transition in which the initial vacancy and one of the electrons that fills the vacancy are in a shell with the same principal quantum number is called a Coster-Kronig transition. Coster-Kronig transitions have a large (by an order of magnitude) transition probability due to the large overlap of the wave functions.

Another nonradiative process is the production of *secondary electrons*. The Auger electrons or the emitted fluorescence X-ray photons can eject electrons from the outer shells as they leave the atom. These secondary electrons, including the "cascade" electrons, are less defined in terms of their kinetic energies but have the particular feature of abundant quantity.

While *primary* photoelectrons or Auger electrons have mean free paths of less than *ca* 20 Å, the *secondary* electrons caused by repeated scattering events may have mean free paths on the order of 50-100 Å for the energy range of interest.

Roughly speaking, the total nonradiative yield ϵ_n is just the complement of the radiative fluorescent yield ϵ_f

$$\epsilon_n = 1 - \epsilon_f \qquad (1.22)$$

To summarize, we note that photoelectric absorption and photoionization are *one-step* processes; X-ray fluorescence and Auger electron are *two-step* processes; and secondary electrons are *multi-step* processes.

1.6.2 Scattering

The scattering processes can be categorized into *coherent* and *incoherent* scattering. Coherent scattering results from the interference between the elastically scattered waves due to individual atoms; there is no loss of energy, that is, no change in wavelength, and the process is called unmodified or Rayleigh scattering. For the incoherently or inelastically scattered radiation, the wavelength is shifted slightly to a longer value and the process is called modified or Compton scattering.

Elastic (Rayleigh) scattering is a process where the X-ray photons are scattered by atomic electrons that are so tightly bound that no ionization or excitation take place. Since the effective photon mass is negligible compared with that of the atom, the elastically scattered photon carries away all its initial energy, giving rise to no loss in energy (or change in wavelength). The Rayleigh scattering cross-section for atoms is proportional to the square of the atomic scattering factor integrated over all directions and hence is proportional to Z^2. However, since the photoelectric absorption increases as Z^4, Rayleigh scattering is in general less important than true absorption except at low photon energies and/or for light elements. Rayleigh scattering gives rise to interference effects in which the scattering from various atoms combines coherently. In the case of crystals, interference between the waves coherently scattered by individual atoms in a periodic lattice gives rise to Bragg diffraction.

Inelastic (Compton) scattering arises from the collision between a photon and a "loosely bound" electron. Part of the incident energy E_0 is transferred to the electron as the photon changes its direction of propagation by a scattering angle θ. The new photon energy is

$$E' = E_0 \frac{1}{1 + \dfrac{E_0}{mc^2}(1 - \cos\theta)} \tag{1.23}$$

It is obvious that the energy loss $E_0 - E'$ increases with increasing incident beam energy E_0 as well as the scattering angle θ. For short wavelengths, the atomic Compton scattering cross-section is roughly proportional to Z since as Z increases, the number of "relatively loosely" bound electrons increases. It also changes much more slowly with λ than either the photoelectric or the Rayleigh cross-section. The Klein-Nishina expression for the Compton scattering cross-section at short wavelengths is given in the "International Tables for X-Ray Crystallography," Vol. III(d).

The Rayleigh and Compton scattering cross-sections are compared with the true absorption cross-section in Fig. 1.6. It is apparent that the former (scattering) are in general less important than the latter (absorption) by orders of magnitude except at low photon energies and/or for light elements.

1.6.3 Electrons

Electrons interact with matter in a similar but not identical fashion to photons. The various excitation and relaxation processes are depicted schematically in Fig. 1.9. White radiation in the X-ray region can result from

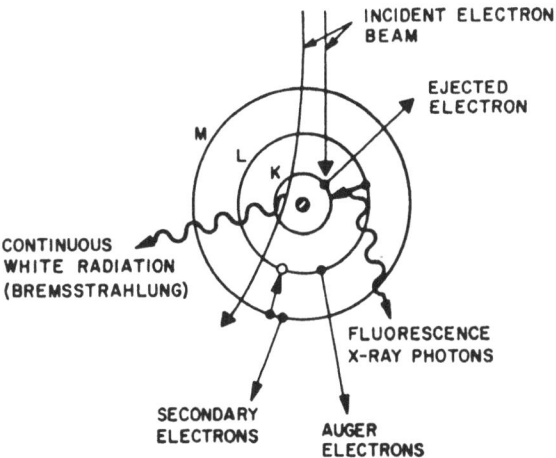

Fig. 1.9 Interaction of electrons with a simplified model of the atom. Incident electrons must have sufficient energy to remove an electron from its shell, leaving a core hole. Electrons falling into the core hole emit their excess energy as fluorescence X-ray photons (characteristic lines). Auger electrons can also be produced by filling the core hole with an electron and simultaneously ejecting another electron, usually from the same shell. The Auger electrons or the emitted fluorescence X-ray photons can eject "secondary electrons" from the outer shells as they leave the atom. Multiple processes involving core-hole filling and electron ejection produce the "cascade electrons." Continuous white radiation (bremsstrahlung) is produced when the incident electron is decelerated in the field of the nucleus.

bremsstrahlung. Core holes can be produced via ionization processes (*viz.*, atomic electrons ejected by the incoming electron beam), which can subsequently give rise to characteristic fluorescent X-rays or Auger and secondary electrons through various relaxation channels as described in the previous section. Inelastic scattering of electrons results in electron energy (or momentum) loss as in electron energy loss spectroscopy (EELS) whereas elastic scattering gives rise to electron diffraction (for gas) and low energy electron diffraction (LEED for crystals).

The contrast between X-rays and electrons is their penetration power. High energy X-ray photons can pass through thick samples with little attenuation whereas electrons have only a small mean free path in condensed matter.

Chapter 2

Extended X-ray Absorption Fine Structure

(EXAFS) Spectroscopy

Extended X-ray absorption fine structure (EXAFS) refers to the oscillatory variation of the X-ray absorption as a function of photon energy beyond an absorption edge. The absorption, normally expressed in terms of absorption coefficient (μ), can be determined from a measurement of the attenuation of X-rays upon their passage through a material. When the X-ray photon energy (E) is tuned to the binding energy of some core level of an atom in the material, an abrupt increase in the absorption coefficient, known as the absorption edge, occurs. For isolated atoms, the absorption coefficient decreases monotonically as a function of energy beyond the edge. For atoms either in a molecule or embedded in a condensed phase, the variation of absorption coefficient at energies above the absorption edge displays a fine structure called EXAFS. Such fine structure may extend up to 1000 eV above the absorption edge and may have an amplitude of up to a few tenths (normally 1-20%) of the edge jump.

Although the extended fine structure has been known for a long time (Kronig, 1931, 1932), its structural content was not fully recognized until the recent work of Stern, Lytle, and Sayers (1974, 1975). In addition, the recent availability of synchrotron radiation has resulted in the establishment of EXAFS as a practical structural tool particularly through the work of

Eisenberger (1975) and Kincaid (1975). This technique is especially valuable for structural analyses of chemical or biological systems where conventional diffraction methods are not applicable.

In this chapter, a brief and qualitative description of various aspects of this powerful structural method will be given. For further details on synchrotron radiation research, readers are referred to excellent reviews by Winick and Bienenstock (1978), Lindau and Winick (1978), Watson and Perlman (1978), Batterman and Ashcroft (1979), as well as the excellent book "Synchrotron Radiation Research" edited by Winick and Doniach (1980). Reviews on different aspects of EXAFS have also appeared (see, for example, Stern, 1978; Eisenberger and Kincaid, 1978; Shulman, *et al*, 1978; Hayes, 1978; Sandstrom and Lytle, 1979; Cramer and Hodgson, 1979; Teo, 1980, 1981; Wong, 1980; Lee, *et al*, 1981; and the book "EXAFS Spectroscopy: Techniques and Applications" edited by Teo and Joy, 1981).

2.1 EXAFS Spectroscopy

EXAFS spectroscopy refers to the measurement of the X-ray absorption coefficient μ as a function of photon energy E above the threshold of an absorption edge. Figure 2.1 shows schematically one edge of an absorber. In a

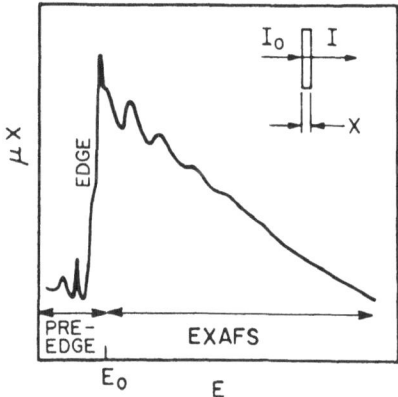

Fig. 2.1. Schematic representation of the transmission experiment and the resulting X-ray absorption spectrum μx *vs* E for an absorption edge of the absorbing atom (e.g. K edge of iron which corresponds to the ejection of a $1s$ electron by absorption of an X-ray photon with $E \geqslant E_0$, where E_0 is the energy threshold). (Reproduced from Teo, 1980).

transmission experiment, μ or μx (x is the sample thickness) is calculated by

$$\mu x = \ln I_o/I \qquad (2.1)$$

where I_o and I are the intensities of the incident and transmitted beams, respectively.

EXAFS spectra generally refer to the region 40-1000 eV above the absorption edge. Near or below the edge, there generally appear absorption peaks due to excitation of core electrons to some bound states ($1s$ to nd, $(n + 1)s$, or $(n + 1)p$ orbitals for K edge, and $2s$ for L_I edge, $2p$ for L_{II}, L_{III} edges to the same set of vacant orbitals, etc.). The selection rules for atomic excitations are described in the previous chapter. This pre-edge region contains valuable bonding information such as the energetics of virtual orbitals, the electronic configuration, and the site symmetry. The edge position also contains information about the charge on the absorber. In between the pre-edge and the EXAFS regions is the X-ray Absorption Near Edge Structure (XANES) which is only recently beginning to be understood. It arises from effects such as many-body interactions, multiple scatterings, distortion of the excited state wavefunction by the Coulomb field, band structures, etc.

Transmission is just one of several modes of EXAFS measurements. The fluorescence technique involves the measurement of the fluorescence radiation (over some solid angle) at right angle to the incident beam. For dilute biological systems, this method removes the 'background' absorption due to other constituents, thereby improving the sensitivity by orders of magnitude. Other more specialized methods include: (1) surface EXAFS (SEXAFS) studies which involve measurements of either the Auger electrons or the inelastically scattered electrons (partial or total electron yield) produced during the relaxation of an atom following photoionization and (2) electron energy loss (inelastic electron scattering) spectroscopy (EELS). These latter methods, which require high vacuum, are useful for light atom EXAFS with edge energies up to a few keV.

It should be emphasized that in the transmission or fluorescence mode, EXAFS spectroscopy involves only X-ray measurements. For the other experimental modes, on the other hand, electrons may be involved in the experimentation. Details of various experimental techniques will be discussed elsewhere.

2.2 Theory

EXAFS is a final state interference effect involving scattering of the outgoing photoelectron from the neighboring atoms. From a qualitative viewpoint, the probability that an X-ray photon will be absorbed by a core electron depends on both the initial and the final states of the electron. The initial state is the localized core level corresponding to the absorption edge. The final state is that of the ejected photoelectron which can be represented as an outgoing spherical wave originating from the X-ray absorbing atom. If the absorbing atom has a neighboring atom, the outgoing photoelectron wave will be backscattered by the neighboring atom, thereby producing an incoming electron wave. The final state is then the sum of the outgoing and all the incoming waves, one from each neighboring atom. It is the interference between the outgoing and the incoming waves that gives rise to the sinusoidal variation of μ vs E known as EXAFS.

Figure 2.2 attempts to convey, pictorially, this current view of EXAFS. For a monatomic gas such as Kr (Fig. 2.2a and c) with no neighboring atoms, a photoelectron ejected by absorption of an X-ray photon will travel as a spherical wave with a wavelength $\lambda = \dfrac{2\pi}{k}$ where

$$k = \sqrt{\frac{2m}{\hbar^2} (E - E_o)} \qquad (2.2)$$

Here E is the incident photon energy and E_o is the threshold energy of that particular absorption edge. The μ vs E curve follows the usual smooth λ^3 decay (cf. Fig. 2.2a). In the presence of neighboring atoms (e.g., in Br_2, Fig. 2.2b and d), the outgoing photoelectron can be backscattered from the neighboring atoms thereby producing an incoming wave which can interfere either constructively or destructively with the outgoing wave near the origin, resulting in the oscillatory behavior of the absorption rate (cf. Fig. 2.2b). The amplitude and frequency of this sinusoidal modulation of μ vs E depends on the type (and bonding) of the neighboring atoms and their distances away from the absorber, respectively.

This simple picture of EXAFS has been formulated into the generally accepted *short-range single-electron single-scattering* theory (Stern, 1974; Stern, Sayers, and Lytle, 1975; Ashley and Doniach, 1975; Lee, *et al*, 1975, 1977). For reasonably high energy ($\geq$ 60 eV) and moderate thermal or static disorders, the modulation of the absorption rate in EXAFS, normalized to the

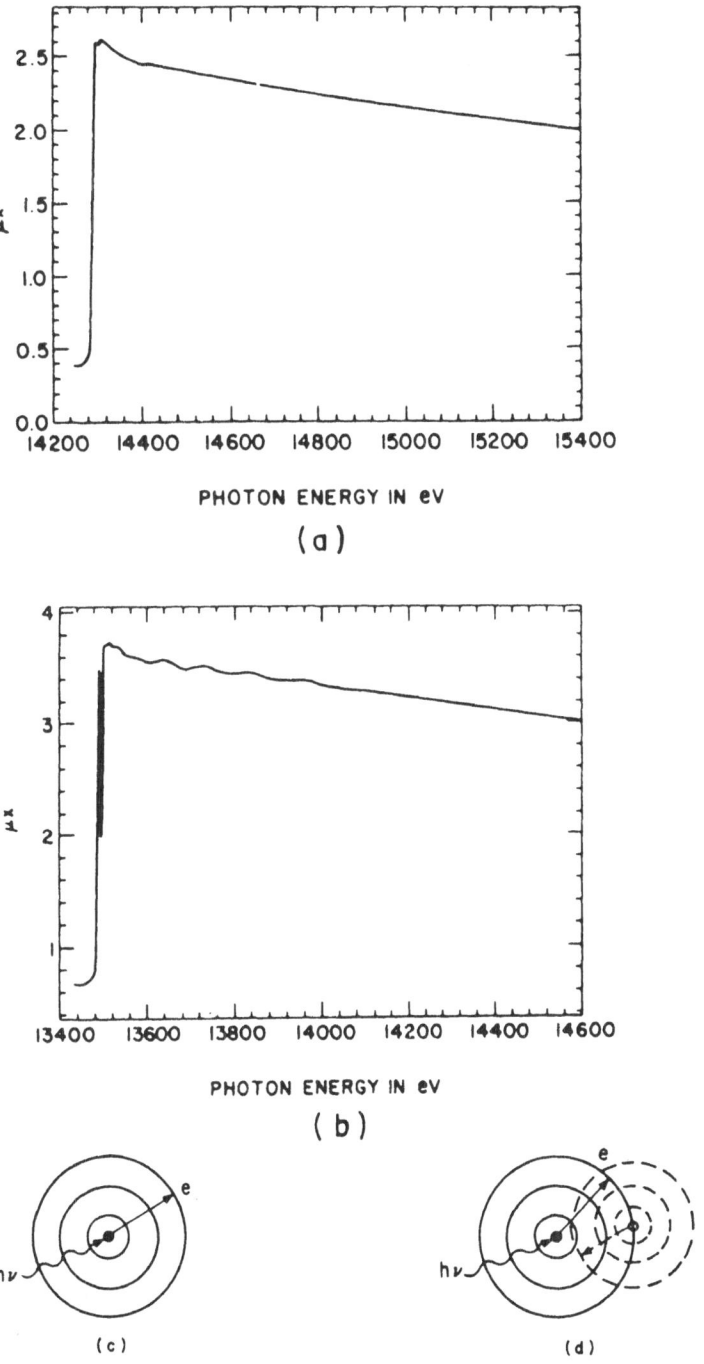

Fig. 2.2. Qualitative rationalization of the absence and presence, respectively, of EXAFS in a monatomic gas such as Kr (a and c) and a diatomic gas such as Br$_2$ (b and d). Figures (a) and (b) are taken from the work of Kincaid and Eisenberger (1975).

'background' absorptions (μ_o) is given by

$$\chi(E) = \frac{\mu(E) - \mu_o(E)}{\mu_o(E)} \qquad (2.3)$$

In order to relate $\chi(E)$ to structural parameters, it is necessary to convert the energy E into the photoelectron wavevector k via Eq. 2.2. This transformation of $\chi(E)$ in E space gives rise to $\chi(k)$ in k space where

$$\chi(k) = \sum_j N_j S_i(k) F_j(k) e^{-2\sigma_j^2 k^2} e^{-2r_j/\lambda_j(k)} \frac{\sin(2kr_j + \phi_{ij}(k))}{kr_j^2} \qquad (2.4)$$

Here $F_j(k)$ is the backscattering amplitude from each of the N_j neighboring atoms of the jth type with a Debye-Waller factor of σ_j (to account for thermal vibration (assuming harmonic vibration) and static disorder (assuming Gaussian pair distribution) and at a distance r_j away). $\phi_{ij}(k)$ is the total phase shift experienced by the photoelectron. The term e^{-2r_j/λ_j} is due to inelastic losses in the scattering process (due to neighboring atoms and the medium in between) with λ_j being the electron mean free path. $S_i(k)$ is the amplitude reduction factor due to many-body effects such as shake up/off processes at the central atom (denoted by i).

It is clear that each EXAFS wave is determined by the backscattering amplitude $(N_j F_j(k))$, modified by the reduction factors $S_i(k)$, $e^{-2\sigma_j^2 k^2}$, and e^{-2r_j/λ_j}, and the $1/kr_j^2$ distance dependence, and the sinusoidal oscillation which is a function of interatomic distances $(2kr_j)$ and the phase shift $(\phi_{ij}(k))$.

The effects of some of these variables or functions are illustrated in Fig. 2.3. In Fig. 2.3(a) we show a backscattering amplitude function $F(k)$ weighted by k^2 for one scatterer (curve A), $NF(k)k^2$ for N scatterers (curve B) and the effect of amplitude damping by the Debye-Waller factor σ in $NF(k)k^2 e^{-2\sigma^2 k^2}$ (curve C). Note that N is a linear term whereas $e^{-2\sigma^2 k^2}$ is an exponential damping which reduces the amplitude at high k regions more than at low k values. In curve D of Fig. 2.3(b) we show the effect of multiplying the amplitude envelope by a weighting factor k^2 and the inclusion of the effects due to the interatomic distance r giving rise to $NF(k)k^2 e^{-2\sigma^2 k^2} \frac{\sin(2kr)}{r^2}$. The sinusoidal EXAFS oscillation is caused by the interference $\sin(2kr)$ term with a frequency $2r$ in k space. The larger the distance r, the more rapid will be the oscillation (higher frequency). Distance r also reduces the EXAFS amplitude by $1/r^2$ which implies that the larger the distance, the weaker will be the

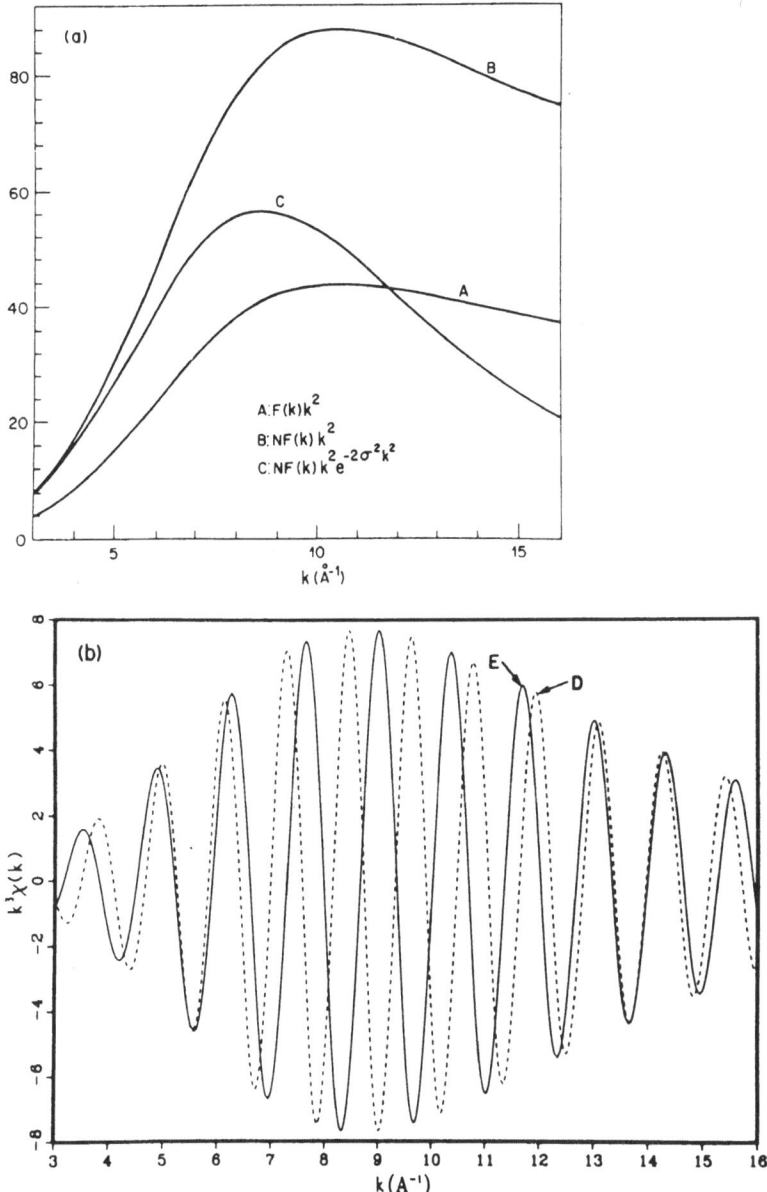

Fig. 2.3. Physical significance of the variables in EXAFS spectroscopy; (a) curve A: $F(k)k^2$, the backscattering amplitude function, curve B: $NF(k)k^2$ for N atoms; curve C: $NF(k)k^2e^{-2\sigma^2k^2}$ which shows the damping effect of the Debye-Waller factor σ; (b) curve D: $NF(k)k^2e^{-2\sigma^2k^2}\dfrac{\sin(2kr)}{r^2}$, the sinusoidal oscillation due to the interference term containing the distance r; curve E: $NF(k)k^2e^{-2\sigma^2k^2}\dfrac{\sin(2kr+\phi(k))}{r^2}$ showing the effect of the total phase shift $\phi(k)$.

EXAFS signal (assuming other things are equal). Finally in curve E we show the effect of the total phase shift $\phi(k)$ in shifting the sine wave in k space: $NF(k)k^2e^{-2\sigma^2k^2}\dfrac{\sin(2kr + \phi(k))}{r^2}$. Since $\phi(k)$ depends on k, it not only changes the origin, but also the frequency, of the sine wave. It has relatively little effect on the amplitude.

It should be emphasized that while the amplitude function $F_j(k)$ depends only on the type of the backscatterers (except, perhaps, the reduction factor $S_i(k)$ which is mainly a function of the absorber), the phase function contains contributions from both the absorber and the backscatterer:

$$\phi_{ij}^l(k) = \phi_i^l(k) + \phi_j(k) - l\pi \qquad (2.5)$$

where $l = 1$ for K and L_I edges and $l = 2$ or 0 for $L_{II,III}$. Here ϕ_i and ϕ_j correspond to Lee and Beni (1977)'s $2\delta'_l$, the l phase shift of the absorber, and θ, the phase of the backscattering amplitude, respectively. Qualitatively, the physical origin of this dependence is that the photoelectron experiences the central atom phase shift twice, once going out and once coming back, but experiences the neighboring atom phase shift once by propagating from the absorber to the neighboring atoms and back to the absorber.

The Debye-Waller factor σ plays an important role in EXAFS spectroscopy. It contains important structural and chemical information which is otherwise difficult to obtain, yet it comes as a bonus in the EXAFS determination of interatomic distance. Generally speaking, the Debye-Waller factor σ has two components σ_{stat} and σ_{vib} due to static disorder and thermal vibrations, respectively. In principle, these two factors can only be separated by a temperature dependent study of $\sigma(T)$. However, if σ_{vib} can be estimated from vibrational spectroscopy or if σ_{stat} is known from other studies, the other term can be calculated from the experimentally determined σ. We shall further discuss the Debye-Waller factor in Chapters 4 to 6.

Broadly speaking, there are two categories of inelastic scattering processes which tend to reduce the EXAFS amplitude. The first is caused by multiple excitations at the central atom whereas the second is associated with excitation of the neighboring environment, including the neighboring atoms and the intervening medium, by the photoelectron. In Equation 2.4 the inelastic losses due to multiple excitations (many-body effects such as shake up/off processes) at the absorber are approximated by an amplitude reduction factor $S_i(k) \leq 1$

whereas, the inelastic losses due to excitation of the neighboring environment is approximated by $e^{-2r/\lambda(k)}$ where $\lambda(k)$, which depends on k, is the electron inelastic mean free path. These phenomena will be discussed in Chapters 4 and 5.

The single-electron single-scattering theory of EXAFS (*cf.* Eq. 2.4) makes use of the fact that in most cases multiple scattering is not important. This assumption is generally valid if one considers that multiple scattering processes can be accounted for by adding all scattering paths that originate and terminate at the central atom (absorber). Each of these processes then behaves like $\sin(2kr_{eff})$ where $2r_{eff}$ is the total scattering path length which is much larger than that of the direct backscattering from the nearest neighbors. Thus, multiple scattering will give rise to rapidly oscillatory waves in k space which tend to cancel out. The amplitude of these waves is also significantly attenuated by the large scattering path lengths, making it relatively unimportant in comparison with the direct backscattering.

On the other hand, multiple scattering in EXAFS can become important when atoms are arranged in an approximately colinear array. In such cases, the outgoing photoelectron is strongly *forward-scattered* by the intervening atom, resulting in a significant amplitude enhancement. In fact, both the amplitude and the phase are modified by the intervening atom(s) for bond angles ranging from 180° to ~75°. The effect, however, drops off very rapidly for bond angles below *ca* 150°. For these systems, it is necessary to rewrite Eq. 2.4 to take into account multiple scattering involving the intervening atom(s). The theory and the use of multiple scattering formalism in bond angle determination will be discussed in Chapter 8.

Finally, it should be emphasized that the phase shifts are unique only if the energy thresholds, E_0, are specified (*cf.* Chapter 7). Changing E_0 will change the momentum k and hence the phase shift function $\phi(k)$. It is clear then that, in order to fit experimental data based upon some empirical E_0 with theoretical phase shifts or phase shifts derived from some model compounds, we must allow E_0 to vary.

The E_0 variation also helps remove the small but significant bonding effects such as electronic configurations and atomic charges.

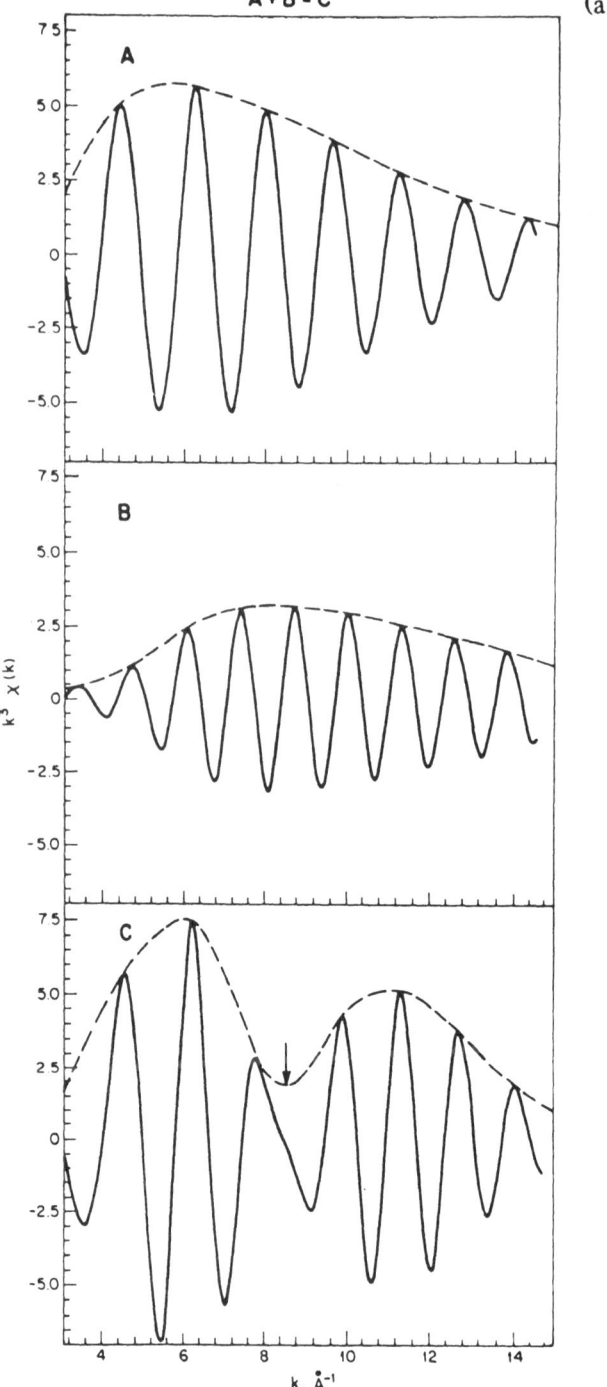

A + B = C

(a)

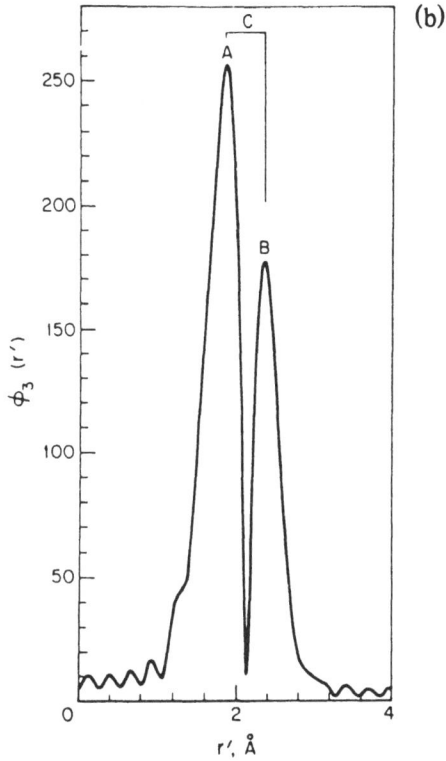

Fig. 2.4. (a) The sum of two individual waves A and B in k space gives rise to a composite wave C with a "beat node" (*i.e.*, a dip in the amplitude function marked by the arrow). (b) Fourier transform of the composite wave C in (a).

2.3 Data Analysis

EXAFS is a sum of individual waves due to the different types of neighboring atoms or different distances of the same type of neighbors:

$$\chi(k) = \sum_{j} A_j(k) \sin(2kr_j + \phi_{ij}(k)) \qquad (2.6)$$

In k space, each wave is characterized by an amplitude

$$A_j(k) = N_j S_i(k) F_j(k) e^{-2\sigma_j^2 k^2} e^{-2r_j/\lambda(k)}/(kr_j^2) \qquad (2.7)$$

and a sinusoidal oscillation with a frequency $2r_j$ and a phase shift $\phi_{ij}(k)$ given by Eq. 2.5. This is shown schematically in Fig. 2.4(a) for two types of neighboring atoms located at different distances from the absorber. Each EXAFS wave has its characteristic amplitude envelope (dashed curves, $A_j(k)$)

and frequency (solid $\sin(2kr_j + \phi_{ij}(k))$ waves). Addition of these two EXAFS waves can result in a dip in the amplitude envelope called a "beat node" and the frequency of the composite wave changing across the "beat node" (*viz.*, the frequency of the composite wave C resembles that of wave A and B in the low and high k regions, respectively. Note that wave A has a lower frequency and an amplitude which peaks at a low k value whereas wave B has a higher frequency and an amplitude which peaks at a high k region. Hence, the low frequency of wave A and the high frequency of wave B dominates at low and high k regions, respectively, in the composite EXAFS wave C).

The frequency of each EXAFS wave depends on the distance between the absorbing atom and the neighboring atom since the photoelectron wave must travel from the absorber to the scatterer and back. During the trip, the photoelectron actually experiences a phase shift (Coulombic interaction) of the absorber twice (i.e., once going out and once coming back) and a phase shfit of the scatterer once (scattering). If one assumes that the phase shifts can be obtained from either model compounds or calculations, one can determine interatomic distances in the vicinity of the absorber. On the other hand, the amplitude of each EXAFS wave depends upon the number and the backscattering power of the neighboring atom, as well as on its bonding to and distance from the absorber (*vide infra*). From an analysis of the scattering profiles, one can quantitatively assess the types and numbers of atoms surrounding the absorber.

Each EXAFS wave contains two sets of highly correlated variables: $\{F(k), \sigma, \lambda, N, S\}$ and $\{\phi(k), E_0, r\}$. Significant correlations can occur both *within* and *between* these two sets of variables as well as *between* different scattering terms. In order to determine N and σ, $F(k)$ must be known reasonable well; similarly, in order to determine r, $\phi(k)$ must be known accurately. In practice, $F(k)$ and $\phi(k)$ can be determined empirically from model compounds of known structure (*cf.* Chapter 6) or calculated theoretically (*cf.* Chapter 7) from first principles.

Structural determinations via EXAFS depend on the feasibility of resolving the data into individual waves corresponding to the different types of neighbors of the absorbing atom. This can be accomplished by either *curve-fitting* or *Fourier transform* techniques. Curve-fitting involves a best fitting of the data with a sum of individual waves modeled by some empirical equations based on Eq. 2.4, each of which contains appropriate structural parameters for each type of neighbor. On the other hand, the Fourier transform technique provides a photoelectron scattering profile as a function of the radial distance from the absorber as shown in Fig. 2.4(b) for the composite wave C depicted in Fig. 2.4(a). In such a radial distribution function, the positions of the peaks are related to the distance between the absorber and the neighboring atoms while the sizes of the peaks are related to the numbers and types of the neighboring atoms. Details of various methods of data analysis will be discussed in Chapter 6.

EXAFS measurements can provide structural information for each type of atom in the sample if one simply tunes the X-ray energy to coincide successively with an absorption edge of each of the atom types in the sample. Such information as the number and kind of neighboring atoms and their distances away from the absorber are contained in the one-dimensional radial distribution function (RDF) centered at the absorber. It is clear that EXAFS is highly specific in that it can focus on the immediate environment around each absorbing species (generally out to *ca* 6Å corresponding to 1-3 coordination shells). Other materials or impurities present in the sample which either do not contain the absorber or are not directly bound to the absorber will not interfere. Furthermore, the technique is highly versatile in that it can be applied with about the same degree of accuracy (0.01 ~ 0.03Å) to matter in the solid (crystalline or amorphous), liquid, solution, or gaseous state.

Chapter 3

EXAFS Parameters

Before describing in detail the various aspects of EXAFS theory and data analysis techniques, it is instructive to first discuss, qualitatively and graphically, the various variables (or parameters) and functions and their effects and correlations.

3.1 Variables and Functions

Figures 3.1-3.10 demonstrate how an EXAFS spectrum in k space is built from various variables and functions. To illustrate their effects on the resulting EXAFS spectrum, two sets of variables or functions are shown in each case.

In Fig. 3.1(a), two sine waves $\sin(2kr)$ of different frequencies $2r < 2r'$ are shown. Note that the shorter distance r (2.3Å, solid curve) gives rise to a lower frequency than the longer distance r' (3.3Å, dashed curve) in k space. The arguments of the sine term, $2kr$, are depicted in Fig. 3.1(b). For a fixed k range (e.g. 3 to 15Å^{-1}), increasing r introduces new waves from the right, thereby generating the illusion that the wave trains are being moved to the left. A better analogy would be a spring fixed on one end at $k = 0$ and free to move on the other. Increasing r then implies moving the free end to the left, i.e., toward the fixed end. It is apparent that changing r affects mainly the high k region since the movement of the waves is linearly proportional to k. The reader will find this somewhat trivial analogy useful in later discussions.

Inclusion of a phase function $\phi(k)$ in $\sin(2kr + \phi(k))$ shifts the wave off the origin (viz., it no longer starts at $k = 0$) as shown in Fig. 3.2. Here we assume a distance $r = 2.3$Å and a linear $\phi(k) = p_0 + p_1 k$ where $p_0 = -2.1515$ and $p_1 = -1.3263$. The constant term p_0 in $\phi(k)$ shifts the sine wave as a

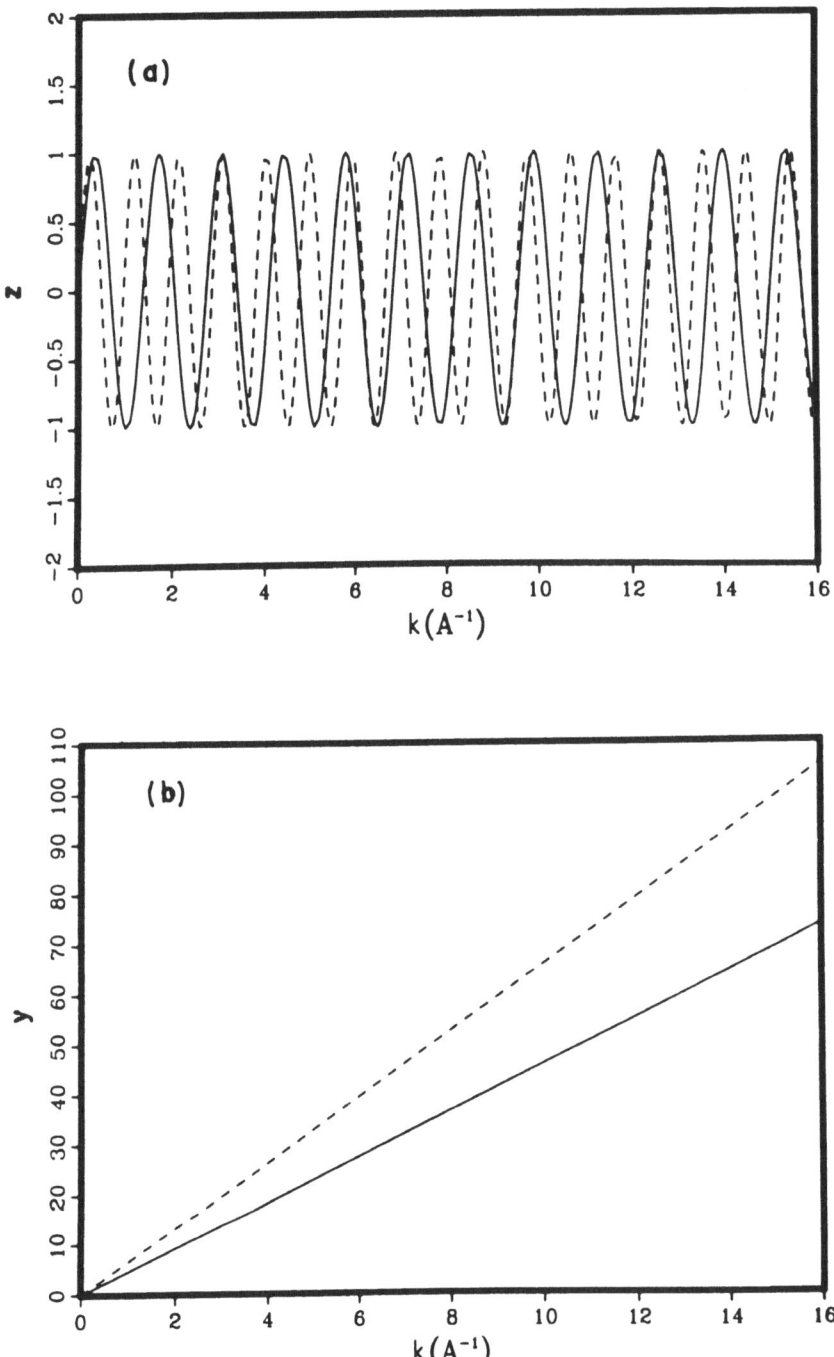

Fig. 3.1. (a) Two sine waves of different frequencies: $z = \sin(2kr)$; (b) the arguments: $y = 2kr$. In (a) and (b), $r = 2.3\text{Å}$ (solid curves) and 3.3Å (dashed curves).

36

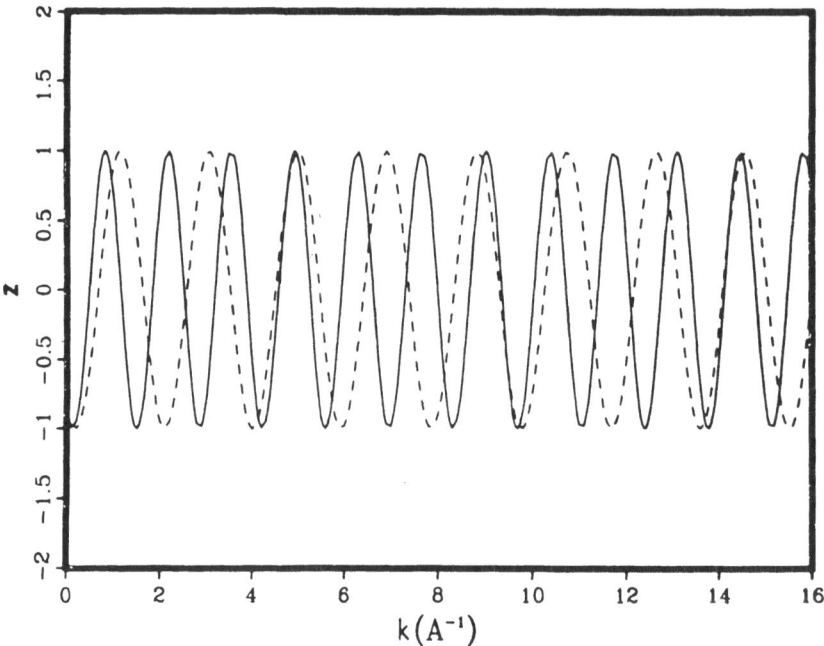

Fig. 3.2. The effect of phase shifts: $z = \sin(2kr + \phi(k))$ where $\phi(k) = p_0$, a constant (solid curve) and $\phi(k) = p_0 + p_1 k$, a linear curve (dashed curve). Here $p_0 = -2.1525$ and $p_1 = -1.3263$ are used.

whole to the left (right) for positive (negative) p_0 without affecting its frequency (solid curve). In contrast, the linear term p_1 affects only the frequency of the wave in much the same way as the distance. Hence, for a linear phase function $\phi(k) = p_0 + p_1 k$, the slope p_1, which normally bears a negative sign, plays an important role in the determination of the distance r since the frequency of the wave is now $(2r + p_1)$. In practice, $\phi(k)$ often has a more complicated k dependence such that higher order terms must be included (*vide infra*).

In Fig. 3.3(a), we show the effect of changing the energy threshold by $\Delta E_0 = E_0' - E_0 = -30$ eV on the function $2kr$ where $r = 2.3$Å. For the practical k range of 3 to 15Å^{-1}, the effect of increasing (decreasing) ΔE_0 is to increase (decrease) the slope of the function $2kr$, thereby resulting in a higher (lower) frequency for the sine wave $\sin(2kr)$ as demonstrated in Fig. 3.3(b). As a result, the wave moves to the right (left) for positive (negative) ΔE_0. Since $k \propto \sqrt{\Delta E}$, the effect of changing ΔE_0 is felt mostly at low k values. Using the same analogy as the distance, the "spring" is now fixed at the high k

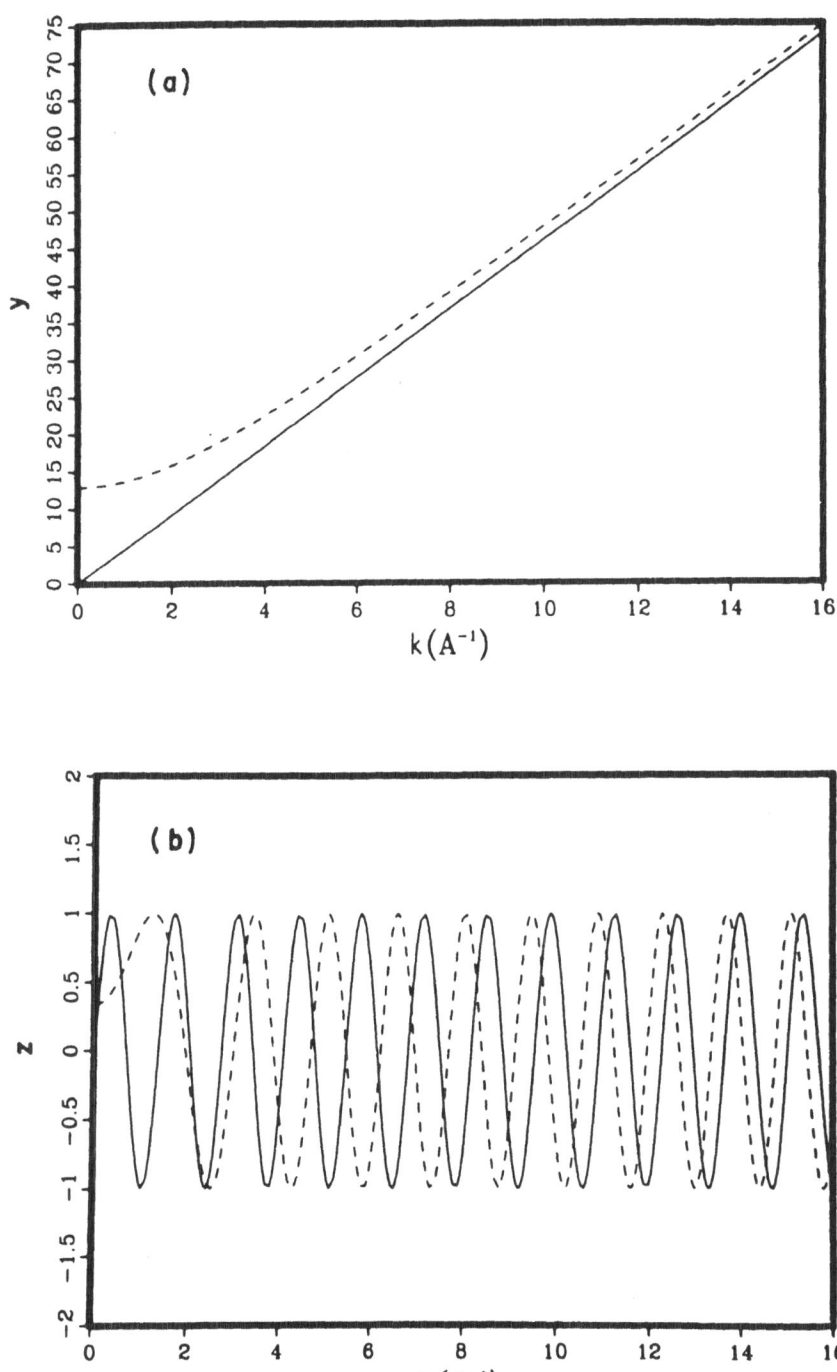

Fig. 3.3. The effect of ΔE_0 on the sine waves: (a) $y = 2k'r$ and (b) $z = \sin(2k'r)$. In (a) and (b), $r = 2.3\text{Å}$ and $k' = (k^2 - 0.2625\,\Delta E_0)^{1/2}$ for $\Delta E_0 = 0$ (solid curve) and $\Delta E_0 = -30$ eV (dashed curve).

end and increasing (decreasing) ΔE_0 causes the "spring" to contract (stretch). The degree of such movement, however, decreases as k increases. Hence, even though both r and ΔE_0 can affect the frequency of the sine wave, they do so in different regions in k space and act in different ways.

In Fig. 3.4(a) we show the effect of $\Delta E_0 = -30\text{eV}$ on a quadratic phase shift function $\phi(k) = p_0 + p_1 k + p_2 k^2$ where $p_0 = -2.1525$, $p_1 = -1.3263$, and $p_2 = 0.0287$. It is seen that a negative ΔE_0 causes the slope of $\phi(k)$ to decrease (from solid curve to dashed curve). The effect, again, is more pronounced at low k region. Inclusion of this phase function in Fig. 3.3(b) gives rise to $\sin(2kr + p_0 + p_1 k + p_2 k^2)$ shown in Fig. 3.4(b). Note the increasing movement of the waves to the left (toward low k values) for $\Delta E_0 = -30\text{eV}$.

In Fig. 3.5(a) we show two parameterized backscattering functions $F(k) = A/(1 + B^2(k-C)^2)$ where $A = 0.783$, 0.656Å; $B = 0.237$, 0.194Å; and $C = 3.4$, 6.4Å^{-1} for the solid and dashed curves, respectively. These functions are in fact the parameterized theoretical backscattering functions for sulfur and iron which peak at $k = 3.4$ and 6.4Å^{-1}, respectively. In Fig. 3.5(b), the sine wave (solid curve) in Fig. 3.4(b) has been multiplied by the backscattering amplitude functions $F(k)$ shown in Fig. 3.5(a). Note that the low Z element has an amplitude envelope which peaks at low k (solid curve) and vice versa.

In Fig. 3.6, the solid curve in Fig. 3.5(b) has been multiplied by the number of atoms $N = 1$ (solid curve) and $N = 2$ (dashed curve). It should be noted that N is a linear scale (which also include all other linear scale factors).

In Fig. 3.7(a) are shown two exponential damping functions $e^{-2\sigma^2 k^2}$ where $\sigma = 0.04\text{Å}$ (solid curve) and 0.09Å (dashed curve) are the Debye-Waller factors. The damping effect of the Debye-Waller factor is demonstrated in Fig. 3.7(b) where the solid curve in Fig. 3.6 has been multiplied by the Debye-Waller functions shown in Fig. 3.7(a). Note that the amplitude attenuation increases exponentially as k increases such that increasing σ can attenuate the EXAFS signal by orders of magnitude at high k.

In Fig. 3.8, another type of exponential damping factor, $e^{-2r/\lambda(k)}$, is shown where $\lambda(k)$ is the inelastic electron mean free path. Since $\lambda(k)$ is approximately proportional to k, we can approximate $e^{-2r/\lambda(k)}$ by $e^{-2\eta r/k}$. In Fig. 3.8(a) we plot the latter function with $\eta = 1\text{Å}^{-2}$ (solid curve) and 2Å^{-2}

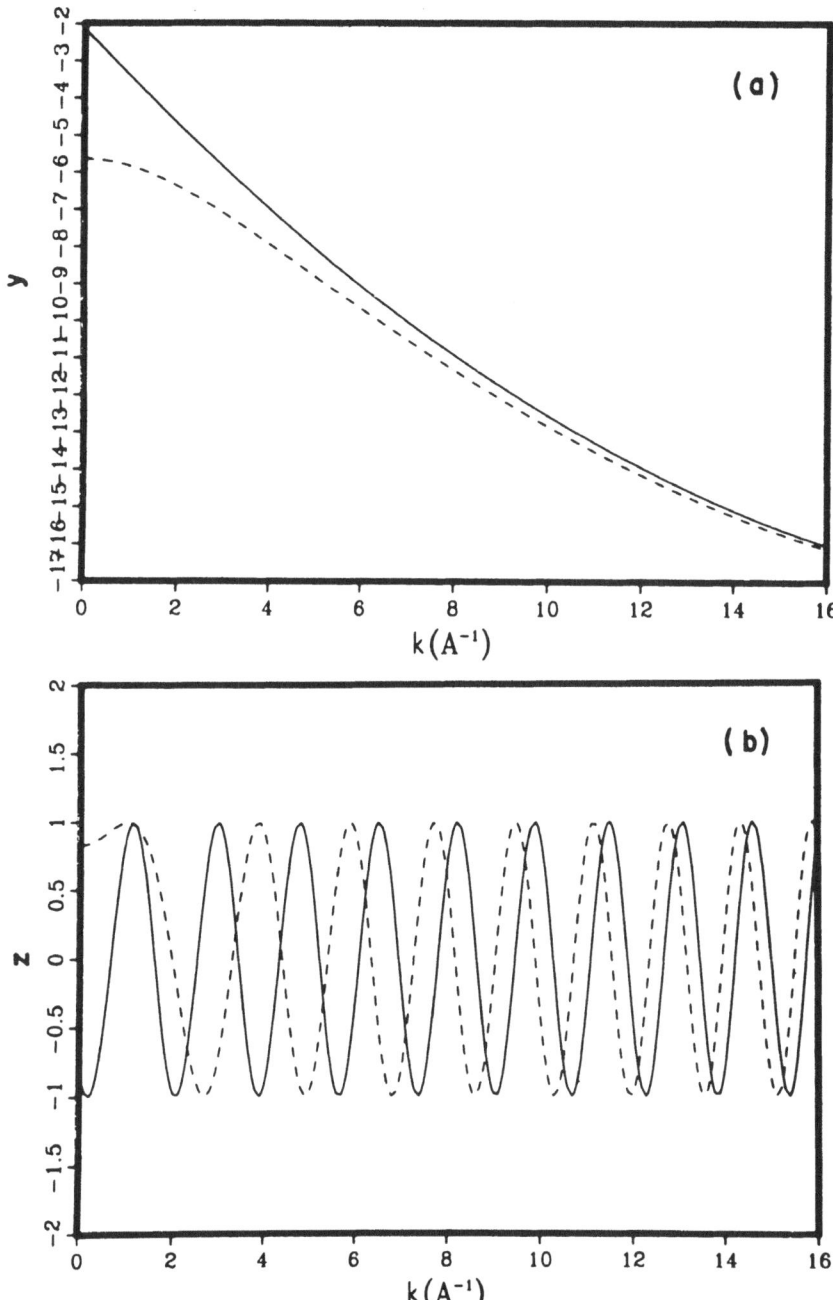

Fig. 3.4. (a) The effect of ΔE_0 on the phase functions: $y = \phi(k) = p_0 + p_1 k' + p_2 k'^2$; (b) The composite effect of ΔE_0 on the sine waves including the phase functions: $z = \sin(2k'r + p_0 + p_1 k' + p_2 k'^2)$, where $r = 2.3\text{Å}$. In both (a) and (b), $p_0 = -2.1525$, $p_1 = -1.3263$, $p_2 = 0.0287$, $k' = (k^2 - 0.2625(\Delta E_0))^{1/2}$ where $\Delta E_0 = 0$ (solid curves) and -30 eV (dashed curves).

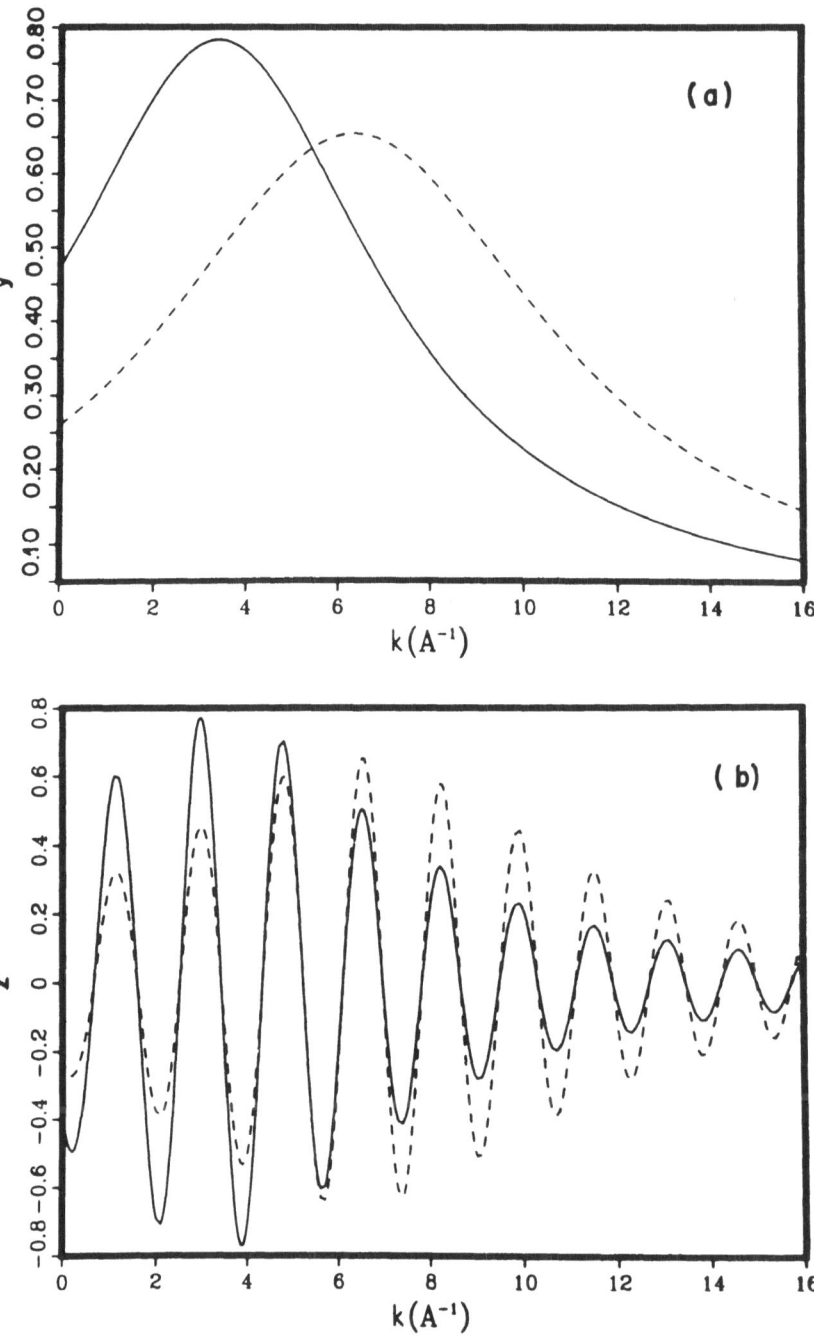

Fig. 3.5. (a) Two backscattering amplitude functions: $y = F(k) = A/(1 + B^2(k - C)^2)$; (b) $z = F(k)\sin(2kr + \phi(k))$ where $r = 2.3\text{Å}$ and $\phi(k)$ is given by the solid curve in Fig. 3.4(a). In both (a) and (b), $A = 0.783$ and 0.656Å, $B = 0.237$ and 0.194Å, and $C = 3.4$ and 6.4Å^{-1} for the solid and dashed curves, respectively.

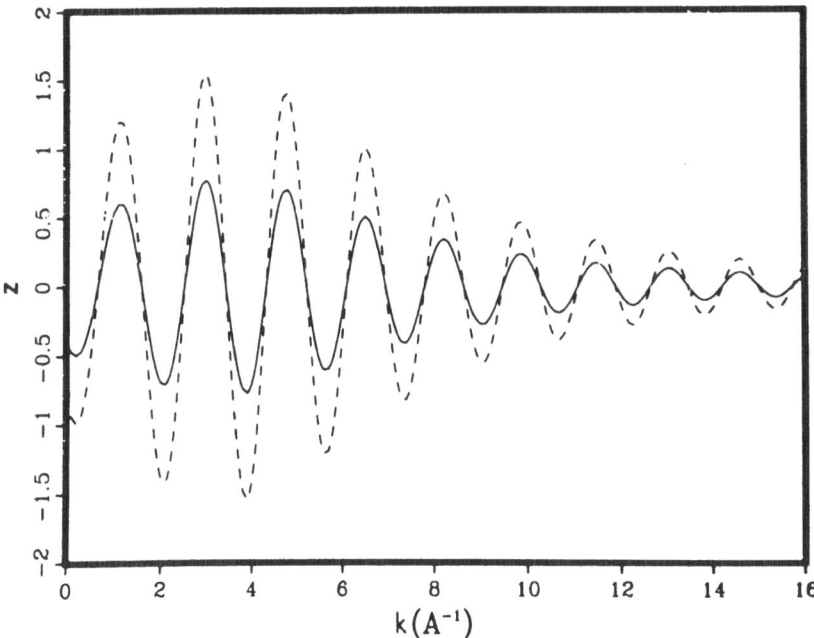

Fig. 3.6. The solid curve in Fig. 3.5(b) has been multiplied by N to give $z = N F(k)\sin(2kr + \phi(k))$ where $N = 1$ (solid curve) and 2 (dashed curve).

(dashed curve). In Fig. 3.8(b) the solid curve in Fig. 3.7(b) has been multiplied by the two functions in Fig. 3.8(a). Note that the amplitude attenuation is felt mostly at low k values, in a way opposite to that of σ. Also, the larger the distance r, the larger will be the attenuation.

In Fig. 3.9 the weighting scheme k^n is applied where $n = 2$ (solid curve) and 1 (dashed curve) in order to compensate for amplitude attenuation at high k. Higher n values put more emphasis on the high k region and vice versa.

Finally, in Fig. 3.10(a) we show the function $(N/r^2)F(k)k^2e^{-2\sigma^2k^2}e^{-2\eta r/k}\sin(2kr + \phi(k))$ where $r = 2.3\text{Å}$ (solid curve) and 3.3Å (dashed curve). The remaining parameters are: $N = 2$, $\sigma = 0.05\text{Å}$, $\eta = 0$, $\Delta E_0 = 0$ while the functions are

$$F(k) = \frac{A}{1 + B^2(k-C)^2}$$

where $A = 0.783\text{Å}$, $B = 0.237\text{Å}$, and $C = 3.4\text{Å}^{-1}$ and $\phi(k) = p_0 + p_1 k + p_2 k^2 + p_3/k^3$ where $p_0 = -2.1525$, $p_1 = -1.3263$,

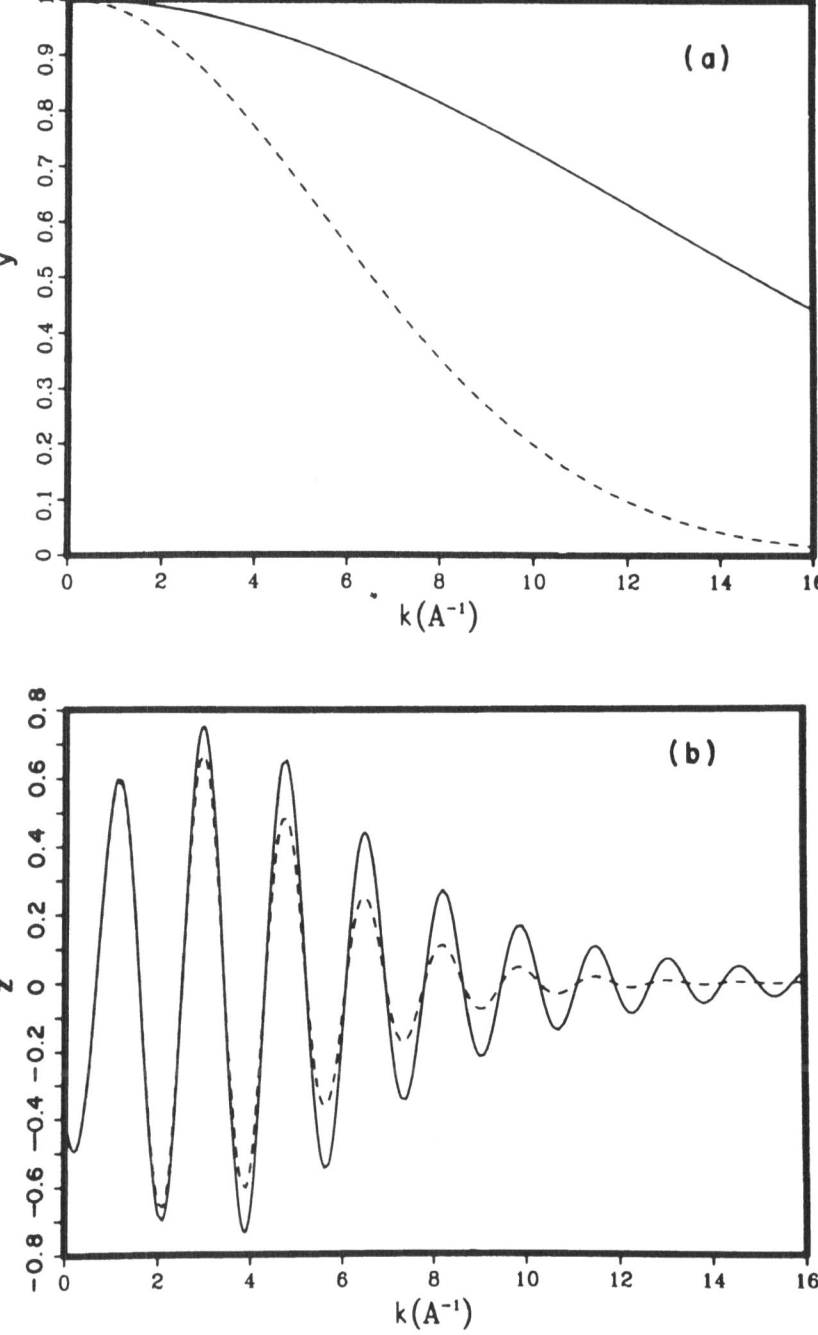

Fig. 3.7. (a) Two exponential damping functions $y = e^{-2\sigma^2 k^2}$ where σ is the Debye-Waller factor; (b) the solid curve in Fig. 3.6 has been multiplied by the Debye-Waller damping factor to give $z = N\,F(k)\,e^{-2\sigma^2 k^2}\sin(2kr + \phi(k))$. In both (a) and (b), $\sigma = 0.04\text{\AA}$ (solid curves), $0.09\text{\AA}$ (dashed curves).

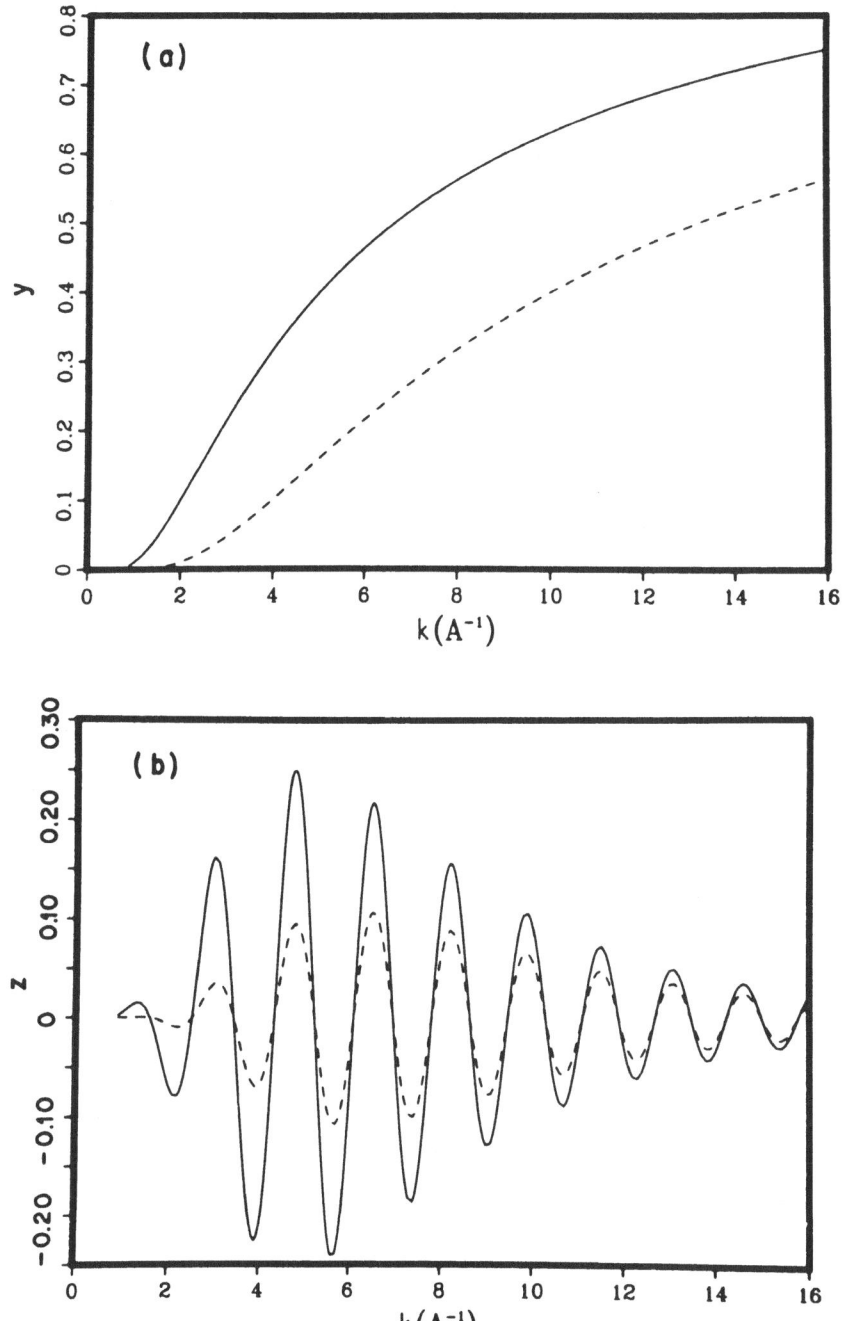

Fig. 3.8. (a) Two exponential damping factors $y = e^{-2\eta r/k}$; (b) the solid curve in Fig. 3.7(b) has been multiplied by y to give $z = N F(k) e^{-2\sigma^2 k^2} e^{-2\eta r/k} \sin(2kr + \phi(k))$. In both (a) and (b), $\eta = 1$ (solid curves) and 2 (dashed curves).

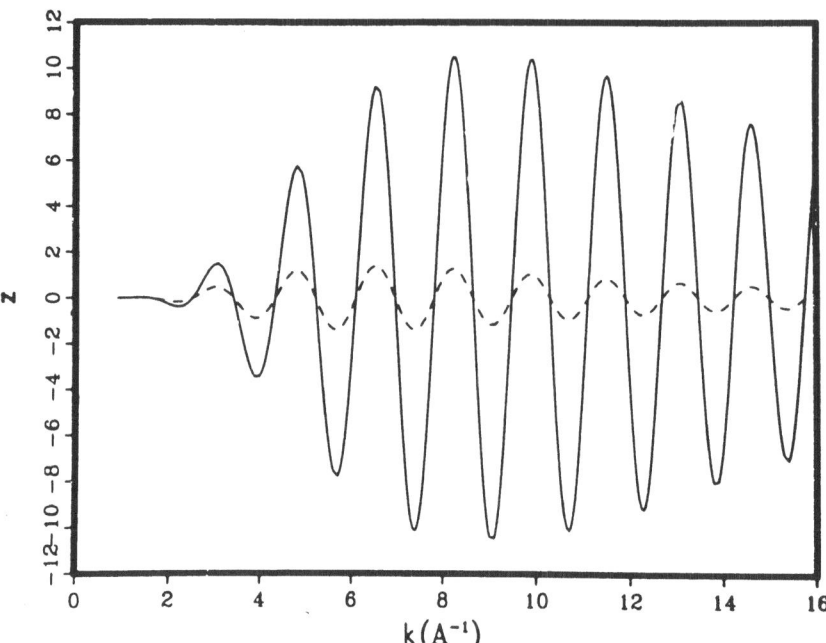

Fig. 3.9. The effect of weighting schemes $z = k^n N \, F(k) \, e^{-2\sigma^2 k^2} e^{-2\eta r/k} \sin(2kr + \phi(k))$ where $n = 2$ (solid curve), 1 (dashed curve). The remaining variables are the same as those used for the solid curve in Fig. 3.8(b).

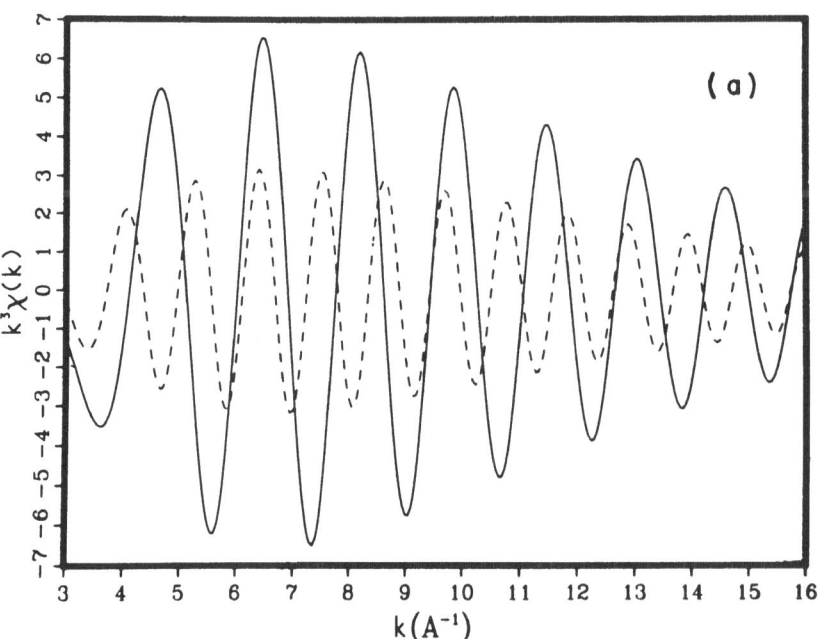

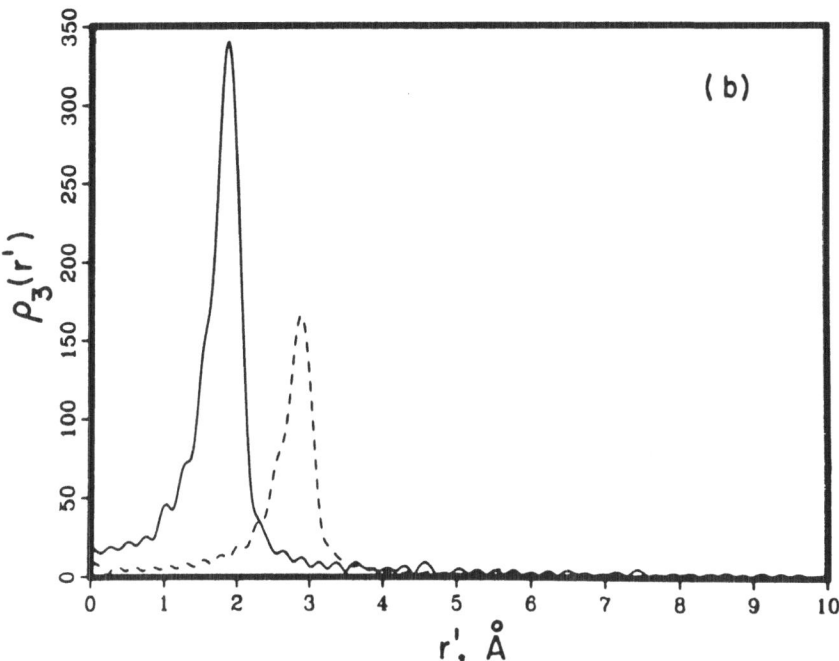

Fig. 3.10. (a) Two EXAFS functions differing only in distances $k^3\chi(k) = (N/r^2)F(k)k^2e^{-2\sigma^2k^2}e^{-2\eta r/k}\sin(2kr + \phi(k))$ where $r = 2.3$Å (solid curve) and 3.3Å (dashed curve). The remaining parameters are: $N = 2$, $\sigma = 0.05$Å, $\eta = 0$, $\Delta E_0 = 0$ while the functions are $F(k) = \dfrac{A}{1 + B^2(k-C)^2}$ where $A = 0.783$Å, $B = 0.237$Å, and $C = 3.4$Å^{-1} and $\phi(k) = p_0 + p_1k + p_2k^2 + p_3/k^3$ where $p_0 = -2.1525$, $p_1 = -1.3263$, $p_2 = 0.0287$, and $p_3 = 54.92$. (b) the corresponding Fourier transforms $\rho_3(r')$ vs r'.

$p_2 = 0.0287$, and $p_3 = 54.92$. The corresponding Fourier transforms are depicted in Fig. 3.10(b). It is apparent that, everything else being equal, increasing the distance r attenuates rapidly the EXAFS signal via the $1/r^2$ factor.

3.2 Effects of Important Parameters

It is clear from the foregoing discussion that the distance r and change in energy threshold, ΔE_0, will affect the frequency of the EXAFS sine wave, but in different ways. Similarly, while both the Debye-Waller factor σ and the electron mean free path λ tend to attenuate the EXAFS magnitude, they do so in different k regions. Finally, the coordination number N is a linear factor.

Figures 3.11-3.15 depict the effects of these important parameters on the EXAFS spectrum in both k and r space. All the parameters and functions for the solid curves in Fig. 3.11-3.15 are the same as those used for the corresponding curves in Fig. 3.10. The dashed curves in Fig. 3.11-3.15 illustrate the effect of changing a particular parameter, one at a time.

In Fig. 3.11, *increasing* the distance r from 2.3Å (solid curve) to 2.4Å (dashed curve) is seen to increase the frequency of EXAFS oscillation in k space (Fig. 3.11(a)) and to shift the peaks to a larger distance in r space (Fig. 3.11(b)). For small increase in distance, as shown in Fig. 3.11(a), the EXAFS waves appear to be shifted to the left (toward low k values); the magnitude of such shifts *increases* with increasing k. Note also that the Fourier transform magnitude is concommitantly attenuated (*cf.* Fig. 3.11(b)).

The effect of changing the energy threshold by $\Delta E_0 = -30$ eV is shown in Fig. 3.12(a) which is seen to shift the EXAFS waves in k space to the left (toward low k values). The magnitude of such shifts, however, *decreases* with increasing k. In r space, a negative ΔE_0 shifts the Fourier transform peak to a shorter distance (*cf.* Fig. 3.12(b)) and *vice versa*. As a rule of thumb, a change of 0.01Å in r corresponds to roughly a change of 3 eV in ΔE_0 (though the latter could vary from 2 to 4 eV depending on the system).

Figure 3.13 shows the effect of the number of neighboring atoms, N (from $N = 2$ for the solid curve to $N = 1$ for the dashed curve). In k space (Fig. 3.13(a)), it simply multiples the spectrum by the linear scale N for all k values while in r space (Fig. 3.13(b)), the peak magnitude is increased by the factor N.

In Fig. 3.14(a), we see that increasing the Debye-Waller factor σ from 0.05Å (solid curve) to 0.09Å (dashed curve) attenuates the EXAFS signal in k space exponentially. The effect increases exponentially as k increases. In r space (Fig. 3.14(b)), the peak broadens and diminishes in an exponential manner. Due to the limited k range used (normally 3 to 15Å^{-1}), the area under the Fourier transform peaks is not conserved.

In Fig. 3.15(a) we show the effect of decreasing the electron mean free path λ, assuming $\lambda \approx k/\eta$ where η goes from 0 in the solid curve to 1Å^{-2} in the

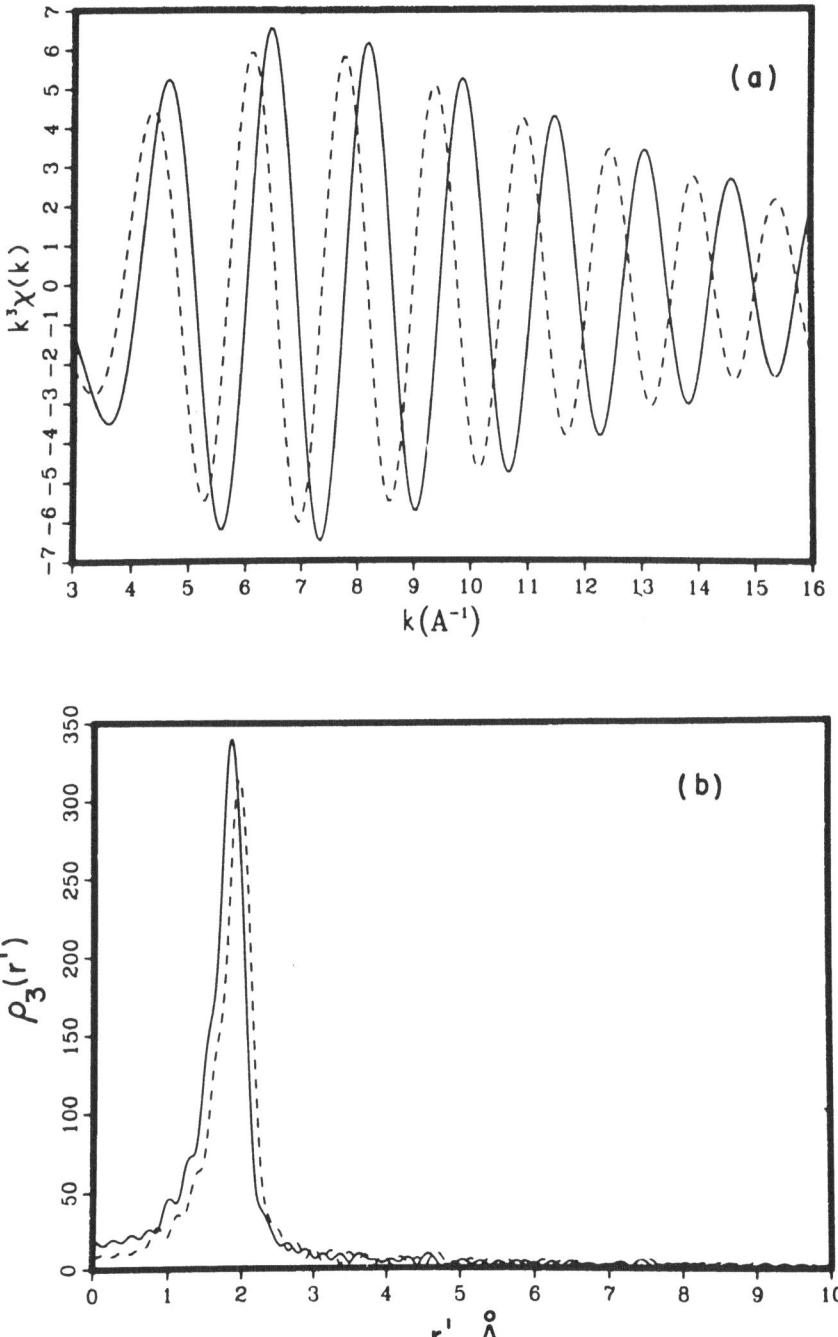

Fig. 3.11. The effect of the distance r on the EXAFS data in k (a) and r (b) space. In both (a) and (b), $r = 2.3\text{Å}$ (solid curve) and 2.4Å (dashed curves); the remaining variables are the same as those used for the solid curve in Fig. 3.10.

48

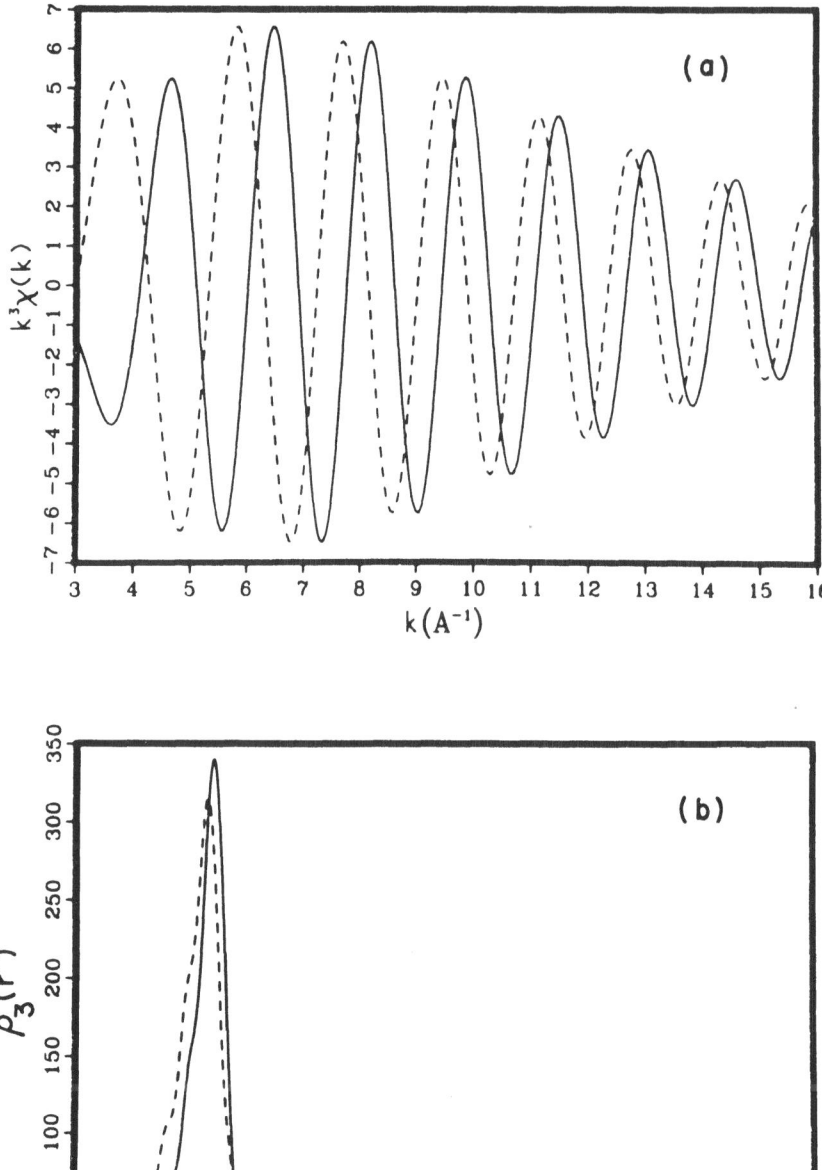

Fig. 3.12. The effect of ΔE_0 on the EXAFS data in k (a) and r (b) space. In both (a) and (b), $\Delta E_0 = 0$ (solid curve) and -30 eV (dashed curve); the remaining variables are the same as those used for the solid curve in Fig. 3.10.

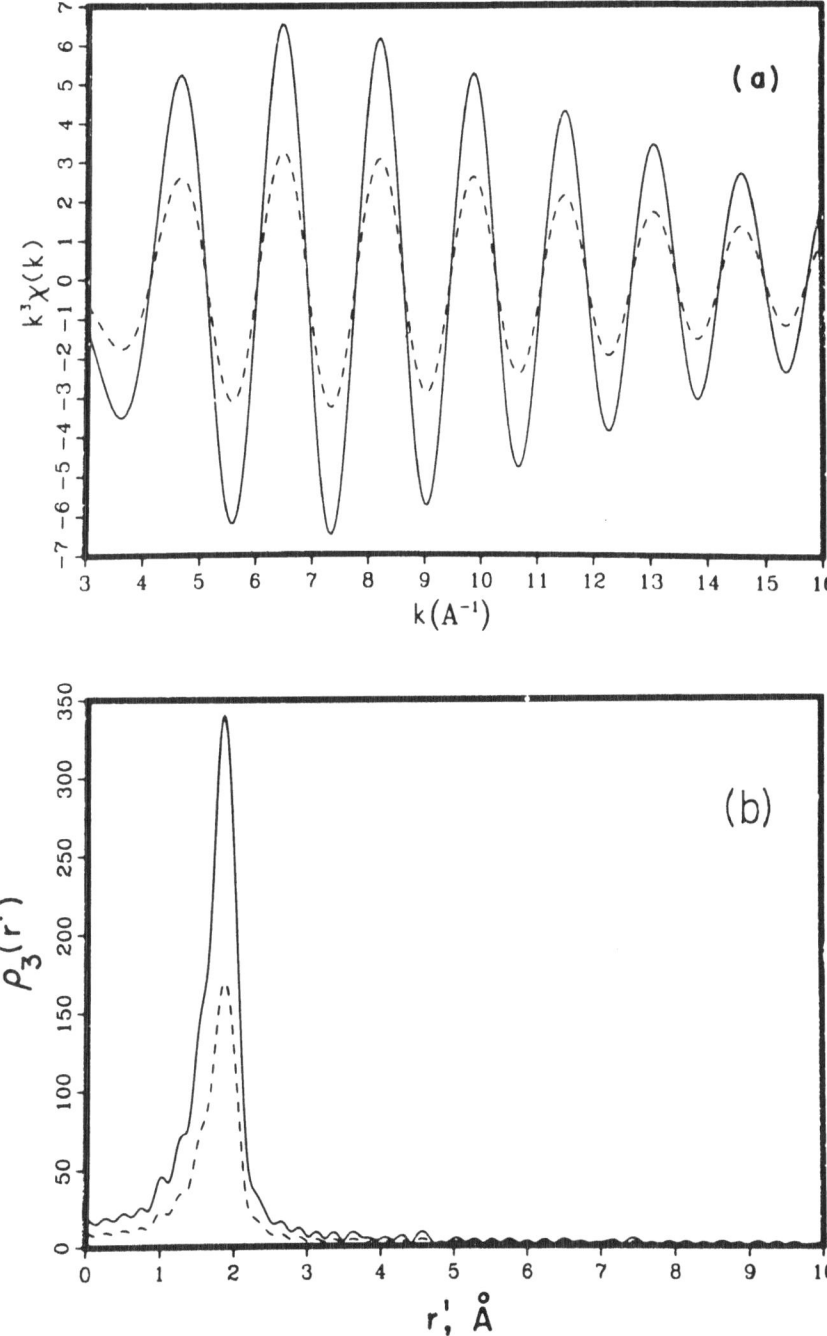

Fig. 3.13. The effect of the coordination number N on the EXAFS data in k (a) and r (b) space. In both (a) and (b), $N = 2$ (solid curve) and 1 (dashed curve); the remaining variables are the same as those used for the solid curve in Fig. 3.10.

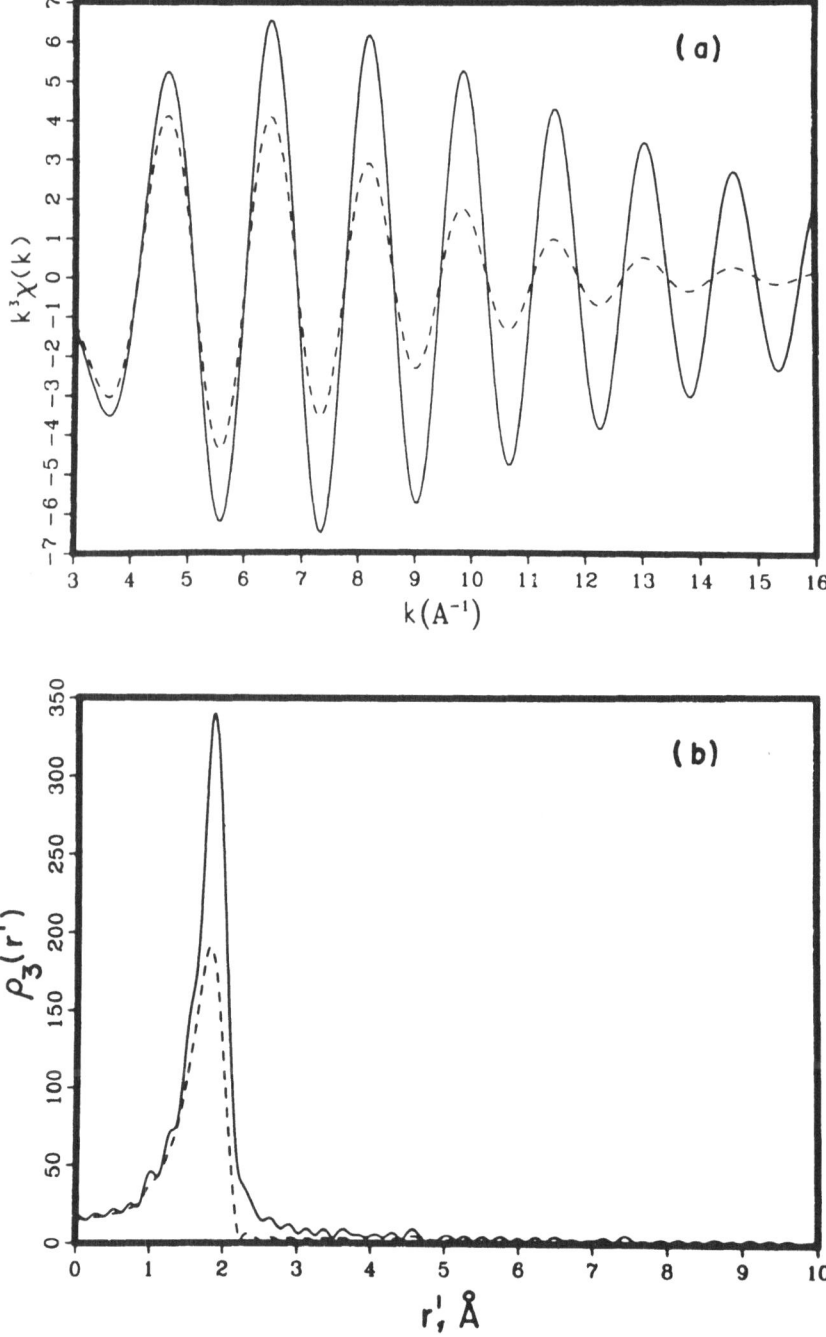

Fig. 3.14. The effect of the Debye-Waller factor σ on the EXAFS data in k (a) and r (b) space. In both (a) and (b), $\sigma = 0.05$ (solid curve) and 0.09 (dashed curve); the remaining variables are the same as those used for the solid curve in Fig. 3.10.

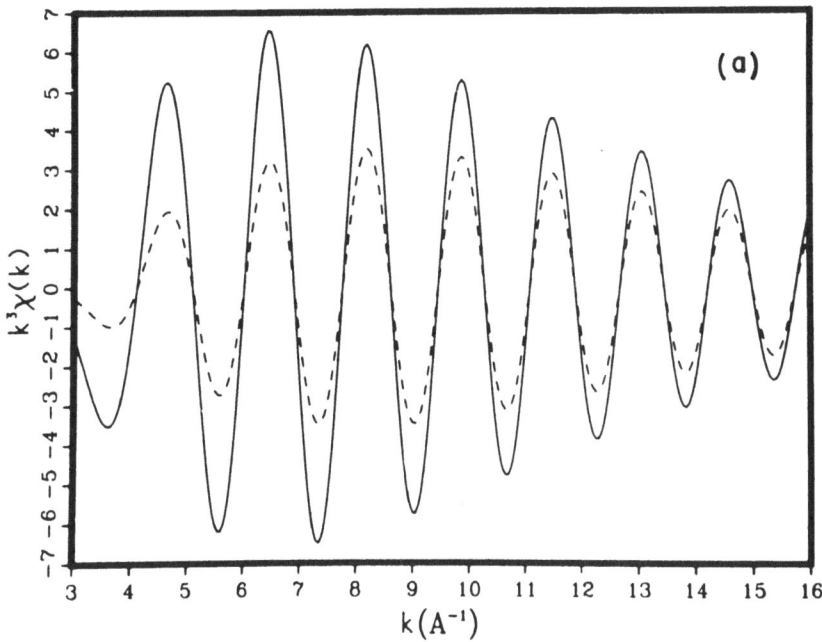

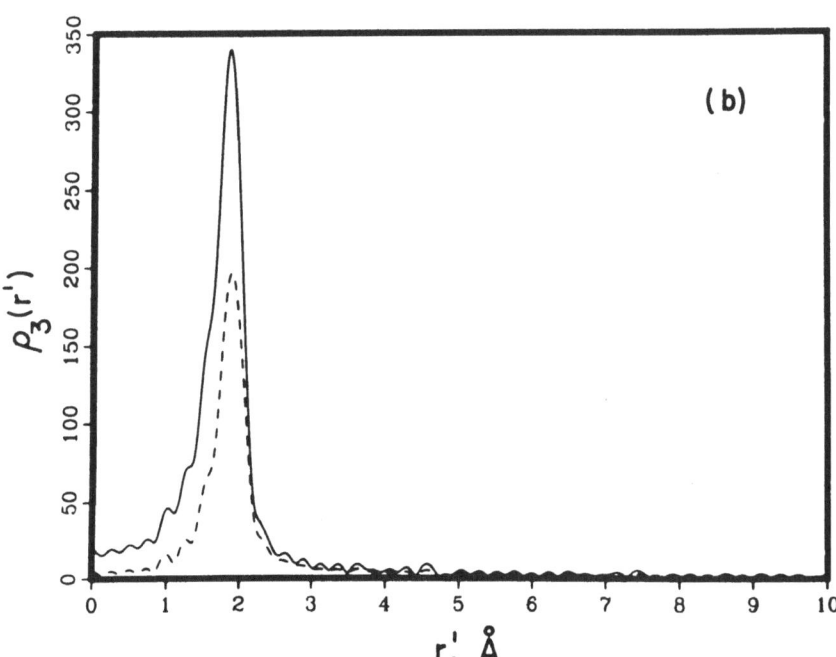

Fig. 3.15. The effect of η on the EXAFS data in k (a) and r (b) space. In both (a) and (b), $\eta = 0$ (solid curve) and 1Å^{-2} (dashed curve); the remaining variables are the same as those used for the solid curve in Fig. 3.10.

dashed curve. Decreasing λ, or equivalently, increasing η tends to attenuate the EXAFS waves in k space. The effect, however, decreases with increasing k. In r space, increasing η attenuates the peak magnitude exponentially (Fig. 3.15(b)). Once again, the area under the peak is not conserved.

3.3 Convention for Changing E_0

A few words of caution are warranted at this point concerning the sign of E_0 change. Changing the energy threshold from E_0 to E_0' causes the electron wave vector to change from k to k' where $\Delta E_0 = E_0' - E_0$. However, there are two ways of plotting any function $y = f(k)$. One can plot (a) the "new" function $y' = f(k')$ vs the "old" variable k via the transformation $k' = \sqrt{k^2 - 0.2625(\Delta E_0)}$ or (b) the "old" function $y = f(k)$ vs the "new" variable k' via the transformation $k = \sqrt{k'^2 + 0.2625(\Delta E_0)}$. Normally when the data is "reprocessed" by choosing a new energy threshold E_0', the data is plotted against the new variable k'. This corresponds to plotting the old function vs the new variable, viz., convention (b). In curve fitting where the experimental data remains the same, it makes more sense to plot the calculated (new) function vs the experimental (old) variable, viz., convention (a). As it turns out, the two methods involve a simple sign change: viz., an *increase* in E_0 within the context of convention (b) is equivalent to a *decrease* in ΔE_0 under convention (a). Throughout this book, convention (a) will be used except specifically stated otherwise.

Chapter 4

Theory of EXAFS

4.1 Introduction

Even though the basic physical explanation of EXAFS was provided by Kronig as being due to modification of the final state of the photoelectron by the crystal (Kronig, 1931) or, in the case of gaseous molecules, by atoms surrounding the excited atom (Kronig, 1932), a great deal of confusion still existed when this field was reviewed by Azaroff in 1963. The question arises as to whether a long-range order theory formulated in terms of Bloch waves (Kronig, 1931) or a short-range order theory in terms of scattering by neighboring atoms (Kronig, 1932, Hartree, et al. 1934, Shiraiwa, et al. 1958) is more appropriate. A major source of the confusion was that quantitative comparison between theory and experiment was difficult if not impossible at the time.

The utilization of synchrotron radiation sources in EXAFS data acquisition since 1975 greatly improved the situation. It is now well established that a single-electron single-scattering short-range theory is adequate under most circumstances. The oscillatory part of the absorption coefficient $\Delta\mu$ normalized to the structureless (atomic-like) background μ_o is given by

$$\chi(k) = \frac{\Delta\mu}{\mu_0}$$

$$= -\sum_j S_i(k) \, \frac{N_j}{kr_j^2} \, |f_j(\pi,k)| e^{-2\sigma_j^2 k^2} e^{-2r_j/\lambda_j(k)}$$

$$\times \sin[2kr_j + 2\delta_1'(k) + \theta_j(k)] \tag{4.1}$$

for excitations of an *s state* (for K or L_I edges) in a system in which the orientation of the sample has been *spherically averaged*. Eq. (4.1) describes the modification of the photoelectron wave function at the origin due to scattering by N_j neighbors, with a backscattering amplitude $|f_j(\pi, k)|$, located at a distance r_j away. The photoelectron wave vector k is related to photon energy E by

$$k = \sqrt{\frac{2m}{\hbar^2}(E - E_0)} \tag{4.2}$$

where E_0 is the threshold energy of the absorption edge and m is the electron mass. It is apparent that the photoelectron wave will be *phase shifted* by $2kr_j$ by the time it travels from the absorbing atom to the neighboring atom j and back. Furthermore, it also experiences two phase shifts: $2\delta_1'(k)$ due to the absorbing atom and $\theta_j(k)$ due to the jth neighboring atom. The factor of 2 in the former term (also called the central atom phase shift) is due to the fact that the photoelectron wave experiences the potential of the central atom twice, once going out and once coming back. The prime denotes that the central atom is photoexcited with a core hole. The smearing effect due to thermal vibration (assumed harmonic) or static disorder (assumed to have a symmetric or gaussian distribution) is accounted for by a Debye-Waller factor $e^{-2\sigma_j^2 k^2}$. Finally, the amplitude losses due to inelastic scattering are crudely approximated by (1) a damping factor $e^{-2r_j/\lambda_j(k)}$ depending upon the ratio of the distance traversed, $2r_j$, to the electron mean free path $\lambda_j(k)$; and (2) an amplitude reduction factor $S_i(k)$ caused by shake up/off processes at the central atom (denoted by i).

4.2 Derivations of EXAFS Theory

The basic EXAFS formula has been derived in the literature by several authors. Sayers, Stern, and Lytle (1971) first derived the EXAFS formula by assuming that the atoms are point scatterers. More formal derivations based on Green's function and generalization to muffin-tin scattering potentials can be found in Ashley and Doniach (1975), Lee and Pendry (1975), and Grosso and Parravicini (1980). Schaich (1973) has made a careful comparison of the short-range and long-range order theories based on Bloch waves in crystalline materials. He concluded that the two are formally identical provided inelastic damping effects are taken into account. It is now clear, as originally pointed out by Stern (1974), that since EXAFS is an interference effect involving the final-state wave function, it is the modulation in the matrix element that is important, and not the density of states.

The following derivations of the EXAFS formula are chosen solely for their simplicity. Other derivations are just as valid and can be found in the literature.

4.2.1 Lee and Pendry (1975)

The X-ray absorption rate μ can be described by the Golden Rule within the *dipole approximation* for the photon-induced transition of an electron from an initial state $|i>$ to a final state $|f>$.

$$\mu \propto |<f | \hat{\epsilon} \cdot \vec{r} | i>|^2 \tag{4.3}$$

where $\hat{\epsilon}$ is the polarization vector of the electric field and $\vec{r}$ is the electron coordinate. We shall first consider the initial state $|i>$ as being a $1s$ core state (e.g. K or L_I edge). The condition for the validity of the dipole approximation is that *the wavelength of the photons must be much greater than the size of the initial state*. Generally this criterion is well satisfied even for light atoms and less tightly bound states.

If we have an *isolated atom* the final state is given by

$$|f_0> = \begin{cases} \frac{1}{2} \left[\frac{ie^{ikr+i\delta_1'}}{kr} - \frac{ie^{-ikr-i\delta_1'}}{kr} \right] Y_{lm}(\Omega) \text{ for } r > r_0 \\ \\ \psi_0 \text{ for } r < r_0 \end{cases} \tag{4.4}$$

where r_0 is the atomic radius. For $r > r_0$ we have written down the asymptotic limit of the $l = 1$ scattering state in terms of a stationary spherical wave. The phase shift δ_1' is the $l = 1$ phase shift which arises due to the potential of the excited central atom (which has a core hole). Inside r_0 the wave function is determined by the condition that its logarithmic derivative is continuous with the scattering state outside r_0, and is denoted by ψ_0. The absorption rate is then given by

$$\mu_0 \propto |M|^2 \tag{4.5}$$

where

$$M = <\psi_0|\vec{r}|i> \tag{4.6}$$

is the dipole matrix element. In the presence of neighboring atoms, the final state is modified by the so-called *final-state interference effect*. The simplest

56

way of visualizing this modification is as follows. The outgoing spherical photoelectron wave originates from the central atom (assumed to be located at the origin) and propagates out toward the neighboring atoms as illustrated schematically in Fig. 4.1(a). Upon its arrival at an atom j located at $\vec{r}_j$ from the origin, it would have acquired a phase shift $(kr_j + \delta_1')$ where kr_j is due to the distance traveled and δ_1' arises from the Coulombic interaction (potential) of the central atom. If we assume that r_j is sufficiently larger than r_0, we can reasonably approximate the outgoing *spherical wave* (solid arcs in Fig. 4.1(a)) by a *plane wave* $e^{i\vec{k}\cdot\vec{r}}$ (dashed lines) in the vicinity of the neighboring atom j. This is the so-called *small-atom* or *plane-wave* approximation. Note that a plane wave does not have the $1/r$ dependence. The scattering of this plane wave $e^{i\vec{k}\cdot\vec{r}}$ by the neighboring atom j produces a spherical wave $f_j(\beta_j,k)e^{ikr_j^*}/r_j^*$ originating from atom j where $f_j(\beta_j,k)$ is the scattering amplitude with a scattering angle β_j at atom j. r_j^* is the distance away from atom j. This is shown schematically in Fig. 4.1(b). If we now assume that multiple scatterings are relatively unimportant and can be neglected, we need only consider the single scattering process called *backscattering* where $\beta_j = \pi$ since the photoelectron wave changes its direction by 180°. We shall call this *backscattered* photoelectron wave the incoming wave since it propagates toward the central atom. Since the *incoming wave* travels in a direction (*viz.*, $-\vec{r}_j$)

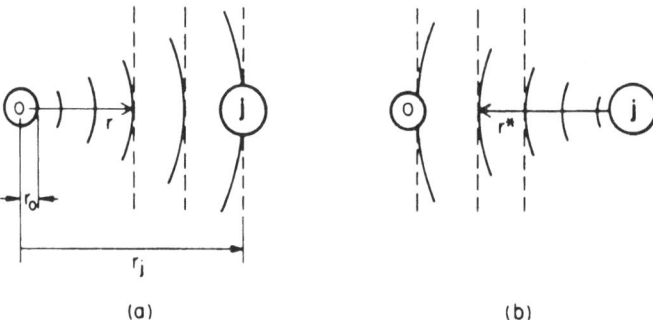

(a)   (b)

Fig. 4.1. Schematic representations of the backscattering process: (a) the *outgoing* spherical photoelectron wave (solid arcs) originating from the central atom (located at the origin) propagates out toward the neighboring atom j (located at $\vec{r}_j$). Assuming $r_j \gg r_0$, one can approximate the outgoing spherical wave by a plane wave (dashed lines) in the vicinity of atom j; (b) the scattering of this plane wave by atom j produces an *incoming* spherical wave (solid arcs) originating from the atom j and propagating toward the central atom.

opposite to that of the *outgoing* wave and originates from a point which is shifted from the origin by $\vec{r}_j$, we can write (*cf.* Fig. 4.1)

$$r_j^* = - \hat{r}_j \cdot (\vec{r} - \vec{r}_j) \tag{4.7}$$

where $\hat{r}_j$ is the unit vector of $\vec{r}_j$. The final state is then the sum of the *outgoing wave* and all the backscattered *incoming waves*:

$$|f> = |f_0> + \sum_j N_j \frac{ie^{ikr_j+i\delta_1'}}{2kr_j} f_j(\pi,k) \frac{e^{-ik\hat{r}_j \cdot (\vec{r}-\vec{r}_j)}}{|\hat{r}_j \cdot (\vec{r}-\vec{r}_j)|} \quad \text{for } r > r_0. \tag{4.8}$$

The sum is over all neighboring atoms with N_j being the number of atoms of the jth type. For the purpose of calculating the absorption rate we need only consider $|f>$ near the *origin* $(\vec{r} \approx 0)$. Again, for an interatomic distance much greater than the size of the atoms, we can approximate the incoming *spherical* wave at the origin by a *plane wave*. We can then expand this incoming plane wave in terms of spherical waves and then continue inside r_0. The net result is that near the origin, $|f>$ is proportional to ψ_0 with a certain complex coefficient

$$|f> = \left[1 + \sum_j N_j \frac{ie^{i2kr_j+i2\delta_1'}}{2kr_j^2} f_j(\pi,k)\right]\psi_0 \quad \text{for } r < r_0 \tag{4.9}$$

In words, the final state wave function is now modified by a term (in bracket) due to the interference between the outgoing wave and all the incoming waves (sum of j over all neighboring atoms). The photoelectron suffers a total phase shift of $(2kr_j + 2\delta_1')$ in the backscattering process simply because it travels the distance r_j twice and experiences the central atom phase shift δ_1' twice, once going out and once coming in.

The absorption rate is then

$$\mu \propto |<f|\vec{r}|i>|^2$$

$$\propto |M|^2\left[1 + \sum_j N_j \frac{ie^{i2kr_j+i2\delta_1'}}{2kr_j^2} f_j(\pi,k) + \text{complex conjugate}\right] \tag{4.10}$$

Here we assume that the backscattered photoelectron wave is sufficiently weak such that the square term due to the sum in the bracket can be neglected. The oscillatory part of the absorption, normalized to the "background" (taken as the absorption due to the "free atom") is then

$$\chi = \frac{\mu - \mu_0}{\mu_0} = - \sum_j \frac{N_j}{kr_j^2} \text{Im}[e^{i2kr_j+i2\delta_1'}f_j(\pi,k)] \tag{4.11}$$

where N_j is the number, r_j is the distance, and $f(\pi,k)$ is the backscattering amplitude of the neighboring atoms of the jth type and $2\delta_1'(k)$ is the central atom (absorber) phase shift. The appearance of the negative sign in Eq. 4.11 is due to the presence of the pre-exponential factor i in Eq. 4.10 (note that $(ie^{ix} - ie^{-ix})/2 = -\sin x = -\operatorname{Im} e^{ix})$.

4.2.2 Lee (1976)

The following alternative derivation of the EXAFS formula, also by Lee (1976), has the advantage of relating the EXAFS phenomenon to the angular resolved photoemission. In fact, it can be shown that the simple EXAFS expression is the result of intricate cancellations upon a 4π spherical averaging of the angular resolved photoemission expression and that the single scattering expression for EXAFS actually includes *second-order scattering processes* in *angular resolved photoemission.*

Consider a situation in which a neighboring atom B_j is located at $\vec{r}_j$ relative to the X-ray absorbing atom A (central atom) located at the origin. The probability for the absorber A to emit an electron in the direction $\vec{k}$ is given by

$$P(\hat{k}) = D\left|\hat{\epsilon}\cdot\hat{k} + \sum_j N_j \frac{f_j(\beta_j,k)}{r_j} e^{ikr_j(1-\cos\beta_j)}\hat{\epsilon}\cdot\hat{r}_j\right|^2 . \qquad (4.12)$$

In this equation, $\hat{\epsilon}$ is the polarization direction of the X-ray and D is a proportionality constant. As illustrated in Fig. 4.2, the second term is due to the *interference* of the original photoelectron wave emitted from the absorber atom A with the wave scattered by neighboring atom B_j through an angle β_j (the angle between $\hat{k}$ and $\hat{r}_j$) with scattering amplitude $f_j(\beta_j,k)$. This is the origin of EXAFS. The difference in the interference path length is simply $r_j(1 - \cos\beta_j)$. Now, if we integrate over all $\hat{k}$ we should get the total absorption coefficient. While this is not immediately apparent, it can be explicitly demonstrated as follows. The nontrivial term in Eq. 4.12 is the cross term (interference), which may be written as

$$D \sum_j N_j \int \frac{1}{4\pi} \left[\hat{\epsilon}\cdot\hat{k} \frac{f_j(\beta_j,k)}{r_j} e^{ikr_j(1-\cos\beta_j)} \hat{\epsilon}\cdot\hat{r}_j\right.$$

$$\left. + \text{ complex conjugate}\right]d\hat{k}$$

$$= D 2Re \sum_j N_j e^{ikr_j} \frac{\hat{\epsilon}\cdot\hat{r}_j l_j}{r_j} \qquad (4.13)$$

where

$$I_j = \int \frac{1}{4\pi} \hat{\epsilon} \cdot \hat{k} \, e^{-ikr_j \cos \beta_j} f_j(\beta_j, k) d\hat{k} \qquad (4.14)$$

We choose $\hat{r}_j$ to be the z-axis and use the identity

$$\hat{\epsilon} \cdot \hat{k} = \sum_m \left[\frac{4\pi}{3} \right] Y_{lm}^*(\hat{k} \cdot \hat{r}_j) Y_{lm}(\hat{\epsilon} \cdot \hat{r}_j) \; . \qquad (4.15)$$

The azimuthal angle integration is straightforward, leaving the $m = 0$ term alone:

$$I_j = \hat{\epsilon} \cdot \hat{r}_j \frac{1}{2} \int \cos \beta_j e^{-ikr_j \cos \beta_j} f_j(\beta_j, k) \, d \cos \beta_j \qquad (4.16)$$

This integration can be done using the standard expansions

$$e^{-ikr_j \cos \beta_j} = \sum_l (2l+1)(-i)^l j_l(kr_j) P_l(\cos \beta_j) \qquad (4.17)$$

and assuming plane wave scattering which allows the partial wave expansion:

$$f_j(\beta_j, k) = \sum_l f_l P_l(\cos \beta_j) \qquad (4.18)$$

where

$$f_l = \frac{1}{i2k} (2l+1)(e^{i2\delta_l} - 1). \qquad (4.19)$$

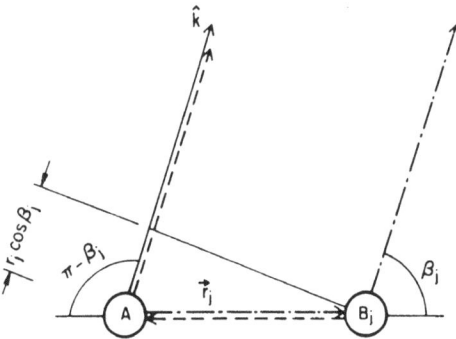

Fig. 4.2. Propagation of the photoelectron wave vector $\hat{k}$ (solid vector) originating from the X-ray absorbing atom A and its scattering by a neighboring atom B_j located at $\vec{r}_j$. The wave can be scattered at an angle β_j (dash-dot vectors) or backscattered at an angle π (dashed vector) by the neighboring atom. The interference of the original and the scattered waves gives rise to EXAFS. The difference in path length is $r_j - r_j \cos \beta_j$. The backscattered wave is responsible for the central atom (absorber) phase shift.

Then

$$I_j = \hat{\epsilon} \cdot \hat{r}_j \sum_l i \frac{\partial}{\partial(kr_j)} [(-i)^l j_l(kr_j)] f_l . \qquad (4.20)$$

where $j_l(kr_j)$ are Bessel functions. For large kr_j, we can use the asymptotic form

$$j_l(kr_j) \rightarrow \frac{1}{kr_j} \sin \left[kr_j - \frac{1}{2} l\pi \right] \qquad (4.21)$$

to obtain

$$I_j = \left[\frac{\hat{\epsilon} \cdot \hat{r}_j}{kr_j} \right] \frac{1}{2} i [f_j(\pi,k) e^{ikr_j} + f_j(0,k) e^{-ikr_j}]. \qquad (4.22)$$

The first term produces $f_j(\pi,k) e^{2ikr_j}$ in the EXAFS formula. Interestingly, the second term is exactly cancelled by the square of the second term in Eq. (4.12) using the optical theorem

$$\mathrm{Im}\, f_j(0,k) = \frac{k}{4\pi} \int |f_j(\beta_j,k)|^2 d\phi d \cos \beta_j , \qquad (4.23)$$

with the final result that

$$\chi(k) = \frac{\Delta\mu}{\mu_0} = \sum_j -3N_j \frac{(\hat{\epsilon} \cdot \hat{r}_j)^2}{kr_j^2} \mathrm{Im}[f_j(\pi,k) e^{i2kr_j}] . \qquad (4.24)$$

The factor of 3 arises from the fact that $\chi(k)$ is defined as the ratio of $\Delta\mu/\mu_0$ with the numerator and the denominator coming from the second and the first terms of Eq. 4.12. Integration of the square of the first term gives rise to 1/3. Note that the cancellation of the square term is important for the following reason. If we have more than one neighbor, the square term can in principle produce cross terms like $\exp(i2k(r_i - r_j))$, etc. If these terms exist the EXAFS spectra will be enormously complicated. Fortunately, all these cross terms exactly vanish as has been shown more generally using the Green's function technique (Ashley and Doniach, 1975, Lee and Pendry, 1975). One way of understanding this cancellation is to note that the terms in question are second order in the scattering amplitude and that there are corrections to the same order in the normalization of the final-state wave function.

The above derivation of Eq. 4.24 can be found in Massey (1969), and is equivalent to the formula originally written down by Kronig (1932) and used by Shiraiwa, *et al.* (1958). However, we now know that it is in error because the central-atom phase sift $2\delta'$ is missing. The importance of the central-atom phase shift was recognized by Kostarev (1941) and Kozlenkov (1961) and was incorporated in a point-scattering theory by Sayers, *et al.* (1971). In our present way of looking at the problem it is clear in the above derivation that the original wave and the scattered wave suffer the same phase shift δ_l, which is *cancelled* in calculating the interference path length. We can regain this term *only* by going to a *higher order* scattering process. Specifically, it is the process shown in Fig. 4.2, whereby the electron goes out to a neighboring atom at $\vec{r}_j$, is backscattered by the neighbor and is scattered again by the central atom in the direction $\hat{k}$ (dashed lines). This probability amplitude is given by

$$A_2(\hat{k}) = \sum_j N_j \ (\hat{\epsilon} \cdot \hat{r}_j) \ \frac{e^{i2kr_j}}{r_j^2} \ f_j(\pi,k) f'(\pi - \beta_j, k) , \qquad (4.25)$$

where f' is the scattering amplitude by the *central atom*. This term must be added to the amplitude in Eq. 4.12 and then squared. The leading term involves a product of $\hat{\epsilon} \cdot \hat{k}$ with A_2, which upon integration over all $\hat{k}$ directions projects out only the $l = 1$ component of the scattering amplitude f'. Using Eq. 4.19 we obtain the following additional contribution to the probability

$$P_2 = - \ \text{Im} \ \sum_j N_j \ \left\{ (e^{i2\delta_1'} - 1) f_j(\pi,k) \ \frac{e^{i2kr_j}}{kr_j^2} \right\} . \qquad (4.26)$$

This term combines with Eq. 4.24 to give us back exactly the central-atom phase shift, so that

$$\chi(k) = - \ \sum_j 3N_j \ \frac{(\hat{\epsilon} \cdot \hat{r}_j)^2}{kr_j^2} \ \text{Im}[e^{i2kr_j + i2\delta_1'} f_j(\pi,k)] . \qquad (4.27)$$

Next we consider the effect of atomic vibrations. Since the EXAFS phenomenon takes place in a time scale much shorter than the time scale for the motion of the atoms, it is like an instantaneous snapshot of the atomic configuration. However, EXAFS experiments generally require much longer time than molecular vibrations which means that we must perform an average over such configurations. As shown in Fig. 4.3, let $\vec{r}_j$ be the equilibrium position of a neighboring atom B_j relative to that of the central atom A which is taken as the origin. The *instantaneous displacements* of the atoms from the equilibrium positions are

$$\vec{u}_j = \vec{r}_j' - \vec{r}_j \qquad (4.28a)$$

$$\vec{u}_0 = \vec{r}_0' \qquad (4.28b)$$

where the prime denotes the instantaneous coordinates and the subscripts j and 0 denote the neighboring atom and the central atom, respectively. Hence, $\vec{r}_j'$ and $\vec{r}_0'$ are the instantaneous positions of the neighboring atom j and the central atom, respectively. The *instantaneous* position of atom j *relative* to that of the central atom is then $(\vec{r}_j' - \vec{r}_0')$. The e^{i2kr_j} factor in Eq. 4.27 must therefore be replaced by the average over a vibration cycle $<e^{i2k|\vec{r}_j' - \vec{r}_0'|}>$. In EXAFS only the component of the relative displacement which lies *along the equilibrium bond direction* needs to be considered:

$$<e^{i2k\hat{r}_j \cdot (\vec{r}_j' - \vec{r}_o')}> \qquad (4.29a)$$

$$= <e^{i2k\hat{r}_j \cdot (\vec{r}_j + \vec{u}_j - \vec{u}_o)}> \qquad (4.29b)$$

$$= e^{i2kr_j} <e^{i2k\hat{r}_j \cdot (\vec{u}_j - \vec{u}_o)}> \qquad (4.29c)$$

Here we make use of Eq. 4.28. Assuming harmonic vibration, Eq. 4.29c becomes

$$e^{i2kr_j} e^{-2k^2\sigma_j^2} \qquad (4.30)$$

where

$$\sigma_j = <[\hat{r}_j \cdot (\vec{u}_j - \vec{u}_o)]^2>^{1/2} \qquad (4.31)$$

is the Debye-Waller factor.

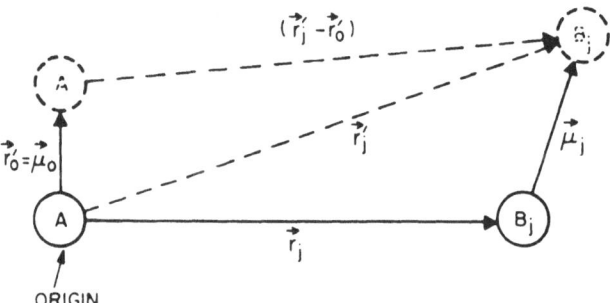

Fig. 4.3. Schematic representation of the equilibrium (solid circles) and instantaneous (broken circles) positions of the X-ray absorbing atom A and a neighboring atom B_j. $\vec{r}_j$ denotes the relative equilibrium position of B_j with respect to that of A which is taken as the origin. The instantaneous positions of atoms A and B_j are $\vec{r}_o'$ and $\vec{r}_j'$, respectively. The corresponding instantaneous displacements of atoms A and B_j from their equilibrium positions are $\vec{u}_o$ and $\vec{u}_j$, resulting in a relative instantaneous position of $\vec{r}_j' - \vec{r}_o' = \vec{r}_j + \vec{u}_j - \vec{u}_o$.

It should be noted that this Debye-Waller factor differs from the Debye-Waller factor for X-ray diffraction in that it is the root-mean-square average of the difference of displacements along the equilibrium bond direction and may in general be different from the root-mean-square displacement. Indeed

$$\sigma_j^2 = <(\hat{r}_j \cdot \vec{u}_j)^2> + <(\hat{r}_j \cdot \vec{u}_0)^2> - 2<(\hat{r}_j \cdot \vec{u}_0)(\hat{r}_j \cdot \vec{u}_j)> \qquad (4.32)$$

The last term is the correlation term and vanishes if the central atom and the scattering atom move independently. In covalently bonded systems the nearest neighbor is strongly bonded and such coherence effects are very important. Indeed, one generally finds that the first shell vibrational motions are significantly correlated. Such correlation diminishes for higher shells. More detailed treatments of vibrational effects in EXAFS can be found in the literature (see, for example, Beni and Platzman, 1976; Greegor and Lytle, 1979; Sevillano, Meuth, and Rehr, 1979; Rabe, *et al.*, 1979).

A static Debye-Waller factor identical in form to that of Eq. 4.30 (viz., $e^{-2\sigma^2 k^2}$) can also be obtained for a Gaussian distribution of distances about the mean value r

$$g(r') = \frac{1}{\sqrt{2\pi}\sigma} e^{-\frac{(r'-r)^2}{2\sigma^2}} \qquad (4.33)$$

As we shall see in the next chapter, two more terms must be added to Eq. 4.27 to account for inelastic scattering processes (losses) the photoelectron experiences as it travels from the X-ray absorbing atom to the neighboring atoms, and back. The first is caused by the so-called *shake-up* or *shake-off* processes due to multiple excitations of the "other" passive (or "bystander") electrons within the X-ray absorbing atom. This can be approximated by an amplitude reduction factor $S_i(k)$ which depends only on the type of central atom involved (designated by the subscript i). The second is the inelastic losses due to excitation of the neighboring environment, including the neighboring atoms and the intervening medium. We shall approximate these losses by an exponential damping term $e^{-2r_j/\lambda_j(k)}$ where $\lambda_j(k)$ is the inelastic electron mean free path.

If we now make use of the *plane wave* or *small-atom* approximation (in this approximation, it is assumed that the atomic sizes are very much smaller than the interatomic distances such that the spherical photoelectron waves can be approximated by plane waves).

$$f(\pi,k) - |f(\pi,k)|e^{i\theta(k)} \qquad (4.34a)$$

$$- F(k)e^{i\theta(k)} \qquad (4.34b)$$

where $F(k)$ and $\theta(k)$ are the backscattering amplitude and phase functions of the neighboring atoms, the EXAFS Eq. 4.27 becomes,

$$\chi(k) - - \sum_j S_i(k)N_j \frac{3(\hat{\epsilon}\cdot\hat{r}_j)^2}{kr_j^2} F_j(k)\mathrm{Im}[e^{i(2kr_j+2\delta_1'(k)+\theta_j(k))}]$$

$$\times e^{-2\sigma_j^2 k^2}e^{-2r_j/\lambda_j(k)} \qquad (4.35)$$

or

$$\chi(k) - - \sum_j S_i(k)N_j \frac{3\cos^2\Theta_j}{kr_j^2} \sin[2kr_j + 2\delta_1'(k) + \theta_j(k)]$$

$$\times F_j(k)\, e^{-2\sigma_j^2 k^2}\, e^{-2r_j/\lambda_j(k)} \qquad (4.36)$$

where Θ_j is the angle between $\hat{r}_j$ and the polarization direction $\hat{\epsilon}$. For polycrystalline samples, or if Eq. 4.36 is spherically averaged, $3\cos^2\Theta_j$ averages out to unity and we have

$$\chi(k) - - \sum_j S_i(k) \frac{N_j}{kr_j^2} \sin[2kr_j + 2\delta_1'(k) + \theta_j(k)]$$

$$\times F_j(k)e^{-2\sigma_j^2 k^2}e^{-2r_j/\lambda_j(k)} \qquad (4.37)$$

If we now define $\phi_i(k) - 2\delta_1'(k)$, $\phi_j(k) - \theta_j(k)$ and $\phi_{ij}(k) - \phi_i(k) + \phi_j(k) + \pi$ we obtain

$$\chi(k) - \sum_j S_i(k)N_jF_j(k)e^{-2\sigma_j^2 k^2}e^{-2r_j/\lambda_j(k)} \frac{\sin(2kr_j + \phi_{ij}(k))}{kr_j^2} \qquad (4.38)$$

Here i and j denote the central and the jth neighboring atoms, respectively. The additional factor of π in the phase shift is to take care of the overall minus sign in the EXAFS expression for K or L_I edges.

4.2.3 Boland, Crane, and Baldeschwieler (1982)

Recently, a very elegant derivation of single and multiple scattering formalisms of EXAFS has been given by Boland, Crane, and Baldeschwieler

(1982). We shall describe briefly their derivation for the single scattering theory and quote their results for the multiple scattering theory. The latter will be discussed in more detail in Chapter 8.

The X-ray absorption cross section in the one-electron and dipole approximations is given by:

$$\sigma = 4\pi^2\alpha\hbar\omega|\langle f|\hat{\epsilon}\cdot\vec{r}|i\rangle|^2 N(\omega), \tag{4.39}$$

where α is the hyperfine structure constant, ω is the photon frequency, and $N(\omega)$ is the density of final states of the photoelectron. The initial and final states of the system (i and f) are both eigenfunctions of an approximate unperturbed Hamiltonian H:

$$H = -\frac{\hbar^2}{2m}\nabla_r^2 - \frac{Ze^2}{r} + V, \tag{4.40}$$

where V is the total potential seen by the final-state photoelectron. V is represented as a sum of nonoverlapping, spherically symmetric, finite-range potentials centered around each atomic site in the system, including the absorbing atom.

Two factors influence the nature of the final state: the potentials of the neighboring atoms, and that of the central atom. For photoelectrons of sufficiently high energy (approximately three times the plasma frequency and higher), the attractive potential of the central atom's nucleus, together with the influence of the other bound electrons (though these are not considered explicitly here), becomes negligible, and the Schrödinger equation reduces to

$$(E - H^0)|f\pm\rangle = V|f\pm\rangle, \tag{4.41}$$

where H^0 is the free-particle Hamiltonian. This equation may be inverted to give the Lippmann-Schwinger equation

$$|f\pm\rangle = |k\rangle + G_0^{\pm}V|f\pm\rangle = |k\rangle + G_0^{\pm}T^{\pm}|k\rangle, \tag{4.42}$$

where $\langle r|k\rangle$ are the normalized eigenfunctions of H^0. We shall use the minus form of the free-particle Green's and T operators, so that $\langle r|k\rangle$ corresponds to the outgoing asymptote of the scattering process described by $\langle r|f\rangle$. The description of the EXAFS phenomenon is thus expressed in terms of the state of the photoelectron after the scattering processes. Furthermore, this choice of asymptote most clearly illustrates the relationship between EXAFS and the modulations observed in electron yield-type experiments.

The full T operator may now be expanded in terms of the operators t_j associated with the individual scattering centers at $\vec{r} - \vec{r}_j$

$$T = \sum_j t_j + \sum_{j \neq m} t_j G_0 t_m + \sum_{j \neq m \neq n} t_j G_0 t_m G_0 t_n + \ldots \qquad (4.43)$$

where $j \neq m \neq n$ run through all neighboring atoms. The first-order terms in the expansion correspond to single scatterings, second-order terms to double-scatterings, and so on.

For a three-atom ABC fragment shown in Fig. 4.4, the important scattering pathways are as follows: (a) represents the unperturbed (i.e. unscattered)

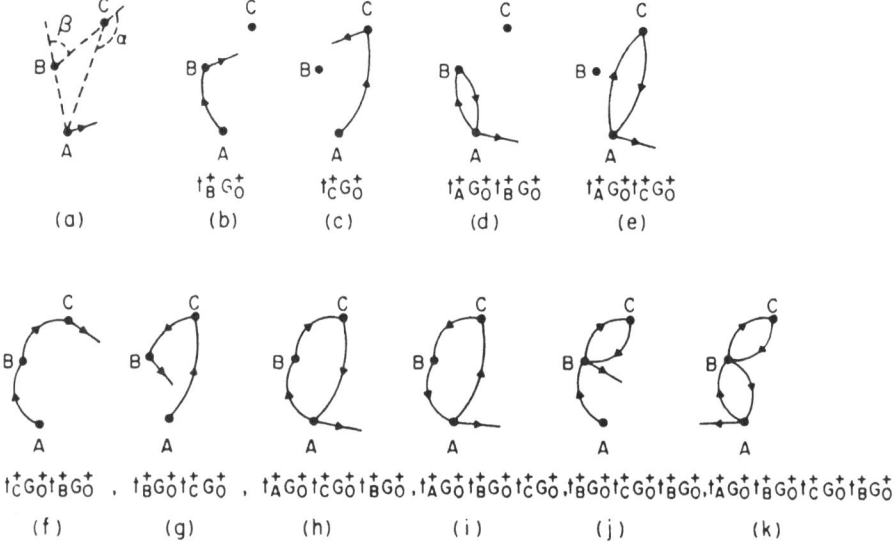

Fig. 4.4. Diagrammatic representation of the unperturbed photoelectron wave (a) and the various scattering processes for a three-atom ABC system where A is the absorbing (central) atom and B and C are two of the neighboring atoms: the single scattering processes (b and c); the double scattering processes involving the absorbing atom (d and e); the double scatterings at neighboring atoms B and C (f and g); the triple scatterings involving the absorbing atom A and neighboring atoms B and C (h and i); the triple scattering process $A \rightarrow B \rightarrow C \rightarrow B$ (j); and the quadruple scattering process $A \rightarrow B \rightarrow C \rightarrow B \rightarrow A$ (k). The operator shown with each diagram is the corresponding term in the expansion of the full T operator in the Lippmann-Schwinger equation (see text). (After Boland, Crane, and Baldeschwieler, 1982).

outgoing photoelectron; (b) and (c) represent the *single* scattering processes involving neighboring atoms B and C, respectively; (d) and (e) represent special types of *double* scattering processes involving neighboring atoms B and C, respectively, and the absorbing atom A. The scattering path lengths of (d) and (e) are identical with those of the single scattering processes (b) and (c), respectively, and hence must be considered in the single scattering theory. In fact it is the double scattering processes (d) and (e) that introduce the central atom phase shift.

The two single scattering terms of Eq. (4.43) may now be substituted into the matrix element in Eq. (4.39):

$$\langle f -|\hat{\epsilon}\cdot\bar{r}|i\rangle - \langle k|\hat{\epsilon}\cdot\bar{r}|i\rangle$$

$$+ \sum_j \langle k|t_j^+ G_0^+ \hat{\epsilon}\cdot\bar{r}|i\rangle$$

$$+ \sum_j \langle k|t_A^+ G_0^+ t_j^+ G_0^+ \hat{\epsilon}\cdot\bar{r}|i\rangle, \qquad (4.44)$$

where $j - B, C$.

The first matrix element on the right-hand side of Eq. 4.44 is responsible for the usual unperturbed photoelectric effect and is given by

$$\langle k|\hat{\epsilon}\cdot r|i\rangle - M(k,Z)\hat{k}\cdot\hat{\epsilon},$$

where $M(k,Z)$, the dipole matrix element, is related to the *atomic* absorption coefficient μ_0 by

$$\mu_0 \propto \frac{1}{3}|M(k,Z)|^2 \qquad (4.45)$$

The EXAFS effect may be viewed as arising from a difference in phase *at the origin* between the *unperturbed* photoelectron and one that has *scattered* off of a neighboring atom. The simplest description of such a phenomenon is one in which all matrix elements are expressed in terms of their effective values *at the origin*. Assuming that the core electrons of neighboring atoms are responsible for most of the scattering at the energies of interest in EXAFS, an approximate Green's function can be used to evaluate the first- and second-order matrix elements in Eq. 4.44. For elastic scatterings, these matrix elements may be expressed in terms of their respective scattering amplitudes $f(\beta_j, k)$ where β_j is the scattering angle. The net results are:

$$\sum_j \langle k | t_j^+ G_0^+ \hat{\epsilon}\cdot\vec{r} | i \rangle \sim \sum_j M(k,Z) \, \frac{1}{r_j} \, (\hat{\epsilon}\cdot\hat{r}_j) f_j(\beta_j,k)$$

$$\times \exp[ikr_j(1 - \cos\beta_j)], \qquad (4.46)$$

$$\sum_j \langle k | t_A^+ G_0^+ t_j^+ G_0^+ \hat{\epsilon}\cdot\vec{r} | i \rangle \sim \sum_j M(k,Z) \, \frac{1}{r_j^2} (\hat{\epsilon}\cdot\hat{r}_j) f_j(\pi,k)$$

$$\times f_A(\pi - \beta_j, k)\exp(2ikr_j) \qquad (4.47)$$

where $\cos(\pi - \beta_j) = \hat{k}\cdot\hat{k}_j' = -\hat{k}\cdot\hat{r}_j$.

The complete matrix element in Eq. 4.44 is the sum of three terms: the unperturbed photoelectric effect (Eq. 4.45); single scattering by the atom at $\vec{r}_j$ (Eq. 4.46); and double scattering by the absorber (Eq. 4.47). Thus,

$$\sigma(k) \propto |\langle f - |\hat{\epsilon}\cdot\vec{r}|i\rangle|^2 = |M(k,Z)\hat{k}\cdot\hat{\epsilon} + (4.46) + (4.47)|^2 . \qquad (4.48)$$

The development above treats the absorption of a single X-ray photon by an absorber-scatterer system. In an EXAFS experiment, however, a large number of such absorptions will occur, and the ejected photoelectrons will be scattered in many directions $\hat{k}$. In order to compute the average absorption cross section of such a macroscopic system, it is necessary to average over all such directions $\hat{k}$ in Eq. (4.48):

$$\int |\langle f - |\hat{\epsilon}\cdot\vec{r}|i\rangle|^2 \frac{d\Omega_k}{4\pi} \sim \int |M(k,Z)(\hat{k}\cdot\hat{\epsilon}) + (4.46) + (4.47)|^2 \frac{d\Omega_k}{4\pi} . \qquad (4.49)$$

The four lowest order terms in r_j in this spherical average are evaluated and given in the original literature and will not be repeated here. As in Lee's derivation (1976), the forward-scattering term $f_j(0,k)$ in one of the cross terms cancels with the second term by virtue of the optical theorem (cf. Eq. 4.23)

$$\text{Im}[f_j(0,k)] \sim \frac{k}{4\pi} \int |f_j(\beta_j,k)|^2 d\Omega_k . \qquad (4.50)$$

The macroscopic absorption coefficient $\mu = n\sigma$ is proportional to

$$\mu = n\sigma \propto \frac{1}{3}|M|^2 - \sum_j |M|^2(\hat{\epsilon}\cdot\hat{r}_j)^2 \, \frac{1}{kr_j^2} \, \text{Im}[e^{i(2kr_j + 2\delta_1')} f_j(\pi,k)] \qquad (4.51)$$

where n is the number of absorbing atoms. By convention, the oscillatory part of the absorption coefficient is normalized to $\mu_0 \propto \frac{1}{3}|M(k,Z)|^2$:

$$\chi(k) = \frac{\mu - \mu_0}{\mu_0} = -\sum_j 3N_j \frac{(\hat{\epsilon}\cdot\hat{r}_j)^2}{kr_j^2} \text{Im}[e^{i(2kr_j + 2\delta_1')}f_j(\pi,k)] \qquad (4.52)$$

or, equivalently,

$$\chi(k) = -\sum_j \frac{3N_j}{kr_j^2}(\hat{\epsilon}\cdot\hat{r}_j)^2|f_j(\pi,k)|\sin[2kr_j + 2\delta_1'(k) + \theta_j(\pi,k)], \qquad (4.53)$$

where $f_j(\pi,k) = |f_j(\pi,k)|e^{i\theta_j(\pi,k)}$ and N_j is the number of atoms of the jth type.

The above treatment can readily be extended to multiple scattering processes. Consider again the three-atom assembly shown in Fig. 4.4. The important scattering pathways involving the shortest path lengths and corresponding to the low order terms in Eq. 4.43 are as follows: (f) and (g) represent double scatterings at atoms B,C and C,B respectively; (h) and (i) represent triple scatterings involving neighboring atoms B and C and the absorbing atom A in either clockwise or counterclockwise directions, respectively; (j) represents the triple scattering pathway from A to B to C to B; and finally (k) represents the quadruple scattering pathway from A to B to C to B to A. With the trivial exception of the unperturbed wave, each pathway in Fig. 4.4 corresponds to a term in the expansion of the full T operator in the Lippman-Schwinger equation: (b) and (c) are first order terms, (d) through (g) are second order terms, (h) through (j) are third order terms, and (k) is a fourth order term. These scattering pathways adequately describe the EXAFS of a three-atom fragment. Other scattering pathways involving longer path lengths or larger scattering angles are relatively unimportant and can be ignored.

Using the same method utilized in the derivation of the single scattering theory, the following EXAFS expression is obtained by summing all the scattering pathways shown in Fig. 4.4.

$$\chi(k) = -\sum_{j=B,C} \frac{3(\hat{\epsilon}\cdot\hat{r}_j)^2}{kr_j^2}|f_j(\pi,k)|\sin[2kr_{Aj} + 2\delta_1'(k) + \theta_j(\pi,k)]$$

$$-\frac{6(\hat{\epsilon}\cdot\hat{r}_{AB})(\hat{\epsilon}\cdot\hat{r}_{AC})}{kr_{AB}r_{BC}r_{AC}}|f_B(\beta,k)||f_C(\gamma,k)|\sin[k(r_{AB} + r_{BC} + r_{AC})$$

$$+ 2\delta_1'(k) + \theta_B(\beta,k) + \theta_C(\gamma,k)]$$

$$- \frac{3(\hat{\epsilon}\cdot\hat{r}_{AB})^2}{kr_{AB}^2 r_{BC}^2} |f_B(\beta,k)|^2 |f_C(\pi,k)| \sin[2k(r_{AB} + r_{BC})$$

$$+ 2\delta_1'(k) + 2\theta_B(\beta,k) + \theta_C(\pi,k)] \tag{4.54}$$

In the case of absorption by polycrystalline samples, an average over all possible polarization directions must be computed, with the results:

$$\int (\hat{\epsilon}\cdot\hat{r}_j)^2 \frac{d\Omega_e}{4\pi} = \frac{1}{3}, \tag{4.55a}$$

$$\int (\hat{\epsilon}\cdot\hat{r}_{AB})(\hat{\epsilon}\cdot\hat{r}_{AC}) \frac{d\Omega_e}{4\pi} = \frac{\hat{r}_{AB}\cdot\hat{r}_{AC}}{3} \tag{4.55b}$$

(Note that Eq. 4.54 differs somewhat from the equation originally given by Boland, et al (1982), which neglects the dependence of scattering phases on the scattering angles, *cf.* Chapter 8.)

The first term in Eq. 4.54 is just the single scattering formula for pathways $A \rightarrow B \rightarrow A$ (see Fig. 4.5(a)) and $A \rightarrow C \rightarrow A$ (Fig. 4.5(b)). The second term is the multiple scattering pathway $A \rightarrow B \rightarrow C \rightarrow A$ (Fig. 4.5(c)). This term should be taken twice since it can also be $A \rightarrow C \rightarrow B \rightarrow A$. The third term is the multiple scattering pathway $A \rightarrow B \rightarrow C \rightarrow B \rightarrow A$ (Fig. 4.5(d)).

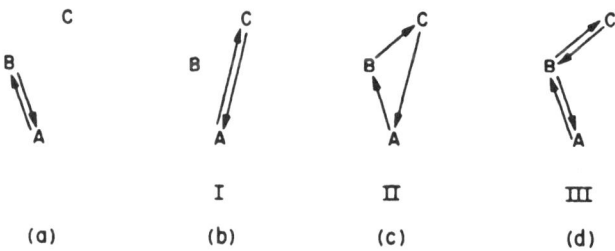

Fig. 4.5. Schematic representation of the terms in the EXAFS expression for a three-atom *ABC* system. Pathways a and b can be described adequately by the single scattering theory whereas pathways c and d must be described by the multiple scattering formalism. Pathways b, c, and d are three important scattering processes involving the more distant neighbor C and correspond to pathways I, II, and III described by Teo (1981).

4.2.4 Curve-Wave Theory

Thus far we have derived the single-scattering EXAFS theory by making the *plane-wave* approximation which assumes that the atomic radii are much smaller than the interatomic distance such that we can approximate the outgoing or the incoming *spherical* electron waves by *plane* waves. In other words, the plane-wave approximation assumes that the curvature of the electron wave at the scattering atom can be neglected (*cf.* Fig. 4.1), thereby greatly simplify the resulting EXAFS expression. This approximate formalism is sufficient for most EXAFS data analysis, especially if one utilizes only the high energy data ($E \geq 60$ eV or $k \gtrsim 4\text{Å}^{-1}$). In reality, however, the plane-wave approximation is only accurate at sufficiently large kr which means high k (or energy) and/or large r (distance). Hence the *plane-wave* approximation may also be called *high-energy* or *small-atom* approximation.

At lower energies the plane-wave approximation breaks down. Use of the low photoelectron energies leads to errors in the calculated phase of $\chi(k)$, which in turn can lead to erroneous determinations of the interatomic distances (r), and to a lesser extent in the amplutide of $\chi(k)$. This is particularly troublesome for systems in which low k data are important in the analysis. The obvious examples include systems with (1) large Debye-Waller factor or higher shells where the EXAFS amplitude in the high k region is exponentially attenuated due to disorder and/or loss mechanisms (2) multiple scatterings where low k data are crucially needed; (3) light atom scatterers which scatter strongly only the low-energy electrons. For these systems, it may be necessary to use the exact EXAFS theory or the so-called *spherical-wave* or *curved-wave* approach.

The exact curve-wave theory can be found in, e.g., Lee and Pendry (1975) and Gurman and Pettifer (1979). Unfortunately it is mathematically quite complicated and difficult to apply in data analysis. More recently, Gurman, Binsted, and Ross (1984) have devised a so-called rapid curved-wave theory which they claim can save up to forty times in computation time. They found, by performing the angle-averaging analytically (for unoriented or polycrystalline samples), major simplifications can be made in the theoretical form of the exact, curve-wave EXAFS expression. This approach may also be extended to higher-order scatterings.

With increasingly more interests in the understanding and the exploitation of the X-ray absorption near edge structure (XANES), the exact curve-wave theory will undoubtedly become more important and popular.

4.3 EXAFS of L Edges

The EXAFS formula described thus far is applicable to photoexcitations of an s state electron which include K and L_I edges. For the excitation of a p or any other states, the following general formula must be used.

$$\chi(k) = \sum_j \sum_{m_o LL''} S_i(k) N_j 4\pi < L_0 |\hat{\epsilon} \cdot \vec{r}| L'' > Y_{L''}^* (-\hat{r}_j) Y_L(\hat{r}_j) < L |\hat{\epsilon} \cdot \vec{r}| L_0 >$$

$$\times \sin(2kr_j + \sigma_L'(k) + \sigma_{L''}'(k) + \theta_j(k)) |f_j(\pi,k)| \frac{e^{-2\sigma_j^2 k^2} e^{-2r_j/\lambda_j(k)}}{kr_j^2}$$

$$\times \left[\sum_{m_o} \sum_L |< L_0 |\hat{\epsilon} \cdot \vec{r}| L >|^2 \right]^{-1} \tag{4.56}$$

This equation describes the excitations of a level $L_o = (l_o, m_o)$ to a state with angular momentum L which propagates in the direction $\vec{r}_j$ with probability amplitude $Y_L(\vec{r}_j)$. It is backscattered by the jth neighboring atom with amplitude $|f_j(\pi,k)| e^{i\theta_j(k)}$. The reflected wave has an angular component L'' as it propagates towards the central atom according to $Y_{L''}^*(-\vec{r}_j)$ and is connected back to L_o by the dipole matrix element. The central atom phase shifts δ_L' and $\delta_{L''}'$ describe the propagation of the outgoing and incoming wave in the central atom potential. The reflected wave is expanded in spherical harmonics about the origin with amplitude $Y_L''(-\vec{r}_j)$. Eq. 4.56 can be simplified using the identity

$$Y_{L''}(-\vec{r}) = (-1)^{l''} Y_{L''}(\vec{r}) = (-1)^{l_o+1} Y_{L''}(\vec{r}). \tag{4.57}$$

For $l_0 = 0$ this accounts for the negative sign in front of Eq. 4.37 for K or L_I edge.

EXAFS of L_{II} or L_{III} edges correspond to the excitation of a core level of p symmetry. The EXAFS formula is more complex owing to the fact that the initial p state can go to a final state of s or d symmetry (i.e., both s and d outgoing electron waves are allowed by the dipole selection rule). If we choose $\hat{\epsilon}$ to be the z-axis then $m' = m'' = m_o$. The matrix element $<L |\hat{\epsilon} \cdot \vec{r}| L_0>$ has an angular part which integrates to a simple number and a radial part which depends on the particular atomic level. [We should remark that in many cases spin-orbit coupling is important and the initial states are split according to $j = 3/2$ and $j = 1/2$. However, if we assume that the photoelectron spin is not changed by the scattering process with the atom it can be shown that summing over the j_z initial state for either $j = 3/2$ or $j = 1/2$ is equivalent to taking $l_0 = 1$ and summing over the initial m_o values.] Heald and Stern (1977) noted that $\chi(k)$ is by definition a second rank tensor and the angular dependence cannot be more complicated than $\cos^2\Theta$. By performing the m sum they obtain

$$\chi(k) = \sum_j S_i(k)\, N_j \frac{|f_j(\pi,k)|e^{-2\sigma_j^2 k^2}}{kr_j^2}\, e^{-2r_j/\Lambda_j(k)}$$

$$\times \left\{ \frac{1}{2}(1+3\,\cos^2\Theta_j)M_{21}^2\sin(2kr_j+2\delta_2'(k)+\theta_j(k)) \right.$$

$$+ \frac{1}{2}M_{01}^2\sin(2kr_j+2\delta_0'(k)+\theta_j(k))$$

$$\left. + M_{01}M_{21}(1-3\,\cos^2\Theta_j)\sin(2kr_j+\delta_0'(k)+\delta_2'(k)+\theta_j(k)) \right\}$$

$$\times \left[M_{21}^2 + \frac{1}{2}M_{01}^2 \right]^{-1} \tag{4.58}$$

Instead of the single term we have three terms in which the central atom phase ϕ_a is given by $2\delta_2'$, $2\delta_0'$, and $\delta_0' + \delta_2'$ where δ_l' is the phase shift of an outgoing wave with angular momentum l (Lee and Pendry, 1977; Heald and Stern, 1977). The matrix elements M_{21} and M_{01} are radial dipole matrix elements between the $2p$ ($l = 1$) atomic wave function and the $l = 2$ and $l = 0$ final states, respectively. The first and second terms in Eq. 4.58 correspond to transitions to $l = 2$ and $l = 0$ final states, respectively. The third term is a cross term originating from the possibility of having an outgoing $l = 2$ state and an incoming $l = 0$ state and vice versa. Unlike the K edge, the dipole matrix elements do not cancel and in theory three sets of phase shifts are involved. Fortunately some simplifications are possible.

For polycrystalline samples the cross term involving $\delta_0' + \delta_2'$ (i.e., the third term in braces in Eq. 4.58) vanishes by spherical averaging. We are still left with a complicated expression which requires two sets of central atom phase shifts and the ratio between the dipole matrix elements.

The matrix elements M_{21} and M_{01} have been calculated for various elements by Teo and Lee (1979). The ratio M_{21}/M_{01} is plotted in Figure 4.6 as a function of photoelectron momentum k for Ti, Zr, and W. We see that the ratio is of the order of 5 and is relatively independent of k. For light atoms, however, M_{21}/M_{01} shows some k dependence and can vary by as much as a factor of 2 as in, for example, chlorine. From Eq. 4.58 we see that transitions to the d final state are favored by a factor of 50 (i.e., $M_{21}^2/(1/2M_{01}^2) = 2(M_{21}/M_{01})^2 \approx 2 \times 5^2$). The smallness of the ratio M_{01}/M_{21} means that the $l = 0$ contribution is practically unobservable and thus for all practical purposes M_{01} can be ignored and the $L_{II,III}$ edge EXAFS can be analyzed in the same way as K and L_I edges with the use of the $l = 2$ phase shift $\delta_2'(k)$ in place of $\delta_1'(k)$ and the removal of the overall minus sign in Eq. 4.37 or, equivalently, the removal of the factor π in Eq. 4.38.

It is interesting to note that the calculated ratio (Fig. 4.6) is in excellent agreement with the value $M_{01}/M_{21} \approx 0.2 \pm 0.06$ obtained by Heald and Stern (1977) from the angular dependence of the tungsten L_{III} edge spectrum. The physical reason that M_{01} is smaller is that the $l = 0$ final state must be orthogonal to the $1s$ core state and therefore is much more rapidly oscillatory

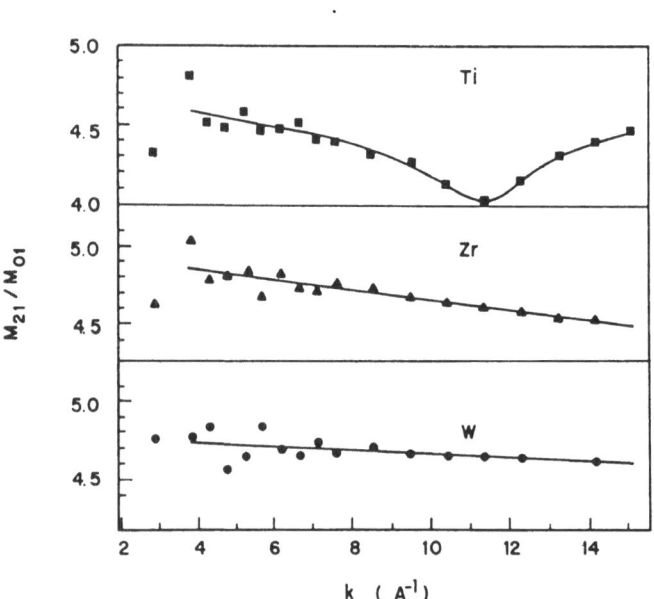

Fig. 4.6. The ratio of the radial dipole matrix elements M_{21}/M_{01} (where M_{21} and M_{01} correspond to the $l = 1$ initial atomic state and the $l = 2$ and $l = 0$ final states, respectively) as a function of electron wave vector k for Ti, Zr, and W. (Reproduced from Teo and Lee, 1979).

than the $l = 2$ final state in the region of the $2p$ wave function. It should also be mentioned that there is a considerable amount of literature dealing with M_{21}/M_{01} for light elements or outer shells (see, for example, Kennedy and Manson, 1972; Codling, *et al.*, 1976).

4.4 The Photoelectron and the Excited Atom

The photoelectric effect takes place in a time scale on the order of 10^{-15} sec. Immediately after the ejection of the photoelectron, the excited atom starts to relax. Since the photoelectron must travel to the neighboring atoms and back, the potential experienced by the photoelectron due to the relaxing excited central atom is therefore a dynamic (time-dependent) one and in theory can be very complex. We shall now consider, in qualitative terms, various events and their time scales which follow the photoelectric effect pertinent to EXAFS.

4.4.1 Lifetime of the Core Hole

The ejection of a photoelectron from a core level creates a *core hole* or equivalently, an inner-shell vacancy, in the excited atom. The excited atom with a core hole has a finite lifetime before it relaxes via various relaxation mechanisms.

As already discussed in Chapter 1, the channel(s) for relaxation of light elements (low atomic number Z) are mainly nonradiative, producing Auger electrons, whereas those of heavy elements are mostly radiative, resulting in the emission of fluorescence photons. The photoabsorption cross sections can therefore be obtained by convoluting the core hole production cross sections with a Lorentzian line shape function with a total width given by the sum of the partial rates for the various relaxation mechanisms. For hydrogen-like wave functions, the radiative line widths are proportional to Z^4 which makes fluorescence the dominating relaxation channel for high Z elements.

In EXAFS, the lifetime of such a core hole must be longer than the transit time for the photoelectron to travel from the central atom to the neighboring atom(s) and back. In other words, the core hole must last long enough for the backscattered electron to interfere with the direct beam. In theory, the finite lifetime of the core hole will smear the EXAFS signal (by convoluting the EXAFS formula with a Lorentzian of width Γ where Γ is the hole-state width). In practice, however, this has not been a serious problem since the hole-state width Γ is on the order of eV's for most currently accessible edges as shown in

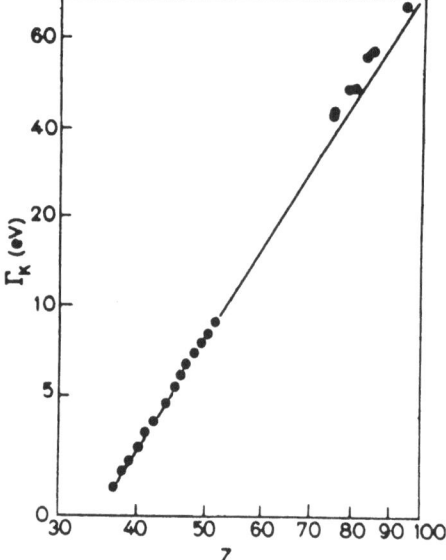

Fig. 4.7. K level width Γ_K as a function of atomic number Z (After Agarwal, 1979).

Fig. 4.7. For example, the full core hole widths for P, Fe, Ge, and Mo are 0.5, 1.15, 1.83, and 4.29 eV, respectively. These values are much smaller than the energy spacing ΔE between EXAFS waves (≥ 40 eV). Since ΔE increases with the photoelectron energy $E - E_o$, the criterion $\Gamma \ll \Delta E$ is easily satisfied and the smearing effect decreases with increasing energy. On the other hand, as shown in Fig. 4.7, Γ_K increases roughly as Z^4; the core hole lifetime broadening problem becomes more serious for deep core levels such as the K edge of platinum and gold where $\Gamma \sim 40$ eV may approach or exceed ΔE.

For a core-state width of Γ, the lifetime t is given by

$$\Gamma \times t \approx \hbar \tag{4.59}$$

For $\Gamma = 1$ eV, $t \approx 7 \times 10^{-16}$ sec. That is, the core hole will take 7×10^{-16} sec to decay. This lifetime is to be compared with the transit time τ for a photoelectron with energy $E - E_o$ to travel a distance of $2r$ where r is the distance between the atoms:

$$\tau = \frac{2r}{v} = \frac{2mr}{\hbar k} \tag{4.60}$$

Here v is the velocity and $k = \sqrt{\dfrac{2m}{\hbar^2}(E-E_o)}$ is the wavevector of the electron. For $r \approx 3\text{Å}$ and $E-E_o \approx 60$ eV, $\tau \approx 4 \times 10^{-17}$ sec. For larger $E-E_o$ or larger k, τ is even smaller. Thus, the criterion $t > \tau$ is easily satisfied.

Ekardt and Tran Thoai (1981) examined the life-time effects of both the excited electron and the core hole left behind. They found that the exponential damping term $e^{-2r/\lambda}$ where λ is the inelastic electron mean free path should be replaced by $e^{-t(2r-\Delta)-g}$ where t includes the life-time of the core hole, Δ is the derivative of the phase shifts involved and e^{-g} is the no-loss probability with respect to various inelastic channels including wave function relaxation effects.

4.4.2 Core Hole Relaxation

When an X-ray photon is absorbed by an atom, a *core hole* is created to which the other electrons in the atom will respond in the *relaxation* process.

The potential experienced by the photoelectron is therefore a time-dependent one, since as the photoelectron is ejected from the central atom, the wave function evolves from a completely unrelaxed state immediately after the photoexcitation (the sudden approximation) to the completely relaxed one some time later. The relaxation rate for each atomic level is roughly given by its binding energy. One must therefore compare these rates with the transit time for the photoelectron to travel from the central atom to the neighboring atoms and back. Or equivalently, one can estimate the binding energy E which corresponds to this transit time (*cf* Eqs. 4.59 and 4.60)

$$ E \approx \frac{\hbar}{\tau} = \frac{\hbar^2 k}{2mr} \tag{4.61} $$

For practical values of $r \sim 4\text{Å}$ and $k \sim 16\text{Å}^{-1}$, one obtains $E \sim 30$ eV. This implies that all the atomic levels with binding energies greater than this value, which includes practically all the core electrons, will have enough time to relax. The relaxation of the valence electrons and the screening by conduction electrons in metals are more complicated problems. However, it can be argued that the errors due to these outer electrons are on the order of chemical bonding (several eV), which will affect only phase shifts in the low k region and can be partially compensated for by varying E_o.

The relaxation of the conduction electrons after absorption of an X-ray photon in a simple metal has been considered by Noguera and Spanjaard (1981) using a dynamic screening theory.

4.4.3 Potential Experienced by Photoelectron

From the previous section, it is obvious that the photoelectron sees practically a completely relaxed central atom with an electron configuration $1s\,2s^2\,2p^6\cdots$ for K-edge, $1s^2\,2s^1\,2p^6\cdots$ for L_I-edge, and $1s^2\,2s^2 2p^5$ for L_{II}, L_{III} edges. Since these core levels are tightly bound, the photoelectron as well as the outer electrons experience a potential very similar to that of the $Z+1$ ion with one electron missing from the valence shell.

It is clear from this discussion that the central atom potential can be calculated from a relaxed ion. One approximation is to use, for the central atom phase shift of atomic number Z, the $Z+1$ ion with an outer electron missing. Another possibility is to solve the self-consistent field potential for the central atom with a core-hole (i.e. with one electron missing from the $1s$, $2s$ or $2p$ orbital). Both of these approximations have been used in the calculation of central atom phase shifts as we shall see in Chapter 7. In theory, one should use a screened $Z+1$ atom (completely relaxed case) at the low kinetic energy limit and an unscreened Z ion (completely unrelaxed) at the high kinetic energy extreme. In practice, however, for the range of $4 \lesssim k \lesssim 16\text{Å}^{-1}$, the photoelectron basically sees a completely relaxed central atom with a core hole.

4.4.4 Multi-electron Excitations

Evidence of *one-photon two-electron* excitations was observed by Salem, Dev, Lee (1980) and Salem, Kuman, Schiessel, Lee (1982) as well as by Madden, Codling (1963), Bonnelle, Wuilleumier (1963), Wuilleumier (1966, 1970). These authors found discontinuities corresponding to simultaneous creation of two electronic vacancies at photon energies approximately equal to the energy required for the sum of two independent excitations.

Two-electron one-photon transitions (two electrons filling a double vacancy (e.g. two core holes in the K shell) have been observed by Wolfli, *et al* (1975, 1976), Briand, *et al* (1976), and Stoller, *et al* (1976), and theoretically examined by Khristenko (1976) and Gavrila and Hensels (1978). For review on multi-electron transitions, see Åberg (1975).

Chapter 5

Improvement of EXAFS

Theory

The single-electron single-scattering theory of EXAFS described in the previous chapter forms the basis of data analysis in EXAFS spectroscopy. In many cases, all it requires is *parameterization* of the relevant variables or functions. These variables or functions can then be determined from model compounds or from theoretical calculations. For some specialized applications or in some particular systems, however, this phenomenological theory is inadequate and *modification* or *generalization* of the theory is required. In this chapter, possible breakdowns of the single scattering theory of EXAFS will be discussed:

1. *Phase* transferability problem caused by uncertainty in *energy threshold;*

2. *Amplitude* transferability problem caused by *inelastic scattering* processes at (a) the central atom and (b) the neighboring environment;

3. *Smearing* effects caused by *static* and *thermal* disorders;

4. *Multiple scattering* effects caused by *intervening atoms.*

We shall discuss the physical significance of these effects with emphasis on the possible problems in data analysis along with proven or suggested remedies. It should be emphasized that some of these problems are minor which require only slight modifications such as the addition of new parameters or functions while others are more serious which may require a generalization of the EXAFS formulation or further developments in EXAFS theory and experimentation.

5.1 Energy Threshold — The Phase Problem.

5.1.1 Choosing E_0

Since EXAFS $\chi(k)$ is a function of the photoelectron wave vector k, the first step in data analysis is to convert the photon energy E to k using the following equation

$$k = \sqrt{\frac{2m}{\hbar^2} (E - E_0)} . \tag{5.1}$$

This relation requires the knowledge of the energy threshold, E_0, which is the minimum energy required to free the electron. E_0 is generally believed to be in the vicinity (say, *ca.* 30 eV) of the edge. Unfortunately, there is no simple way to determine E_0 from the observed experimental spectrum; nor is there any universally characteristic or identifiable feature in the edge spectrum which will allow unequivocal determination of E_0. Though in theory E_0 can be determined by elaborate first-principle calculation, it has been done only for small and simple systems. Even then the accuracy is questionable.

Most EXAFS practitioners choose to use certain features near the edge as a mean to identify E_0 in a consistent manner. This is exemplified schematically in Fig. 5.1 where the onset of the edge (a), the inflection point (b), the "edge position" (c) defined as the energy at half the edge jump, the first absorption peak (d), the "start" of the oscillation (e), or some value based on theoretical or other experimental evidence (f) may be used. While these procedures or prescriptions can be used to provide a first approximation to E_0, they are somewhat arbitrary. Furthermore, E_0 is known to be affected by chemical bonding effects such as the charge distribution within the cluster, covalency or ionicity of the bonds, as well as central atom effects such as many-body excitations, core relaxations, etc.

5.1.2 Phase Transferability

Fortunately it turns out that it is not necessary to know the absolute value of E_0. If there is a chemically similar model compound of known structure, one can arbitrarily choose a reasonable E_0 for either the unknown or the model compound via any of the schemes described above, so long as they are consistent with each other. One can then determine the unknown distance r_u using the

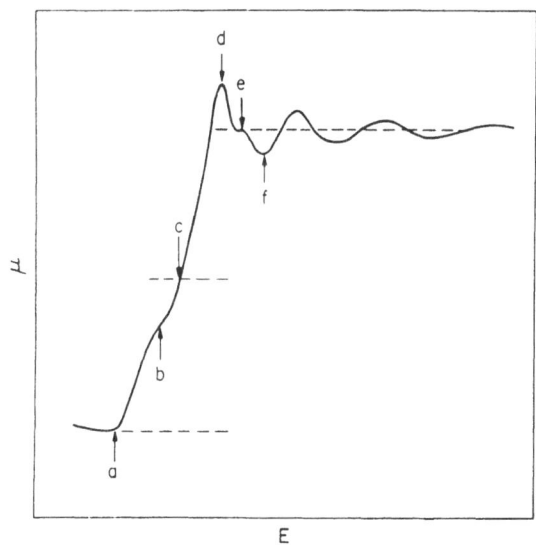

Fig. 5.1. Different ways of choosing the energy threshold, E_0.

phase function $\phi(k)$ deduced from the model compound with distance r_m based on the concept of *phase transferability*. This assumption states that at sufficiently high photoelectron kinetic energies, say 60 eV above the absorption edge, where EXAFS scattering processes are dominated by the core electrons, the phase function is relatively insensitive to chemical bonding effects such that once determined for a known system, they can be applied to unknown systems which contain the same corresponding pair of elements.

5.1.3 Varying E_0

Since determination of distance r_{AB} depends upon the phase shift function $\phi_{AB}(k)$ which in turn depend on k, the uncertainty in E_0 implies possible error in r_{AB} if E_0 is not chosen properly. Even if the E_0 values for the model compound and the unknown systems are chosen in a consistent manner, holding it fixed in the determination of the unknown distance will cause some error. In other words, there is no guarantee that the chosen E_0 values will give rise to identical phase functions for different systems unless the two systems happen to be chemically and structurally virtually identical or very similar. This is particularly problematic in lieu of the fact that E_0 is somewhat sensitive to chemical effects.

To see that the phase shifts are unique only if the energy thresholds, E_0, are specified, we note that changing E_0 by $\Delta E_0 = E'_0 - E_0$ will change the momentum k to

$$k' = \sqrt{k^2 - \frac{2m}{\hbar^2}(E'_0 - E_0)} \qquad (5.2)$$

For k in Å^{-1} and ΔE_0 in eV,

$$k' = \sqrt{k^2 - 0.2625(\Delta E_0)} \qquad (5.3)$$

If we denote the phase functions as $\phi'(k')$ and $\phi(k)$ based on E'_0 and E_0, respectively, we can write

$$\phi'(k') + 2k'r' = \phi(k) + 2kr \qquad (5.4)$$

since the argument of the sine term in the EXAFS $\chi(k)$ expression, which corresponds to the experimentally observed sinoisodal oscillation, is invariant to the choice of E_0. If we further define $\Delta\phi = \phi'(k') - \phi(k)$, $\Delta r = r' - r$, and using the approximation

$$\Delta k = k' - k = k\left[1 - \frac{0.2625(\Delta E_0)}{k^2}\right]^{1/2} - k \qquad (5.5a)$$

$$\approx k\left[1 - \frac{0.2625(\Delta E_0)}{2k^2}\right] - k \qquad (5.5b)$$

$$= -\frac{0.2625(\Delta E_0)}{2k} \qquad (5.5c)$$

for $0.2625(\Delta E_0) \ll k^2$, we obtain the phase shift modification as

$$\Delta\phi = -2(\Delta k)r - 2k(\Delta r) \qquad (5.6a)$$

$$\approx \frac{0.2625(\Delta E_0)}{k}r - 2k(\Delta r) \qquad (5.6b)$$

For $\Delta r = 0$ (viz., if we demand that r is independent of the choice of E_0), Eq. 5.6b reduces to

$$\Delta\phi \approx 0.2625r(\Delta E_0)/k \qquad (5.7)$$

It is obvious from Eq. 5.7 that the difference $\Delta\phi(k) = \phi'(k') - \phi(k)$ decreases with increasing k, indicating that phase shifts are more sensitive to a change in E_0 at small k than at large k (Lee and Beni, 1977; Teo and Lee, 1979). This effect is shown in Fig. 5.2 where $\Delta E_0 = 4$ eV is applied to a phase function

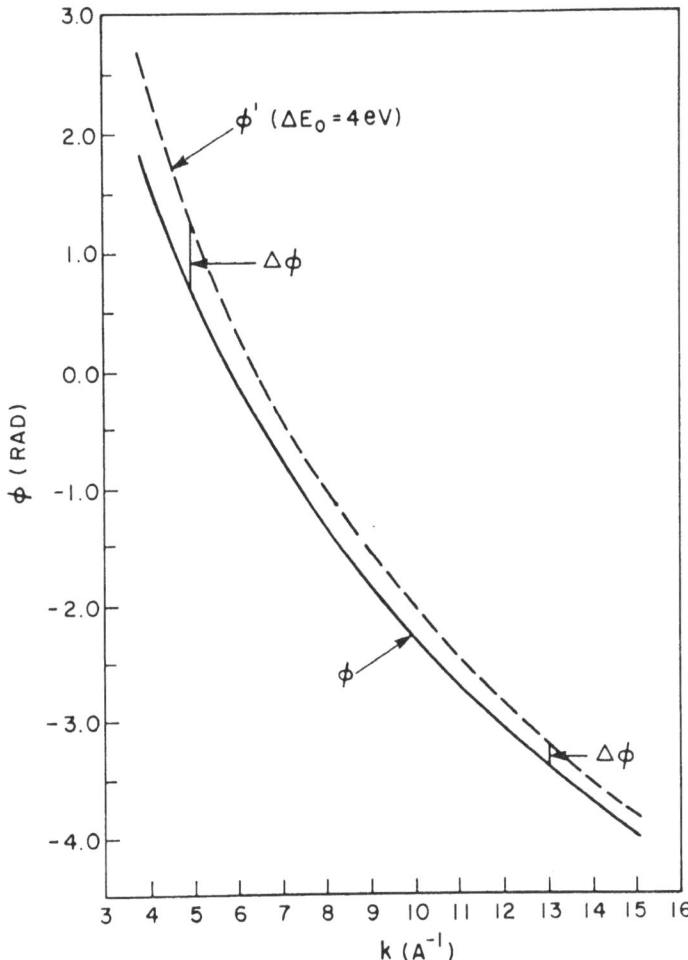

Fig. 5.2. The effect of $\Delta E_0 = 4$ eV on a phase function $\phi(k)$, resulting in $\phi'(k')$, for a system with $r = 2.5$Å.

$\phi(k)$, resulting in $\phi'(k')$ for $r = 2.5$Å. Note that the apparent constant $\Delta\phi$ is an optical illusion. It is also important to emphasize that $2(\Delta k)r$ is the dominant term in changing E_0. Changing the k scale alone would require a much larger ΔE_0 for the same $\Delta\phi$.

It is clear then that, in order to analyze the EXAFS of an unknown compound with the phase function derived from the EXAFS of a model compound, E_0 for the unknown system must be allowed to vary $(\Delta E_0 = E'_0 - E_0)$ in Eq. 5.3 where E_0 and E'_0 are the initially chosen and the refined values, respectively.

The same line of argument applies to EXAFS data analysis using theoretically calculated phase functions (see, for example, Teo and Lee, 1979). That is, in order to analyze experimental data based upon some empirical E_0 with the theoretical phase shifts, we must allow ΔE_0 to vary (*i.e.*, refine) where $\Delta E_0 = E_0^{\text{th}} - E_0^{\text{exp}}$ with E_0^{th} and E_0^{exp} denoting the "theoretical" and "experimental" energy thresholds, respectively.

Since the determination of interatomic distance r depends on the precise knowledge of $\phi(k)$, the nonuniqueness of phase shifts naturally causes concern about the uniqueness of the distance determination. Fortunately, it can be seen from Eq. 5.6b that by adjusting E_0 it is not possible to produce an artificially good fit with an incorrect distance r, simply because changing E_0 will affect $\phi(k)$ mainly at low k values by $\sim 0.2625 \, (\Delta E_0)/k$ (first term) whereas changing r will affect $\phi(k)$ mostly at high k values by $-2k \, (\Delta r)$ (second term). That is, the first term, with a $1/k$ dependence, dominates at low k while the second term, with a linear k dependence, dominates at high k. The extent of correlation between ΔE_0 and Δr will be discussed in a later chapter.

The E_0 variation also helps remove the small but significant bonding effects such as electronic configurations and atomic charges as well as effects such as many-body phenomena, core relaxations, multiple scatterings, etc. In essence, the E_0 variation preserves the otherwise less reliable phase transferability in EXAFS data analysis.

5.2 Inelastic Scatterings — The Amplitude Problem

In order to determine the number of each type of neighboring atoms N (the so-called coordination number), one must invoke the *amplitude transferability* approximation. As in the case of phase transferability, this concept assumes that at sufficiently high energies above the absorption edge, the EXAFS backscattering processes are dominated by the core electrons of the neighboring atoms such that the backscattering amplitude $F(k)$ is relatively insensitive to chemical effects. Once the $F(k)$ is determined for a particular type of neighboring atom, either from EXAFS analysis of a model compound of known structure or from theoretical calculations, it can be transferred to an unknown system in order to determine N. However, there are two processes which tend to complicate the situation.

The two processes are the so-called inelastic scattering processes which the photoelectron experiences as it leaves the X-ray absorbing atom, travels to the neighboring atoms, and back. The first is caused by multiple excitations at the central atom whereas the second is associated with excitation of the neighboring environment, including the neighboring atoms and the intervening medium. Both of these inelastic scattering processes tend to reduce the EXAFS amplitude and are often referred to as inelastic losses. These losses tend to limit the usefulness of the concept of amplitude transferability.

The limitations of amplitude transferability have been addressed by several groups. Eisenberger and Lengeler (1980) concluded from extensive measurements at both ambient and liquid nitrogen temperatures that there is a significant sensitivity of EXAFS amplitude to the chemical environment and the asymmetric pair distribution function for the same atom pair.

5.2.1 Central Atom: Shake up/off Processes

In the previous chapter, the EXAFS formula was derived from the *photon-induced "elastic" dipole* transition of a single electron (from a core state to continuum) in the presence of a fully relaxed atom with the remaining $(Z-1)$ electrons. Here Z is the total number of electrons in the central atom. In reality, the $(Z-1)$ "other" *"passive"* (or "bystander") electrons may also be excited along with the photoelectron in the so-called *shake-up* (excitation to a bound state) and *shake-off* (excitation to continuum) processes. These may be referred to as "inelastic" transitions in which the final state consists of an excited photoelectron and a "partially relaxed" ion with $(Z-1)$ electrons. Since the "dipole sum rule" states that the total absorption rate must remain the same regardless of multi-electron effects, these absorption processes imply a loss of intensity in the primary channel — viz., a loss in EXAFS amplitude. Shake up/off processes are multi-excitations or many-body (correlation) effects at the X-ray absorbing atom.

The physical origin of this loss mechanism is that the excess energy $(E - E_0)$ in the photoionization process (where E is the photon energy and E_0 is the threshold energy) can excite (shake-up) or ionize (shake-off) other low-binding (outer or valence) electrons within the central atom as shown schematically in Fig. 5.3. Since the total energy must be conserved, a natural consequence is that the photoelectron has a kinetic energy less than $(E - E_0)$. This implies that the resulting EXAFS will be shifted in energy and may have a different phase. Since these multi-electron excitations generally have a broad spectrum, they do not contribute coherently to the EXAFS, i.e., they tend to wash out the

86

EXAFS signal. (One exception is that if the spectrum contains a significant fraction of low energy (≤ 10 eV) excitations, then these channels can contribute significantly, especially at high energies where a small energy shift is insignificant.)

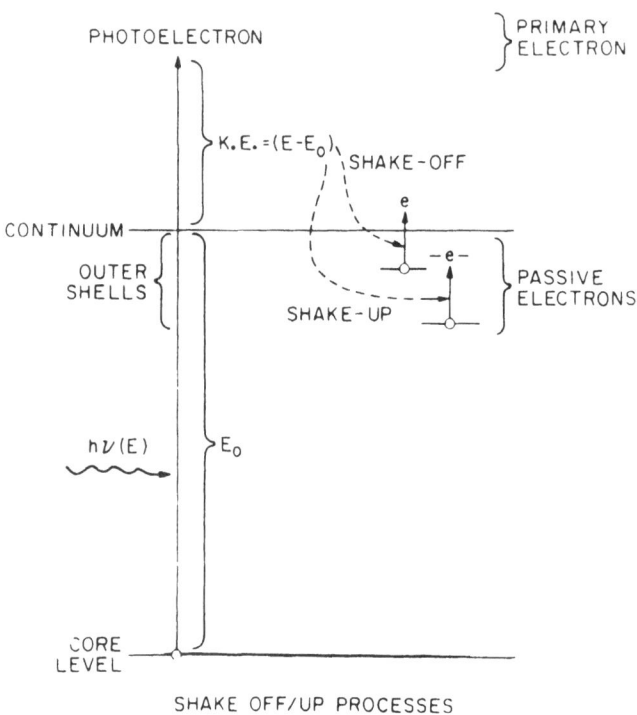

SHAKE OFF/UP PROCESSES

Fig. 5.3. A schematic representation of the shake-up and shake-off processes within the X-ray absorbing atom. Shake-up/off refers to the excitation/ionization of outer (mainly valence) electrons by the excess energy $E - E_0$ of the primary photoelectron. Here E is the photon energy, E_0 is the threshold energy and $(E - E_0)$ is the kinetic energy of the ejected photoelectron. Shake-up/off processes are important for $(E - E_0) \geq 200$ eV which is several times larger than the binding energies (~ 60 eV) of the outer shells.

In Chapter 4, the amplitude reduction due to these multiple excitations at the X-ray absorbing atom is approximated by $S_i(k)$. A better approximation is to use instead the overlap of the initial and the final state wave functions of the $(Z-1)$ passive electrons which are excited along with the photoelectron:

$$s_0^2(k) = |<\phi'_{Z-1}|\phi_{Z-1}>|^2 < 1 \qquad (5.8)$$

where ϕ_{Z-1} and ϕ_{Z-1} are the wave functions of the $Z-1$ electrons before and after the photoexcitation, respectively. That is, since the final state of the X-ray absorbing atom has a core hole induced by the *dipole* transition of the primary (photo)electron, the "other" $(Z-1)$ passive electrons sense a different (more attractive) potential than in the initial state and their wave functions are correspondingly modified. Consequently $s_0^2(k)$ is less than unity since the final state wave functions of the $(Z-1)$ passive electrons no longer have 100% overlap with their initial state wave functions. Since the many-electron matrix element includes a factor $s_0^2(k)$ of the overlap between the initial and the final states of the passive electrons, the net result is that the many-electron matrix element will decrease below that given by the one-electron theory which neglects the effects of the passive electrons for the dipole transition of the primary electron. In this context, the shake up/off processes are sometimes referred to as monopole excitation/ionization. (For monopole transitions, only the principal quantum number changes. The spin and angular momentum remain unchanged.)

The multi-electron excitations generally do not turn on until the excess energy reaches a value several times the binding energies of the outer electrons. For binding energies on the order of 60 eV, this implies that $(E-E_0)$ must be ≥ 200 eV for the shake up/off processes to occur. For most elements, this means that $s_0^2(k) \approx 1$ at low k values and $s_0^2(k) < 1$ for $k \geq 7\text{Å}^{-1}$. It has been shown by Stern, Heald, and Bunker (1979) for Br_2 and by Stern, Bunker, and Heald (1980, 1981) for metal dichlorides ($MnCl_2$, $FeCl_2$, $CoCl_2$) and tetrahedral semiconductors (Ge, GaAs, ZnSe, CuBr) that in general $s_0^2(k)$ decreases from unity at low k to a value somewhere between 0.6 and 0.8 at high k. For $k \geq 7\text{Å}^{-1}$, $s_0^2(k)$ is roughly independent of k. This is shown in Fig. 5.4 for Br_2 in both liquid and vapor phases where the ratio of the measured $F(k)$ divided by the single-electron theoretical calculation is plotted as a function of k. At high k ($\geq 7\text{Å}^{-1}$), the ratio approaches the value of 0.797 predicted by the single-electron calculation of Martin and Davidson (1977).

Table 5.1 lists the EXAFS amplitude reduction factor $s_0^2(k)$ at high k region for some representative elements as approximated by $(1-P)$ where P is the total electron shake-off probability. Since the shake-off probability for a given shell decreases rapidly with increasing atomic number Z and for a given atom it increases with increasing principal quantum number, the electrons most likely to be excited are in the valence shell orbitals. Consequently, the total

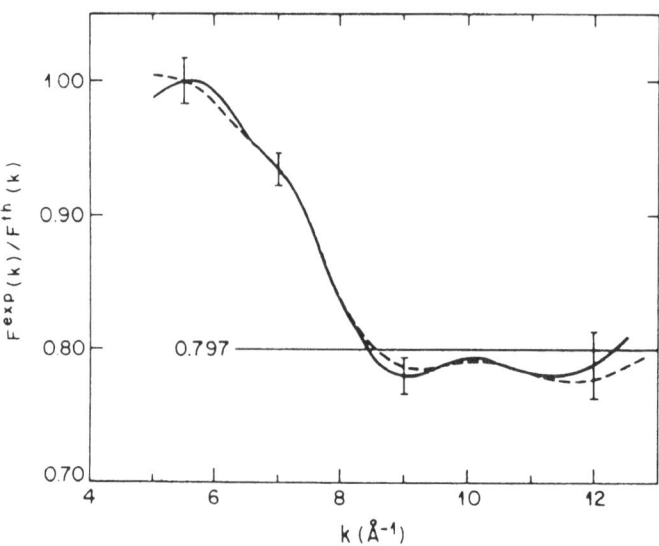

Fig. 5.4. The ratio of the measured $F(k)$ divided by the single-particle theoretical calculation (Teo and Lee, 1979) as a function of k. The solid line and dashed line show this ratio for the Br_2 vapor and Br_2 liquid, respectively. The thin horizontal line shows the multielectron corrections at high k to the single-particle calculation by Martin and Davidson (1977). (Reproduced from Stern, Heald, and Bunker, 1979).

Table 5.1. EXAFS Amplitude Reduction factor s_0^2 Due to Total Electron Shake-off at High k Values ($k \geq 7\text{Å}^{-1}$) for Some Elements.[a]

Z	CHEM	s_0^2
2	He	0.731
10	Ne	0.775
18	Ar	0.747
21	Sc	0.623
26	Fe	0.690
31	Ga	0.703
36	Kr	0.782
46	Pd	0.782
54	Xe	0.787
61	Pm	0.729
70	Yb	0.758
79	Au	0.803
83	Bi	0.774
92	U	0.733

a. Computed from $s_0^2 \approx 1 - P$ where P is the total electron shake-off probability taken from Carlson, et al. (1968).

shake-off probability shows no obvious overall trend with Z (Carlson, 1975). These values can be used to approximate EXAFS amplitude reduction at high k since Rehr, Stern, Martin, and Davidson (1978) have shown that in the high k region, EXAFS is affected only by the relaxation within the central atom and is not appreciably affected by the additional relaxation that occurs in the molecule. We note that further reduction in EXAFS amplitude by about (4-12)% may arise from correlation effects as indicated by X-ray photoemission spectroscopy (Mehta, Fadley, Bagus, 1976; Carlson, 1963).

Finally, it is interesting to note that $s_0^2(k)$ is the only term in the EXAFS amplitude which is significantly dependent upon the central atom. In many cases, it can be treated as a scale factor (independent of k), particularly when the high k data are weighted more heavily in the data analysis. This will facilitate the curve fitting of the data and more importantly, compensate for the breakdown, if any, of the amplitude transferability.

That EXAFS amplitude reduction factor is to a large extent due to the shake up/shake off processes is also demonstrated by Bomben, Bahl, Gimzewski, Chambers, and Thomas (1979) who measured the satellite spectrum accompanying ionization of $3d$ electrons from Br_2 and found approximate agreement between the amplitude attenuation calculated from their photoelectron spectrum and that determined from EXAFS.

Using ab initio Hartree-Fock Roothaan calculations and estimates for correlation effects, Rehr, Stern, Martin, Davidson (1978) calculated the many-electron overlap integral, s_0^2, to be 0.60, 0.64, and 0.64 for F_2, Cl_2 and Br_2, respectively, in good agreement with the amplitude reduction factor found between one-electron theory and experiment. However, the calculated amplitude reduction factors are quite sensitive to the nature of local chemical environment.

5.2.2 Scatterer: Electron Inelastic Mean Free Path

In Eq. 4.37-38, the inelastic losses due to excitation of the neighboring environment is approximated by $e^{-2r/\lambda(k)}$ where $\lambda(k)$, which depends on k, is the electron inelastic mean free path. Strictly speaking, one should instead use the expression $L_i(k)L_j(k)L_{ij}(r,k)$ where $L_i(k)$, $L_j(k)$, and $L_{ij}(r,k)$ are inelastic losses due to excitations of the medium, $L_{ij}(r,k)$, as the electron travels from the central atom, $L_i(k)$, to the neighboring atom j, $L_j(k)$, and back (Eisenberger and Lengeler, 1980). Clearly, the exponential damping term

$e^{-2r_j/\lambda(k)}$ approximates, albeit insufficiently, only $L_{ij}(r,k)$ for higher shells. Furthermore, since inelastic scattering processes are dominated by weakly bound electrons (outer electrons such as conduction and valence electrons) which are highly dependent upon chemical environment, we expect these loss mechanisms to be somewhat chemical sensitive, especially for *light-atom* scatterers where the number of valence electrons is a substantial portion of the total number of electrons (e.g. $Z \leq 10$) and for *higher shells* where the emitted photoelectron must travel a greater distance (and hence more excitable medium in between). As demonstrated by Lee and Beni (1977), the inelastic losses *within the scattering atom* can be accounted for approximately by using a complex effective potential in the scattering process. Stern, Bunker, and Heald (1980, 1981) showed that a damping factor of the form $e^{-2(r-d)/\lambda(k)}$ can be used in conjunction with the backscattering amplitude calculated by the one-electron theory. Here d accounts for inelastic losses already included in the central atom $s_0^2(k)$ and the theoretical backscattering amplitude $F(k)$. In most cases, d amounts to roughly the first coordination shell (covalent) distance.

Granting the exponential damping form $e^{-2r/\lambda(k)}$, the inelastic electron free path can be approximated by (Teo, 1981)

$$\lambda(k) = \frac{1}{\eta}\left[\left(\frac{\xi}{k}\right)^4 + k^n\right] \qquad (5.9a)$$

For $k \geq 5\text{Å}^{-1}$, Eq. 5.9a reduces to

$$\lambda(k) = k^n/\eta \qquad (5.9b)$$

Eq. 5.9 stems from the works of Powell (1974), Penn (1976), Seah and Dench (1979). These authors showed that the electron inelastic mean free path $\lambda(E) \propto E_e$ at high energies ($E_e \geq 500$ eV), $\propto E_e^{1/2}$ at intermediate energies ($100 \leq E_e \leq 500$ eV), and $\propto E^{-2}$ at low energies ($E_e \leq 50$ eV). In fact, there is a *universal* curve $\lambda(E_e)$ vs E_e for each type of material (elements, inorganic or organic compounds, adsorbed gases, etc.):

$$\lambda = \frac{A}{E_e^2} + B\sqrt{E_e} \qquad (5.10a)$$

For $E_e \geq 150$ eV

$$\lambda = B\sqrt{E_e} \qquad (5.10b)$$

The coefficients A, B, η, and ξ for the universal curves are tabulated in Table 5.2. Here λ is in Å, E_e in eV, and $k = \sqrt{0.2625\,E_e}$ in Å^{-1}.

Table 5.2. Universal Curves for Electron Inelastic Mean Free Path (λ in Å) as a Function of Energy (E_e in eV) or Photoelectron Wavevector (k in Å^{-1}):
$\lambda = \dfrac{A}{E_e^2} + B\sqrt{E_e} = \dfrac{1}{\eta}\left[\left[\dfrac{\xi}{k}\right]^4 + k\right]$ where $k = \sqrt{0.2625\,E_e}$ (A and B values were taken from Seah and Dench, 1979).

Material	A	B	η	ξ
Elements	1430	0.54	0.95	3.106
Inorganic	6410	0.96	0.53	3.913
Organic	310	0.87	0.59	1.881
Adsorbed Gases	--	0.64	0.80	--

The universal λ vs E_e curves for elements and inorganic materials are shown in Fig. 5.5(a) and (b), respectively. While significant scattering can occur for individual cases, these universal curves can be used as a first approximation to λ in EXAFS. Fig. 5.6 compares the λ determined by EXAFS (Stern, Bunker, and Heald, 1981) for Ge (solid curve) with that determined by Auger spectroscopy (Powell, 1974) for Cu, Ag, and Al.

In EXAFS data analysis, it has been shown (Teo, 1981) that $n = 1$ and $n = 2$ models give rise to about the same quality fit for data with $k \geq 5$Å^{-1}. However, the latter generally yields values of λ in the range 2-17Å (for $k = 5$-13 Å^{-1}) which is more in line with that predicted by the universal curves for electron inelastic mean free paths. Parameters η and ξ in Eqs. 5.9 can be refined to best fit the EXAFS data.

5.3 Static and Thermal Disorder Effects

The Debye-Waller factor σ plays an important role in EXAFS spectroscopy. It contains important structural and chemical information which is otherwise difficult to obtain, yet it comes as a bonus in the EXAFS determination of interatomic distance. Generally speaking, the Debye-Waller factor σ has two components σ_{stat} and σ_{vib} due to static disorder and thermal vibrations, respectively. Both of these can be derived from the pair distribution function $g(r)$ via the following equation:

$$\chi(k) = \frac{F(k)}{k} \int_0^\infty g(r) e^{-2r/\lambda(k)} \frac{\sin(2kr + \phi(k))}{r^2}\, dr \qquad (5.11)$$

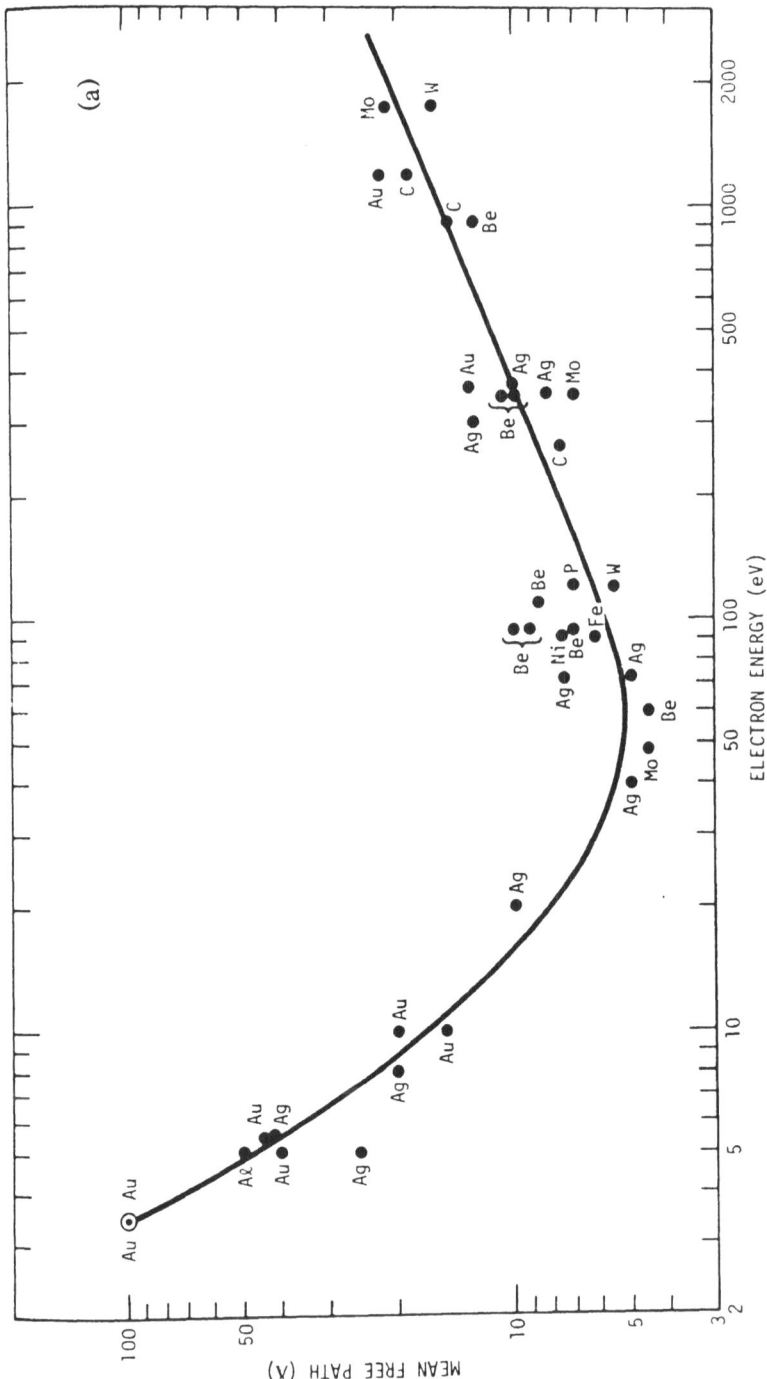

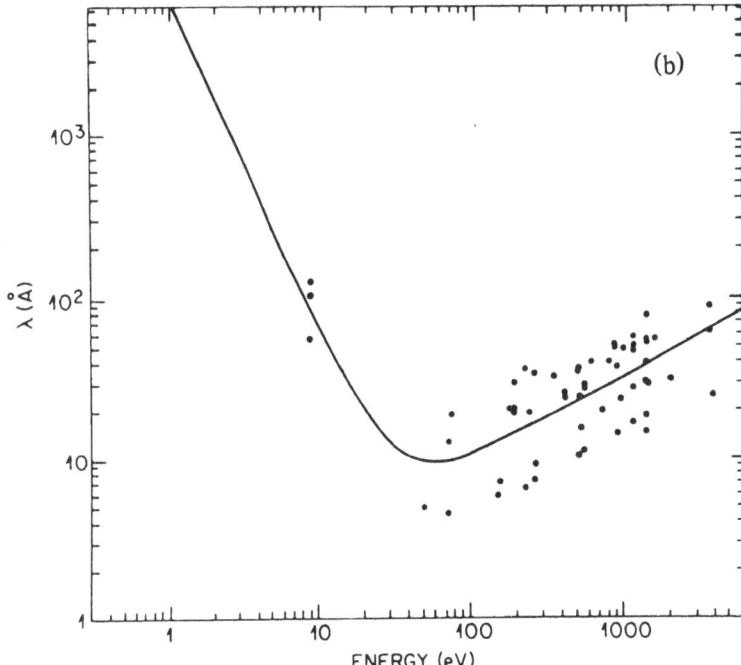

Fig. 5.5. "Universal curve" for the electron mean free path λ (in Å) as a function of electron kinetic energy. Dots indicate individual measurements: (a) elements (reproduced from Somorjai, 1981); (b) inorganic materials (adapted from Seah and Dench, 1979).

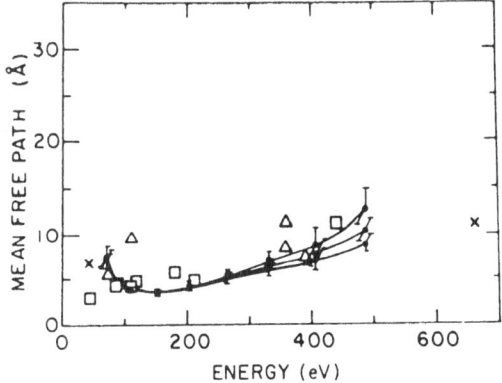

Fig. 5.6. Electron mean free path for Ge as measured by EXAFS. For comparison, results from Auger measurements (from Powell, 1974) are also given: X, Cu; Δ, Ag; $\square$, Al. (Reproduced from Stern, Bunker, and Heald, 1980, 1981).

In general, $g(r)$ depends on the interatomic potential. For σ_{vib}, in particular, it depends on the vibrational frequency and the absolute temperature.

For crystalline or molecular complexes, the coordination shells are often well-defined with discrete bonds or distances ($g(r)$ is in essence a delta function for each shell which integrates to a number equal to the number of bonds) whereas for amorphous or liquid materials, the distances are disordered so that $g(r)$ must be represented by a continuous function and then integrated to yield the EXAFS expression.

5.3.1 Small Disorders

Assuming small disorders with a *symmetric* pair distribution function for static disorder and *harmonic* vibration for thermal disorder, the overall Debye-Waller factor σ is given by

$$\sigma^2 = \sigma_{stat}^2 + \sigma_{vib}^2 \tag{5.12}$$

where the static and thermal disorders give rise to factors of $e^{-2\sigma_{stat}^2 k^2}$ and $e^{-2\sigma_{vib}^2 k^2}$, respectively, in the EXAFS expression. In principle, these two factors can only be separated by a temperature dependent study of $\sigma(T)$. However, if σ_{vib} can be estimated from vibrational spectroscopies or if σ_{stat} is known from other studies, the other term can be calculated from the experimentally determined σ.

5.3.1.1 Symmetric Pair Distribution

For a continuous symmetric Gaussian pair distribution function of distance r

$$g(r) = \frac{1}{\sqrt{2\pi}\sigma_{stat}} e^{-(r-r_0)^2/2\sigma_{stat}^2} \tag{5.13}$$

σ_{stat} is just the *root-mean-square* deviation or displacement from the mean distance r_0. Fig. 5.7 shows two normalized Gaussian distributions with different half-widths, $\sigma_1 = 0.05\text{Å}$ and $\sigma_2 = 0.09\text{Å}$.

The fact that a symmetric Gaussian distribution of distances gives rise to an exponential damping $e^{-2\sigma_{stat}^2 k^2}$ in the EXAFS in k space can be seen by integrating the EXAFS contribution over all distances. Based on Eq. 5.11, we shall use $e^{-2\sigma_{vib}^2 k^2}$ for the vibrational contribution to the Debye-Waller factor. We further assume that this as well as other factors are identical such that they can be taken out of the integral. For small disorder, $r \approx r_0$ such that the $1/r_0^2$

factor can also be taken out of the integral. The integral of interest is then

$$\int_0^\infty g(r) \sin(2kr + \phi(k)) dr$$

$$= \frac{1}{\sqrt{2\pi}\sigma_{stat}} \int_0^\infty e^{-\frac{(r-r_o)^2}{2\sigma_{stat}^2}} \sin(2kr + \phi(k)) dr \qquad (5.14)$$

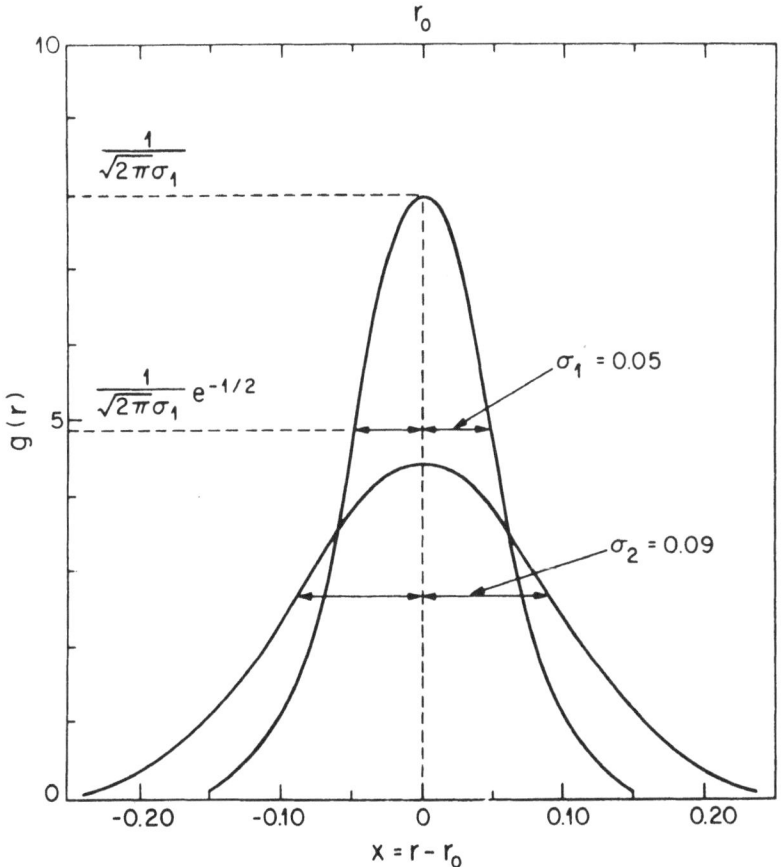

Fig. 5.7. Two normalized Gaussian pair distribution functions with different half-widths: $\sigma_1 = 0.05\text{Å}$, $\sigma_2 = 0.09\text{Å}$.

Substituting $x = r - r_0$, the above equation becomes

$$\frac{1}{\sqrt{2\pi}\sigma_{\text{stat}}} \int_{-\infty}^{\infty} e^{-\frac{x^2}{2\sigma_{\text{stat}}^2}} [\sin(2kr_0 + \phi(k))\cos(2kx + \phi(k))$$

$$+ \cos(2kr_0 + \phi(k))\sin(2kx + \phi(k))]dx \qquad (5.15a)$$

$$- \frac{1}{\sqrt{2\pi}\sigma_{\text{stat}}} \sin(2kr_0 + \phi(k)) \int_{-\infty}^{\infty} e^{-\frac{x^2}{2\sigma_{\text{stat}}^2}} \cos(2kx + \phi(k))dx \qquad (5.15b)$$

$$- \frac{1}{\sqrt{2\pi}\,\sigma_{\text{stat}}} \sin(2kr_0 + \phi(k))\, \sqrt{2\pi}\,\sigma_{\text{stat}}\, e^{-2\sigma_{\text{stat}}^2 k^2} \qquad (5.15c)$$

$$- e^{-2\sigma_{\text{stat}}^2 k^2} \sin(2kr_0 + \phi(k)) \qquad (5.15d)$$

Note that the second term (an odd function of x) in Eq. 5.15a integrates to zero. It is of interest to note that functions of the form $Ae^{-B(x-C)^2}$ (where A, B, and C are constants) are often called Gaussian functions because of their occurrence in the least-squares method of data analysis originated by Gauss. Note that a Gaussian function of r (or x) (Eq. 5.13) transforms to another Gaussian function of k (cf. Eq. 5.15d). The two widths are roughly inverse to each other $\Delta x \Delta k \sim 1$. In fact, Fourier transform of one Gaussian always gives rise to another Gaussian.

5.3.1.2 Discrete Bonds

For *discrete* bonds, σ_{stat} is related to the *root-mean-square* standard deviation δ (keeping terms to second order):

$$\sigma_{\text{stat}} \approx \delta = \sqrt{\sum_{j=1}^{N} \frac{(r_j^2 - r_0)^2}{N}} \qquad (5.16)$$

For a two-distance system with m bonds at a distance r_m and n bonds at a distances r_n, we have

$$N = m + n \qquad (5.17)$$

the average distance,

$$r_0 = \frac{mr_m + nr_n}{m + n} \qquad (5.18)$$

and the distance disparity,

$$\Delta r = r_m - r_n \qquad (5.19)$$

Eq. 5.16 then reduces to

$$\sigma_{stat} \approx \frac{\sqrt{mn}}{m+n} |\Delta r| = \frac{\sqrt{mn}}{m+n} |r_m - r_n| \qquad (5.20)$$

Because of the importance of the interplay between the various components of Debye-Waller factors and the interatomic distances, we shall derive Eq. 5.20 as follows. Form Eq. 5.18 and 5.19,

$$r_m = r_0 + \frac{n}{m+n} \Delta r \qquad (5.21a)$$

$$r_n = r_0 - \frac{m}{m+n} \Delta r \qquad (5.21b)$$

the root-mean-square displacement is then

$$\sigma_{stat} = \sqrt{\frac{m(r_m - r_0)^2 + n(r_n - r_0)^2}{m+n}} \qquad (5.22a)$$

$$= \sqrt{\frac{mn(\Delta r)^2}{(m+n)^2}} \qquad (5.22b)$$

$$= \frac{\sqrt{mn}}{m+n} |\Delta r| \qquad (5.22c)$$

Alternatively, Eq. 5.20 can be derived from the basic EXAFS formula

$$\chi(k) \propto [m \sin(2kr_m + \phi(k)) + n \sin(2kr_n + \phi(k))] \qquad (5.23)$$

Here we assume other things (not shown) are the same or approximately equal such that they can be factored out of the square bracket. We can rewrite Eq. 5.23 in the exponential form

$$\chi(k) \propto \text{Im} \left[me^{i\left[2k\left(r_\bullet + \frac{n}{m+n}\Delta r\right) + \phi(k)\right]} + ne^{i\left[2k\left(r_\bullet - \frac{m}{m+n}\Delta r\right) + \phi(k)\right]} \right]$$

$$= \text{Im} \left[e^{i(2kr_\bullet + \phi(k))} \right] \left[me^{i\left(\frac{2n}{m+n}\right)k\Delta r} + ne^{-i\left(\frac{2m}{m+n}\right)k\Delta r} \right] \qquad (5.24)$$

Using the series expansion

$$e^{ix} = 1 + ix - \frac{x^2}{2!} - i\frac{x^3}{3!} + \cdots \qquad (5.25)$$

We can write

$$me^{i(\frac{2n}{m+n})k\Delta r} + ne^{i(\frac{2m}{m+n})k\Delta r}$$

$$\approx (m+n) - \frac{2mn}{(m+n)} k^2(\Delta r)^2 + i\frac{4mn(m-n)}{3(m+n)^2} k^3(\Delta r)^3 + \cdots \qquad (5.26)$$

Keeping terms up to second order (assuming $\frac{2mn}{(m+n)} k^2(\Delta r)^2 \ll \pi$) and using

$$e^{-x^2} = 1 - x^2 + \frac{x^4}{2} - \cdots \qquad (5.27)$$

Eq. 5.26 becomes

$$(m+n)\left[1 - \frac{2mn}{(m+n)^2} k^2(\Delta r)^2 + \cdots\right] \approx (m+n)e^{-2\sigma_{stat}^2 k^2} \qquad (5.28)$$

where

$$\sigma_{stat} \approx \frac{\sqrt{mn}}{(m+n)} |\Delta r|$$

Combining Eq. 5.24, 5.26, and 5.28, we obtain

$$\chi(k) \propto (m+n)e^{-2\sigma_{stat}^2 k^2} \operatorname{Im}[e^{i(2kr_\bullet + \phi(k))}] \qquad (5.29a)$$

$$= (m+n)e^{-2\sigma_{stat}^2 k^2} \sin(2kr_0 + \phi(k)) \qquad (5.29b)$$

Thus, for small disorder ($\Delta r \leq 0.1\text{Å}$) the two terms in the EXAFS expression, corresponding to the two sets of distances, can be combined to give a single term, corresponding to the average distance, with an additional static contribution to the Debye-Waller factor σ_{stat} given by Eq. 5.22c. It is obvious from Eq. 5.22c that one can only determine the magnitude of the difference in distance, Δr, and not its relative sign from σ_{stat}.

For systems with large difference in discrete distances where the approximation $k(\Delta r) \ll \pi$ no longer holds, particularly those with interference patterns (beat nodes), it is possible to separate distances from Debye-Waller factors as well as to determine the signs of the differences of both. This can be done by keeping higher order terms in Eq. 5.26 or using a beat-node technique to be discussed in the next chapter. It is important to note that, in either case, both the amplitude and the phase of the resulting EXAFS are modified as is evident from the $ik^3(\Delta r)^3$ term in Eq. 5.26 (rather than just the amplitude alone as in Eq. 5.29).

5.3.1.3 Harmonic Vibration

Assuming harmonic motion for a diatomic system, the vibrational contribution to the Debye Waller factor is given by

$$\sigma_{\text{vib}}^2 = \frac{h}{8\pi^2\mu\nu} \coth \frac{h\nu}{2kT} \tag{5.30}$$

where μ is the reduced mass, T is the temperature, and ν is the vibrational frequency (see, for example, Cyrin, 1968). It can be used as a qualitative assessment of bond strength for closely related systems assuming that ν is a qualitative measure of bond strength. For μ in atomic mass units and $\bar{\nu} = \nu/c$ in cm^{-1}, and T in °K, σ_{vib} in Å is given by

$$\sigma_{\text{vib}} = 4.106 \left[\frac{1}{\mu\bar{\nu}} \coth \left(\frac{x}{2} \right) \right]^{1/2} \tag{5.31}$$

where $x = 1.441 \dfrac{\bar{\nu}}{T}$. Fig. 5.8 depicts the function $\sqrt{\coth (x/2)}$ as a function x (dashed curve) as well as the dependence of σ_{vib} on x for $\mu\bar{\nu} = 3{,}000;\ 10{,}000;$ and 30,000. It is apparent that as the vibration frequency $\bar{\nu}$ increases and/or as the temperature T decreases, x increases, $\coth (x/2)$ and hence σ_{vib} decreases exponentially.

For strong bonds or at the low temperature limit $h\nu >> kT, x >> 1$ (typically $x \geq 2$)

$$\sigma \approx 4.106 \left[\frac{1}{\mu\bar{\nu}} \right]^{1/2} = 3.151\times10^{-3} \left[\frac{\bar{\nu}}{K} \right]^{\frac{1}{2}} \tag{5.32}$$

For weak bonds or at the high temperature limit $h\nu << kT, x << 1$ (typically $x \leq 0.3$):

$$\sigma \approx 4.836 \left[\frac{T}{\mu\bar{\nu}^2} \right]^{1/2} = 3.712\times10^{-3} \left[\frac{T}{K} \right]^{\frac{1}{2}} \tag{5.33}$$

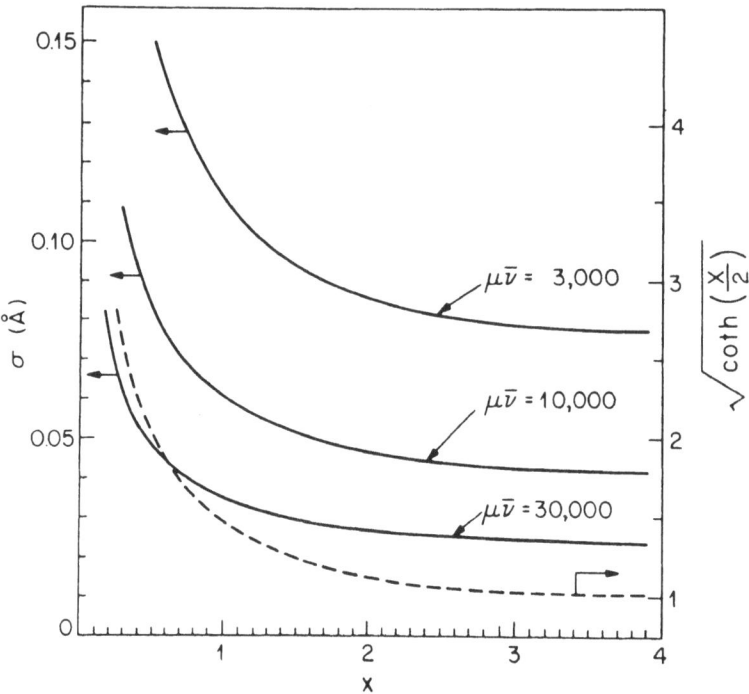

Fig. 5.8. The dependence of vibrational (harmonic) contribution to σ on x where $x = 1.441\ \bar{\nu}/T$. The solid curves are for $\mu\bar{\nu} = 3,000;\ 10,000;$ and $30,000$. Also shown is the $\sqrt{\coth(x/2)}$ function (dashed curve).

Here $K = 4\pi^2\mu c^2\bar{\nu}^2$ in mdyne/Å is the force constant. For weak bonds, it is often necessary to do the experiments at low temperatures so as to reduce σ_{vib}, thereby enhancing the EXAFS signal, especially at high k values.

In crystals, the temperature dependence of the Debye-Waller factor can be fitted with the Einstein model

$$\sigma_E^2 = \frac{\hbar^2}{Mk_B\theta_E}\coth\frac{\theta_E}{2T}$$

where θ_E is the Einstein temperature, M is the atomic mass, and k_B is Boltzmann's constant. For crystalline Ge, θ_E of $315(10)\,°K$ was found (see, for example, Crozier and Seary, 1981).

It is obvious that temperature dependent EXAFS measurements are the only way to separate the vibrational contribution from the static contribution to the

Debye-Waller factor. For further discussions on the determination and calculation of EXAFS σ, including its temperature and symmetry dependence, see papers by Beni and Platzman (1976); Greegor and Lytle (1979); Sevillano, Meuth, Rehr (1979); Rabe, *et al.* (1979), Boland and Baldeschwieler (1984a,b), among many others.

5.3.2 Large Disorders

For systems with large disorders ($\sigma \geq 0.1$Å), the above treatments are inadequate. For large disorder, the EXAFS expression must be averaged over a pair distribution function $g(r)$ characteristic of the system via Eq. 5.11.

Large disorders are often accompanied by either an asymmetric pair correlation function or an anharmonic vibration potential. The former is a static disorder which is an intrinsic property characteristic of the system whereas the latter is a thermal disorder which can be reduced by lowering the temperature. Failure to take into account these two types of large disorder effects will lead to serious errors. Large disorder can lead to a drastic reduction of the EXAFS amplitude, and hence a lowering of the apparent coordination numbers. It can also cause an apparent contraction (or even expansion) in the near neighbor distances which can be as large as 0.15 Å.

The physical reason is that the broad tail(s) of a pair distribution function will contribute to EXAFS only at low k values due to the exponential damping effect. A large part of this information is lost owing to a large $k_{min} \geq 3$ Å^{-1} cut-off of the data necessitated by the edge complication. The truncated EXAFS therefore only contains structural information characteristic of the sharpest feature/edge in $g(r)$. A broad distance distribution function in itself can also lead to a smearing and hence a reduction of the EXAFS signal.

Large disorders are often accompanied by asymmetric pair distributions and/or anharmonic vibrations. For these systems, a generalized phenomenological expression of EXAFS such as the following one (Eisenberger and Brown, 1979) must be used.

$$\chi(k) = \sum_j N_j S_i(k) F_j(k) \sqrt{S_j^2 + A_j^2}$$

$$\times \frac{\sin (2kr_j + \phi_j(k) + \tan^{-1}(A_j/S_j))}{kr_j^2} \qquad (5.34)$$

where

$$S_j = \int_{-\infty}^{\infty} e^{-2r_j/\Lambda_j(k)} \frac{g_j(x,r_j,T)\cos2kx}{(1+x/r_j)^2} \, dx \qquad (5.35)$$

$$A_j = \int_{-\infty}^{\infty} e^{-2r_j/\Lambda_j(k)} \frac{g_j(x,r_j,T)\sin2kx}{(1+x/r_j)^2} \, dx \qquad (5.36)$$

Here $x = r - r_j$ is the deviation of the distance r from the mean distance r_j of the neighboring atom j. S_j and A_j are potential functions of k, r_j (for static disorder), and T (for thermal disorder). S_j and A_j are symmetric and asymmetric transforms of the distribution function $g(r,T)$. Given $g(r,T)$, it is often possible to integrate Eq. 5.35 and 5.36 to yield analytical expressions for S and A in terms of k, r, and T. Alternatively, Eq. 5.35 and 5.36 can be integrated numerically. It is apparent that the effect of large disorders with asymmetric and/or anharmonic distributions is to introduce an amplitude modification factor $\sqrt{S^2+A^2}$ and a phase correction $\tan^{-1}(A/S)$ in the EXAFS expression. For a symmetric Gaussian pair correlation function or a harmonic thermal vibration and assuming $x \ll r_j$, $A_j = 0$ and $S_j = e^{-2\sigma^2k^2}$ where σ is σ_{stat} (cf., Eq. 5.13) or σ_{vib} (cf., Eq. 5.30), respectively.

The absence of the low k information together with the modification of the amplitude and phase function indicated in Eq. 5.34 implies that one cannot assume amplitude and/or phase transferabilities for systems with differing pair distribution functions.

5.3.2.1 Derivation of the Generalized EXAFS Formalism

We shall derive Eq. 5.34 in the following manner. Consider one term with a mean distance r_0 and a phase function $\phi(k)$ and substituting $x = (r-r_0)$ (the possible temperature dependences of $g(r)$, $S(k)$, and $A(k)$ are neglected for clarity).

$$\chi(k) \propto \int_0^{\infty} e^{-2(r_\bullet+x)/\Lambda(k)} \frac{g(r)}{kr^2} \sin(2kr+\phi(k)) \, dr \qquad (5.37a)$$

$$= \operatorname{Im} \int_0^{\infty} e^{-2(r_\bullet+x)/\Lambda(k)} \frac{g(r)}{kr^2} \, e^{i(2kr+\phi(k))} \, dr \qquad (5.37b)$$

$$= \operatorname{Im} \frac{1}{kr_0^2} e^{i(2kr_0+\phi(k))} \int_{-\infty}^{\infty} e^{-2(r_0+x)/\Lambda(k)} \frac{g(r_0+x)}{(1+x/r_0)^2} \, e^{i2kx} \, dx \qquad (5.37c)$$

$$= \operatorname{Im} \frac{1}{kr_0^2} e^{i(2kr_0+\phi(k))} [S(k) + iA(k)] \qquad (5.37d)$$

where

$$S(k) = \int_{-\infty}^{\infty} e^{-2(r_e+x)/\lambda(k)} \frac{g(r_0+x)\cos 2kx}{(1+x/r_0)^2} dx \qquad (5.38)$$

$$A(k) = \int_{-\infty}^{\infty} e^{-2(r_e+x)/\lambda(k)} \frac{g(r_0+x)\sin 2kx}{(1+x/r_0)^2} dx \qquad (5.39)$$

Since $x+iy = \sqrt{x^2+y^2}\, e^{i\theta}$ where $\theta = \tan^{-1} y/x$ Eq. 5.37d can be rewritten as

$$\chi(k) \propto \text{Im} \frac{1}{kr_0^2} e^{i(2kr_e+\phi(k))} \sqrt{S(k)^2+A(k)^2}\, e^{i\Sigma(k)} \qquad (5.40a)$$

$$= \text{Im} \frac{1}{kr_0^2} \sqrt{S(k)^2+A(k)^2}\, e^{i(2kr_e+\phi(k)+\Sigma(k))} \qquad (5.40b)$$

$$= \sqrt{S(k)^2+A(k)^2}\, \frac{\sin(2kr_0+\phi(k)+\Sigma(k))}{kr_0^2} \qquad (5.40c)$$

where

$$\Sigma(k) = \tan^{-1} \frac{A(k)}{S(k)} \qquad (5.41)$$

5.3.2.2 Moments of g(r)

Since the integrals for $S_j(k)$ and $A_j(k)$ may not be straightforward (especially when x/r_j is no longer small), an alternative method is clearly desirable. One way is to simply expand $\sin 2kx$ and $\cos 2kx$ and keep terms up to $1/r_0$, $1/\lambda$, and x^3 to derive expressions for $S_j(k)$ and $A_j(k)$ in terms of the moments of $g_j(r)$. Considering a normalized pair distribution function $g(r)$ and defining $x = r-r_0$, we find

$$S(k) \approx \frac{e^{-2r_e/\lambda}}{r_0^2} e^{-2k^2<x^2>} \qquad (5.42)$$

$$A(k) \approx -\frac{e^{-2r_e/\lambda}}{r_0^2} [4k<x^2>(\frac{1}{\lambda} + \frac{1}{r_0}) + \frac{4}{3}k^3<x^3>] \qquad (5.43)$$

where the nth moment of the distribution $g(r)$ is defined as the mean value of x^n:

$$<x^n> = <(r-r_0)^n> \qquad (5.44a)$$

$$= \int_{-\infty}^{\infty} g(r_0+x)x^n dx \qquad (5.44b)$$

Specifically, $<x> = 0$ and $<x^2> = \sigma^2$ (the so-called *variance of x*) by definition. Hence, given any normalized pair distribution function $g(r)$, $S(k)$ and $A(k)$ can readily be calculated via Eqs. 5.42-44. One interesting observation is that even if $g(r)$ is symmetric such that $<x^{2n+1}> \equiv 0$, there should still be a phase shift correction of $\tan^{-1}[-4k\sigma^2(k)(1/r_0 + 1/\Lambda(k))]$ if r deviates significantly from r_0, the average distance (note that the factorization of $e^{-2\sigma^2 k^2}$ and $e^{-2r/\Lambda(k)}$ in the EXAFS expression is possible only if one assumes $r \approx r_0$ such that the factor $e^{-2r_0/\Lambda(k)}/r_0^2$ can be taken out of the integral). Normally, however, this amounts to only ≤ 0.01Å in distance correction.

Equations 5.42 and 5.43 can be derived in the following manner.

$$S(k) + iA(k) = \int_{-\infty}^{\infty} e^{i2kx} \frac{g(r)e^{-2r/\Lambda}}{r^2} dx \tag{5.45}$$

$$= \frac{e^{-2r_0/\Lambda}}{r_0^2} \int_{-\infty}^{\infty} e^{i2kx} \frac{g(r_0+x)e^{-2x/\Lambda}}{(1+x/r_0)^2} dx$$

$$= \frac{e^{-2r_0/\Lambda}}{r_0^2} \left[1 + i2k <x> - 2k^2 <x^2> - i\frac{4}{3}k^3<x^3> + \cdots \right]$$

$$= \frac{e^{-2r_0/\Lambda}}{r_0^2} \left[1 - i4k <x^2> \left(\frac{1}{\lambda} + \frac{1}{r_0} \right) - 2k^2 <x^2> \right.$$

$$\left. + 4k^2 <x^3> \left(\frac{1}{\lambda} + \frac{1}{r_0} \right) - i\frac{4}{3}k^3<x^3> + \cdots \right]$$

Here we have made use of the relations

$$e^{i2kx} \approx 1 + i2kx - 2k^2x^2 - i\frac{4}{3}k^3x^3 + \cdots \tag{5.46}$$

$$e^{-2x/\Lambda} \approx 1 - \frac{2x}{\lambda} + \cdots \tag{5.47}$$

$$(1 + \frac{x}{r_0})^{-2} \approx 1 - 2\frac{x}{r_0} + \cdots \tag{5.48}$$

Hence

$$S(k) \approx \frac{e^{-2r_o/\lambda}}{r_0^2} \left[1 - 2k^2 <x^2> + 4k^2 <x^3> \left(\frac{1}{\lambda} + \frac{1}{r_0} \right) + \cdots \right] \quad (5.49)$$

$$\approx \left(\frac{e^{-2r_o/\lambda}}{r_0^2} \right) e^{-2k^2<x^2>}$$

$$A(k) \approx - \left(\frac{e^{-2r_o/\lambda}}{r_0^2} \right) \left[4k <x^2> \left(\frac{1}{\lambda} + \frac{1}{r_0} \right) + \frac{4}{3} k^3 <x^3> \right] \quad (5.50)$$

Eisenberger and Brown (1979) suggested a broad classification of materials into three phenomenological classes based on the consideration that the phase correction in the generalized EXAFS formalism amounts to $\tan^{-1}(S(k)/A(k))$ where $S(k)$ and $A(k)$ are given by Equations 5.42 and 5.43, respectively: (1) systems for which $2<x^2>/r_0$ and $2<x^2>/\lambda < 0.01\text{Å}$, $(2k)^{2n+1} <x^{2n+1}> \ll 1$ require no corrections to maintain the 0.01Å accuracy (since $A(k)$ is insignificant); (2) systems for which $2<x^2>/r_0 \geq 0.01\text{Å}$, $8k^3<x^3> \leq 1$ but $(2k)^{2n+1} <x^{2n+1}> \ll 1$, one can apply a simple cubic correction (*cf.* Eq. 5.43); and (iii) systems with $<x^2> \geq 0.05\text{Å}$, $k^3 <x^3>$ and/or higher moments > 1, more correction terms must be added and the information obtainable by EXAFS is limited.

5.3.2.3 Symmetric Pair Distributions

It may seem possible to determine $g(r)$ by Fourier transforming the EXAFS data $\chi(k)$, assuming that $F(k)$ and $\phi(k)$ are known. In practice, however, it has long been recognized that for EXAFS data of $k \leq 3\text{Å}^{-1}$ the simple backscattering picture of EXAFS does not apply. The region of $k \leq 3\text{Å}^{-1}$, which corresponds to the low k cutoff in EXAFS data analysis, actually contains a great deal of information concerning $g(r)$ which is unfortunately unusable.

First let us consider symmetric or harmonic systems with large disorders. It is apparent from Eq. 5.13 that in the normalized form, the area under each Gaussian distribution curve is conserved with larger values of σ_{stat} merely representing a large spread of distances. However, as we shall see later in this section, the slowly-varying tails of the pair distribution function $g(r)$ give rise to only a small contribution to EXAFS data at low k which is often lost owing to the low k cutoff in the data analysis. That is to say that even with symmetric distribution function, large disorders often affect EXAFS amplitude and phase to a point that it is impossible to reconstruct $g(r)$ from the data. This is often manifested in the Fourier transform where the area under the peak with a large Debye-Waller factor is not conserved. In the final analysis, one frequently finds a drastic decrease in the EXAFS amplitude and an additional phase shift correction for systems with large disorders. These effects are shown in Fig. 3.14(b) for the two distributions depicted in Fig. 5.7. Note that the peak areas in the Fourier transforms are not conserved.

5.3.2.4 Asymmetric Pair Distributions

For systems with asymmetric distribution and/or anharmonic potential, the problem is even more troublesome since the EXAFS signal in k space will be dominated by the sharply peaked or rapidly varying of $g(r)$. The broad or slowly varying features will contribute only to small k region (which is often lost due to the low k cutoff) and/or buried in the noise.

In most cases, the asymmetry in $g(r)$ arises from hard-core repulsion which has a sharp rise in $g(r)$ at a low distance and a long tail at larger distances. Examples include liquids, molten salts, metallic glasses, etc. The sharp rise in $g(r)$ gives rise to the dominant feature in EXAFS which persists to high k. In contrast, the broad tail in $g(r)$ contributes to EXAFS only at low k values, much of which is lost due to truncation at $k_{min} \approx 3\text{Å}^{-1}$. It is then not surprising to find an apparent reduction in amplitude and an apparent shrinkage in the distance.

The dramatic effect of the loss of $k \lesssim 3\text{Å}^{-1}$ information has been demonstrated by Eisenberger and Brown (1979). They used the Percus-Yevick solution to the radial distribution function of a hard sphere liquid to generate an EXAFS spectrum. The transform of the resulting $\chi(k)$ over all k is shown in Fig. 5.9 (curve 1). The reconstructed function obtained by discarding data below $k = 3\text{Å}^{-1}$ is also shown in Fig. 5.9 as curve 2. The hard sphere distribution is characterized by a sharp rise at r_d followed by a more slowly decaying tail. The information contained in the sharp rise will be evident at high k, but the information contained in the tail will be present only for small k, and hence will be lost. Furthermore, the second nearest neighbor peak, which has no sharp structure, will be obliterated by the truncation procedure. It thus appears that without resorting to the modeling of the pair distribution function, one will only be able to extract something related to the distance of the closest approach (r_d).

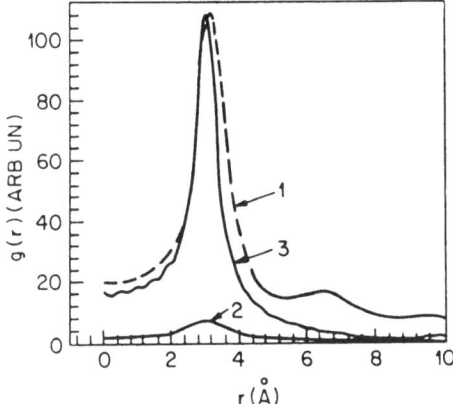

Fig. 5.9. Curve 1, The absolute value of the Fourier transform of the $\chi(k)$ generated by using the Percus-Yevick $g(r)$. Curve 2, the same distribution transformed only for $k > 3$ Å^{-1}. Curve 3, a blown up version of curve 2 to show the absence of the second peak. (Reproduced from Eisenberger and Brown, 1979).

For highly asymmetric distributions, $g(r)$ must be properly modelled. The EXAFS data, in either k or r space, can then be fitted with an appropriate expression based on the model. The results, with appropriate amplitude and phase corrections, are to a significant extent *model dependent* simply because the information loss due to low k truncation cannot be recovered.

As an example, Crozier and Seary (1980) used the following asymmetric distribution function in analyzing EXAFS of liquid metals:

$$g(r) = \frac{1}{2\sigma^3}(r - r_0)^2 e^{-(r-r_0)/\sigma} \quad r \geqslant r_0$$
$$= 0 \qquad\qquad\qquad\qquad r < r_0 \tag{5.51}$$

Here $g(r)$ is normalized and r_0 is the minimum distance of approach which corresponds to the hard sphere contact distance. It can be shown that, for such a distribution function, the maximum $g(r_m) = 2\sigma^{-1}e^{-2}$ occurs at $r_m = r_0 + 2\sigma$. The mean distance $\bar{r}$ occurs at

$$\bar{r} = \int_0^\infty g(r)\, r\, dr$$

$$= r_0 + 3\sigma .$$

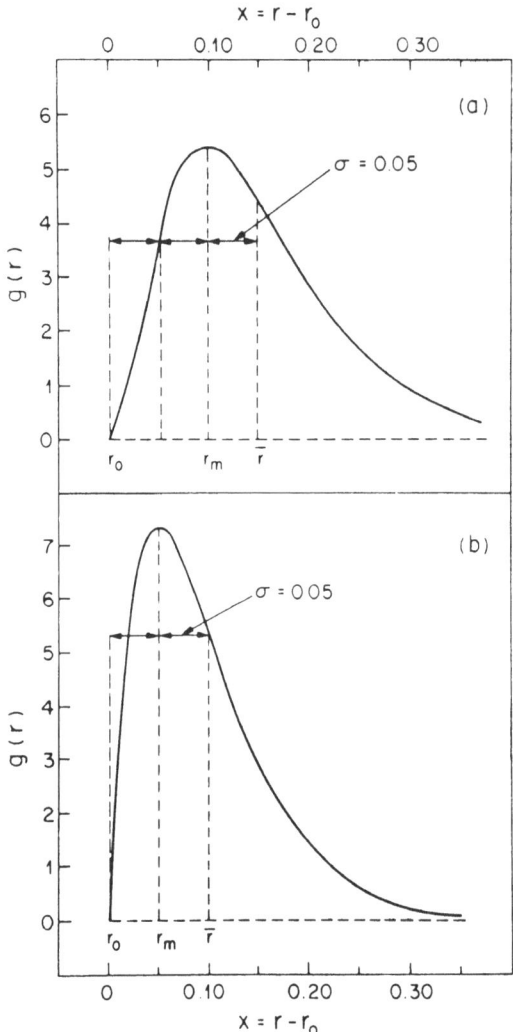

Fig. 5.10. Two asymmetric pair distribution functions: (a) Eq. 5.51 with $\sigma = 0.05\text{Å}$; (b) Eq. 5.54 with $\sigma = 0.05\text{Å}$.

At $r = \bar{r}$, $g(\bar{r}) = (9/2)\sigma^{-1}e^{-3}$. Fig. 5.10(a) depicts one such distribution functions with $\sigma = 0.05\text{Å}$. It is seen that a larger σ value not only broadens the distribution function but also makes it more asymmetrical with a long tail on the high distance side.

Substituting Eq. 5.51 into Eqs. 5.35 and 5.36 and assuming that the $e^{-2r/\lambda(k)}$ and $(1 + x/r)^2$ factors can be taken out of the integration, we obtain

$$\sqrt{S^2 + A^2} = [1 + (2k\sigma)^2]^{-3/2} \qquad (5.52)$$

and

$$\tan^{-1}\frac{A}{S} = \tan^{-1}\left[(2k\sigma)\frac{3 - (2k\sigma)^2}{1 - 3(2k\sigma)^2}\right] \qquad (5.53)$$

Equations 5.51-53 have also been utilized in the EXAFS analysis of metallic glasses (Teo, Chen, Wang, Antonio, 1983).

Other asymmetric distribution functions can also be used. For

$$g(r) = \frac{1}{\sigma^2}(r - r_0)e^{-(r-r_0)/\sigma} \qquad r \geqslant r_0$$
$$= 0 \qquad\qquad\qquad r < r_0 \qquad (5.54)$$

the maximum $g(r_m)$ occurs at $r_m = r_0 + \sigma$ whereas the average distance occurs at $\bar{r} = r_0 + 2\sigma$ (cf. Fig. 10(b), $\sigma = 0.05\text{Å}$). The amplitude and phase modification factors are, respectively:

$$\sqrt{S^2 + A^2} = [1 + (2k\sigma)^2]^{-1} \qquad (5.55)$$

$$\tan^{-1}\frac{A}{S} = \tan^{-1}\left[\frac{4k\sigma}{1 - (2k\sigma)^2}\right] \qquad (5.56)$$

For the pair distribution function

$$g(r) = \frac{1}{\sigma}e^{-(r-r_0)/\sigma} \qquad r \geqslant r_0$$
$$= 0 \qquad\qquad r < r_0 \qquad (5.57)$$

shown in Fig. 11(a) for $\sigma = 0.2$ (solid curve) and 0.4 (dashed curve), it can be shown that the maximum $g(r_m)$ occurs at $r_m = r_0$ and the average distance occurs at $F = r_0 + \sigma$. The amplitude and phase modification factors are:

$$\sqrt{S^2 + A^2} = [1 + (2k\sigma)^2]^{-\frac{1}{2}} \qquad (5.58)$$

and

$$\tan^{-1}\frac{A}{S} = \tan^{-1}(2k\sigma) \qquad (5.59)$$

And for the pair distribution function

$$g(r) = \frac{1}{(\sigma_a - \sigma_b)}(e^{-(r-r_0)/\sigma_a} - e^{-(r-r_0)/\sigma_b}) \qquad r \geqslant r_0$$

$$\qquad\qquad (5.60)$$

$$= 0 \qquad\qquad\qquad\qquad r < r_0$$

where $\sigma_a > \sigma_b$, shown in Fig. 11(b) for $\sigma_a = 0.2$, $\sigma_b = 0.1$ (solid curve) and $\sigma_a = 0.4$, $\sigma_b = 0.2$ (dashed curve), we obtain

$$\sqrt{S^2 + A^2} = \{[1 + (2k\sigma_a)^2][1 + (2k\sigma_b)^2]\}^{-\frac{1}{2}} \tag{5.61}$$

$$\tan^{-1} \frac{A}{S} = \tan^{-1} \frac{2k(\sigma_a + \sigma_b)}{1 - 4k^2\sigma_a\sigma_b} \tag{5.62}$$

One may also introduce the constraint $\sigma_a = 2\sigma_b$ in Eqs. 5.60-62 to reduce the number of parameters as well as to simplify the equations.

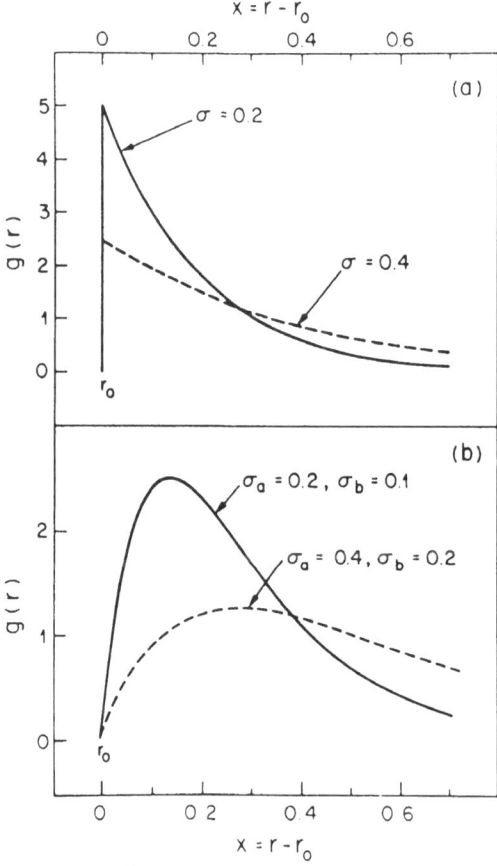

Fig. 5.11. Two asymmetric pair distribution functions: (a) Eq. 5.57 with $\sigma = 0.2$ (solid curve) and 0.4 (dashed curve); (b) Eq. 5.60 with $\sigma_a = 0.2$, $\sigma_b = 0.1$ (solid curve) and $\sigma_a = 0.4$, $\sigma_b = 0.2$ (dashed curve).

5.3.2.5 Anharmonic Vibration Potentials

With an anharmonic potential $U(r)$, serious phase and amplitude corrections can occur. A case in point is the room temperature EXAFS of the c-direction of single crystal Zn where a conventional analysis using the low temperature $(10°K)$ data as a model predicts (1) an "apparent" *contraction* of 0.09Å of the nearest Zn-Zn distance at room temperature, contrary to the known *expansion* of 0.05Å, thereby producing a large error of 0.14Å; and (2) an order of magnitude decrease in peak area in the Fourier transform even though the coordination number is unchanged as the temperature increases (Eisenberger and Brown, 1979). The Fourier transforms of the $k^3\chi(k)$ data at 10 and 280°K are shown as curves 1 and 2, respectively, in Fig. 5.12. It is apparent that the peak area is not conserved. This is due to the fact that the shape of the radial distribution function changes as the temperature (or disorder) is increased. Since EXAFS sees only the more sharply peaked or rapidly varying parts of $g(r)$ and since thermal expansion is due to an increase in width and a distortion of $g(r)$ due to anharmonic terms in the interatomic potential, only the more ordered part of the distribution, which is the part at shorter distances, is observed in the EXAFS spectrum. The area under the $g(r)$ curve in this close-approach region decreases with increasing temperature, so the apparent coordination number determined by EXAFS also decreases. A related study on polycrystalline cadmium was done by Thulke and Rabe (1983).

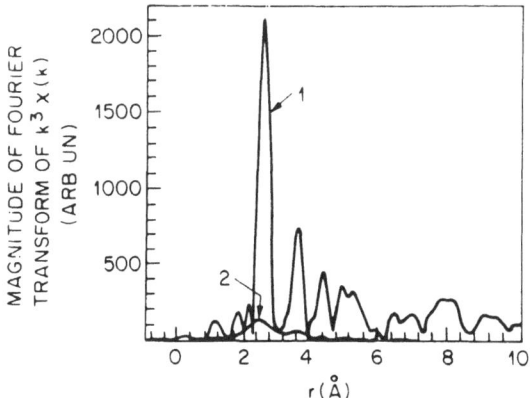

Fig. 5.12. Curves 1 and 2 are the Fourier transform of the $k^3\chi(k)$ for $k > 3Å^{-1}$ for 10°K and 285°K data of single-crystal Zn along the c direction. (Reproduced from Eisenberger and Brown, 1979).

For large thermal disorder, the EXAFS must be *integrated* over the Boltzman distribution

$$g(r) = Ce^{-U(r)/k_{\blacksquare}T} \tag{5.63}$$

where $U(r)$ is the interatomic potential (which may well be anharmonic) and C is a normalization constant.

As an example, Eisenberger and Brown (1979) modelled the interatomic potential with the Lennard-Jones potential

$$U(r) = U^{\bullet}\left[\left(\frac{r_0}{r}\right)^{12} - 2\left(\frac{r_0}{r}\right)^6\right] \tag{5.64}$$

with different values of $U^{\bullet}/kT$. As expected, for small $U^{\bullet}$ (weak potential) or at high temperature T, $g(r)$ is increasingly broadened. Marques, Sandstrom, Lytle, and Greegor (1982) instead used the Morse potential

$$U(r-r_0) = U^{\bullet}[e^{-2\alpha(r-r_e)} - 2e^{-\alpha(r-r_e)}] \tag{5.65}$$

to model the asymmetric pair distribution in supported catalysts. Here $U^{\bullet}$ is the depth of the potential well at $r = r_0$ and α provides a measure of the disorder. Again, the peak area in the Fourier transform is not conserved.

5.3.2.6 Comparison of EXAFS and Diffraction

Recently, detailed comparisons have been made between EXAFS and other diffraction techniques in terms of their ability in providing information concerning the Debye-Waller factor and the radial distribution function (See, for example, Mobilio and Incoccia, 1984). It is now clear that X-ray diffraction is sensitive to the absolute displacement of atoms from their equilibrium position such that its Debye-Waller (DW) factor measures the total density of phonons. On the other hand, EXAFS is sensitive only to the relative distance between atoms, so its DW factor provides a projected density of phonons (Sevillano, Meuth, Rehr, 1979). Both methods provide the static radial distribution function (RDF), but with different emphasis: X-ray diffraction is sensitive to the smooth behavior of the RDF, EXAFS is more sensitive to the sharp feature of the RDF. For highly disordered systems or systems with an anharmonic potential, a comparative study is the only way to fully understand the dynamic and structural properties of the system. The difficult task of calculating the full potential parameters of the system have been attempted only in a few cases (Crozier and Seary, 1980; Crozier in Teo and Joy, 1981; Boyce, Hayes, Mikkelsen, 1981).

5.4 Multiple Scattering EXAFS Formalism

The single-electron single-scattering theory of EXAFS discussed thus far assumes that EXAFS is the sum of terms which correspond to the *single backscattering* processes from the neighbors, one for each type of atom(s). This approximation makes use of the fact that in most cases multiple scatterings are not important. This assumption is generally valid if one considers that multiple scattering processes can be accounted for by adding all scattering paths that originate and terminate at the central atom (absorber). Each of these processes then behaves like $\sin(2kr_{eff})$ where $2r_{eff}$ is the total scattering path length which is much larger than that of the direct backscattering(s) from the nearest neighbors. Thus, multiple scatterings will give rise to rapidly oscillatory waves in k space which tend to cancel out. The amplitude of these waves are also significantly attenuated by the large scattering path lengths, making it relatively unimportant in comparison with the direct backscattering.

On the other hand, multiple scattering in EXAFS can become important when atoms are arranged in an approximately colinear array. In such cases, the outgoing photoelectron is strongly *forward-scattered* by the intervening atom, resulting in a significant amplitude enhancement. In fact, both the amplitude and the phase are modified by the intervening atom(s) for bond angles. The effect, however, drops off very rapidly for bond angles below *ca* 150°. For these systems, it is necessary to take into account multiple scattering involving the intervening atom(s).

For the sake of simplicity, let us consider a three-atom array A-B-C where A is the central atom (absorber), B is the nearest neighbor (the intervening atom), and C is the next nearest neighbor. For such a system, the EXAFS of the absorber A is comprised of two contributions, one from the *backscattering* of B and the other from the scattering of C via the intermediary atom B. We shall designate these two contributions as AB and AC respectively. The former can be described quite adequately by the *backscattering* from the atom B with the *single-electron single-scattering theory*. The latter, which is affected by multiple scattering involving the intervening atom B, must be treated with a modified formulation as will be discussed in Chapter 8. It suffices to note here that the single-scattering theory provides only distance information whereas the multiple scattering theory can provide both distance and angle information. For the three-atom system, the former gives rise to two distances while the latter yields two distances and one angle which completely characterize the ABC triangle.

Chapter 6

Data Analysis in Practice

The first step in EXAFS data analysis is to convert the experimentally measured total absorption data $\mu(E)$ in energy (E) space to the interference function $\chi(k)$ in photoelectron momentum (k) space. This *data reduction* procedure involves background removal, conversion of E to k, normalization, μ_0 correction, weighting scheme, etc. In order to extract structural information from $\chi(k)$, an appropriate function of $\chi(k)$ in terms of the structural parameters must be chosen (or developed in some cases) as described in the previous chapters. The choice of these phenomenological expressions, often quite equivalent or similar, depends largely on the system of interest and the data analysis technique to be employed. Though vary widely in form, there are basically two major approaches to EXAFS data analysis: the Fourier transform (FT) and the curve fitting (CF) techniques. Each of these techniques has its own advantages and disadvantages. A compromise of these two methods is Fourier filtering (FF) followed by curve fitting which is the most commonly used technique. We shall describe these as well as other methods in some detail.

6.1 Data Reduction

6.1.1 Conversion of Experimental Variables

The first step in data reduction is to convert the experimental "variables" into an experimental "spectrum." The Y-axis is the total linear absorption coefficient $\mu(E)$ which is

$$\mu(E)x = \ln \frac{I_0}{I} \qquad (6.1)$$

in transmission experiments but

$$\mu(E)x = \frac{F}{I_0} \qquad (6.2)$$

in fluorescence experiments. Notice the inversion of I_0 and I in the two expressions as well as the absence of the natural log in the latter. Here I_0, I, and F are the incident, transmitted, and fluorescent X-ray intensities, respectively, and x is the sample thickness. For fluorescence measurements with more than one detector, F is the weighted average fluorescence signal (the center detector usually accounts for 50% of the total signal). The abscissa is the photon energy E which can be calculated from the monochromator settings. In the digitized form, each data point then consists of I_0, I or F's and E from which $\mu(E)$ vs E can be computed via Eq. 6.1 or 6.2. Plots of $\mu(E)$ vs E are shown schematically as solid curves in Fig. 6.1(a) and (b) for transmission and fluorescence data, respectively.

6.1.2 Background Removal

The interference function $\chi(k)$ is defined as

$$\chi(E) = \frac{\mu(E) - \mu_0(E)}{\mu_0(E)} \qquad (6.3)$$

above the absorption edge where $\mu(E)$ is the absorption coefficient due to the particular edge of the element of interest in the sample and $\mu_0(E)$ is the absorption coefficient of an isolated atom. Since $\mu_0(E)$ is generally not known, it is often assumed that the smooth part of the measured $\mu(E)$ (i.e. without the wiggles or oscillations) approximates $\mu_0(E)$. With this assumption, the background can be removed from $\mu(E)$ by calculating $\Delta\mu(E) = \mu(E) - \mu_0(E)$ where $\mu(E)$ is the experimental curve and $\mu_0(E)$ is the smooth curve best fit to $\mu(E)$ as illustrated in Fig. 6.1 (dashed curves).

In principle, a distinction should be made between the experimentally determined total absorption $\mu(E)$, either in transmission (solid curve in Fig. 6.1(a)) or fluorescence (solid curve in Fig. 6.1(b)) mode and the absorption due to the particular edge of the element under consideration $\mu(E)$ (solid curve in Fig. 6.1(c)). A procedure has been described by Lytle, Sayers, Stern (1975) with which the "elemental absorption" $\mu(E)$ (solid curve in Fig. 6.1(c)) can be extracted from the experimental absorption curve (solid curves in Fig. 6.1(a) and (b)). The pre-edge absorption curve can be fitted with a polynomial, a line, or simply a constant. The latter methods have been used for both transmission and fluorescence data while a polynomial which corresponds to Victoreen's $C\lambda^3 - D\lambda^4$ formula is commonly used for transmission data. The pre-edge fit (dash-dot curves in Figures 6.1(a) and (b)) is then *extrapolated* beyond the edge and *subtracted* from the total absorption, $\mu(E)$, in Fig. 6.1(a) or (b), to

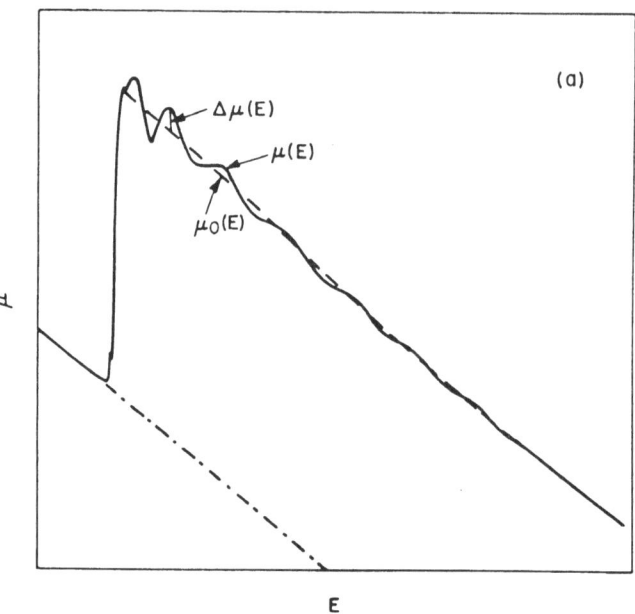

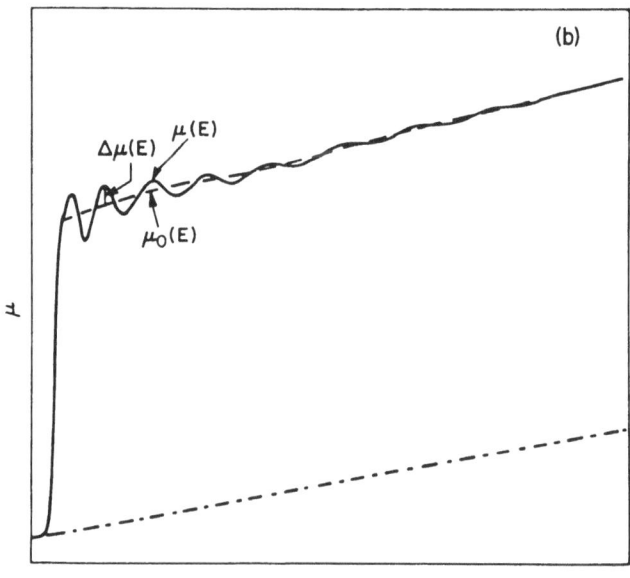

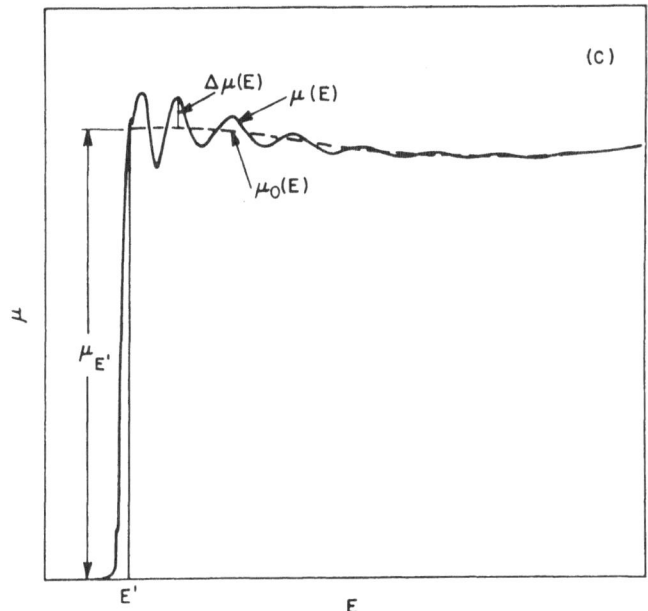

Fig. 6.1. A commonly used procedure for background removal for either transmission (a) or fluorescence (b) absorption data $\mu(E)$ *vs* E. The pre-edge region is first fitted with a polynomial, a line, or simply a constant which is extrapolated beyond the edge. The pre-edge fit (dash-dot curves in (a) and (b)) is then subtracted from the total absorption data (raw data, solid curves) to give the "elemental absorption" $\mu(E)$ *vs* E shown as the solid curve in (c) from which the edge-jump $\mu_{E'}$ and the experimental energy threshold E' can be determined. (This step, however, can be omitted if one chooses to use a theoretical $\mu_0(E)$ curve.) The background removal amounts to first fitting the experimentally measured total absorption curve $\mu(E)$ (solid curves in (a)-(c)) with a smooth "background absorption" curve $\mu_0(E)$ (dashed curves in (a)-(c)) which is subsequently subtracted from $\mu(E)$ to give the $\Delta\mu(E)$ curve. Normalization, μ_0 correction, and conversion of E to k give rise to $\chi(k)$ *vs* k data (see text).

give the "elemental absorption," $\mu(E)$, in Fig. 6.1(c). In practice, however, the "elemental absorption" $\mu(E)$ thus determined still contains "other background factors" such as spectrometer baseline, beam harmonics, elastic scatterings, etc. Hence, this step (from Fig. 6.1(a) or (b) to (c)) is sometimes omitted. The background removal then reduces to first fitting the experimentally measured total absorption curve $\mu(E)$ (solid curves in Fig. 6.1(a) or (b)) with a smooth "background absorption" curve $\mu_0(E)$ (dashed curves) which is subsequently subtracted from $\mu(E)$ to give the $\Delta\mu(E)$ curve.

Various fitting procedures have been developed for background removal with varying degrees of sophistication: e.g., a single polynomial fit over the whole range of the data, iterative low-order polynomial fitting combined with fitting the EXAFS, orthogonal polynomial fitting, Fourier transform filtering, running averages or sliding windows (Wong, 1980), convolution with Gaussian functions of different widths (Boland, Halaka, Baldeschwieler, 1983), etc. While most of these methods are in general quite adequate for background removal, some suffer from being too sensitive to experimental boundary conditions such as end point effects or experimental imperfections such as noise or glitches in the data.

One commonly used method for removing the background to fit $\mu(E)$ with polynomial splines or B-splines (Ahlberg, Nilson, Walsh, 1967; de Boor 1968, 1972; Fox et al.; 1976) using a least squares procedure. A polynomial spline or B-spline is a function defined over a series of intervals with each interval containing a polynomial of some order. The ends or knots of the intervals are tied together such that the function and a specified number of derivatives are continuous across the knots. By specifying the number of intervals (sections) and the order of the polynomials a very flexible function can be defined. A least squares fit with such a spline function readily enables the removal of low frequency background components from $\mu(E)$ without affecting the higher frequency EXAFS oscillations. Since the number of degrees of freedom of the function is controllable, continuous changes in the background removal can be obtained and their effect on the EXAFS part of the data tested. Spline fitting is essentially a local fitting procedure in that the polynomial function within each interval is mainly determined by the local quality of the fit. In some cases only the first derivative is required to be continuous, so the global requirement of continuity across the knots is rather weak. This enables the spline-fitting method to deal with a variety of slowly varying bumps and valleys in the background.

It is obvious that the spline technique is well suited for EXAFS background removal since continuous control of the low-frequency background removed is possible with a few well-behaved parameters. Higher orders or greater number of sections will in general allow better removal of the low-frequency (slow-varying) background. The danger of using too high orders or too many sections is that it will remove part of the EXAFS oscillation as well and hence reduce some part of the EXAFS signal. On the other hand, too low orders or not enough sections will result in a low distance peak (at around 1Å) in the Fourier transform which may distort the real peaks. For most EXAFS data with a typical energy range of 600-1000 eV above the edge, cubic splines (with polynomials to the third order) of 3-5 sections (such that $\Delta k \approx 3\text{Å}^{-1}$ within each section) are often used.

Cook and Sayers (1981) suggest the use of three parameters obtained from the Fourier transform of the k^3 weighted EXAFS data as criteria for choosing smoothing parameters in cubic spline background removal. The three parameters are H_R, the average value of the transform magnitude between 0 and 0.25Å; H_M, the maximum value in the transform magnitude between 1 and 5Å; and H_N, the average value of the transform magnitude between 9 and 10Å. H_R is the measure of any low-frequency (low r) component which remains in the EXAFS data; H_M is a reference value against which H_R and H_N can be judged; and H_N is related to the high-frequency noise in the data. The criteria used may be written as either $H_R - H_N \geqslant 0.05\, H_M$, or if $H_N > 0.1\, H_M$, then $H_R \geqslant 0.1\, H_M$. The latter criterion gives somewhat better results for noisy data.

6.1.3 Normalization and μ_0 Correction

Since $\chi(E) = \Delta\mu(E)/\mu_0(E)$, the background-removed $\Delta\mu(E)$ curve must be normalized with respect to $\mu_0(E)$. Ideally one would like to measure or calculate $\mu_0(E)$ in the absence of neighboring atoms (i.e., with no EXAFS). In practice, however, the "experimental" background $\mu_0(E)$ (dashed curves in Fig. 6.1) fitted to the experimental curve $\mu(E)$ is complicated by many factors including imbalance of the counting efficiencies of the detectors (spectrometer baseline), absorption due to other constituents or media (i.e., residual absorption due to other elements or other edges of the same element under consideration), beam harmonics, sample thickness, elastic scatterings, the λ^3 decay (cf. Chapter 1), etc. For example, in transmission experiments, the background $\mu_0(E)$ generally decreases with increasing energy E; the (negative) slope, however, can vary significantly under different experimental conditions (detecting gases, other materials in the sample, etc.). In contrast, the fluorescence data normally show just the opposite trend. The positive slope of $\mu(E)$ vs E in fluorescence experiments is due to the fact that as the photon energy increases, the X-ray beam is penetrating deeper into the sample thereby producing more fluorescence signals. The spectrometer baseline also depends on the detector gains. It is obvious from these considerations that the "experimental background" cannot be used in the normalization procedure except, perhaps, under ideal conditions.

One commonly used method in this regard is to normalize $\Delta\mu(E)$ by dividing it with the theoretically calculated $\mu_0^{th}(E)$ such as the Victoreen's (1948) true absorption coefficient

$$\mu_0^{th}(E) = C\lambda^3 - D\lambda^4 \qquad (6.4a)$$

$$= C\left(\frac{hc}{E}\right)^3 - D\left(\frac{hc}{E}\right)^4 \qquad (6.4b)$$

or those tabulated by McMaster (1969). A simple way to implement this is to first divide $\Delta\mu(E)$ by the jump $\mu_{E'}$ at the absorption edge (i.e., the increase in absorption at the edge — a constant easily determined from the experimental $\mu(E)$ vs E data; cf. Fig. 6.1) followed by dividing each $\Delta\mu(E)$ data point with the ratios $\mu_0^{th}(E)/\mu_0^{th}(E')$. Thus

$$\chi(E) = \frac{\Delta\mu(E)}{\mu_{E'}[\mu_0^{th}(E)/\mu_0^{th}(E')]} \qquad (6.5)$$

The division of $\Delta\mu(E)$ by $\mu_{E'}$ "normalizes" the spectrum whereas the division by $[\mu_0^{th}(E)/\mu_0^{th}(E')]$ "corrects" for the μ_0 dropoff which amounts to roughly 15% at ca 500 eV above the edge. Obviously the energy E' at which the edge jump $\mu_{E'}$ is determined should be chosen in a consistent manner when comparing different data. It must be above and near the absorption edge. It should be close to, but not necessarily the same as, the energy threshold E_0. For fluorescence data, experience shows that the latter correction is not necessary since the EXAFS signal increases with increasing energy which roughly cancels the μ_0 drop off.

6.1.4 Conversion of E to k

The next step in the data analysis is to convert the photon energy E to the photoelectron wave vector k via

$$k = \sqrt{\frac{2m}{\hbar^2}(E - E_0^{exp})} \qquad (6.6a)$$

where E_0^{exp} is the "experimental" threshold. For E in eV and k in Å^{-1},

$$k = \sqrt{0.2625(E - E_0^{exp})} \qquad (6.6b)$$

E_0^{exp} is normally chosen as some noticeable feature at or near the absorption edge (e.g. the first sharp peak, the edge position, etc.) as described in the previous chapter (cf. Fig. 5.1). The exact choice of E_0^{exp} is not very important since in the final analysis, E_0 will be allowed to vary. However, it should be chosen consistently for compounds within the same series. The resulting spectrum is $\chi(k)$ vs k based on the E_0^{exp} used as shown in Fig. 6.2(a).

6.1.5 Weighting Scheme

In order to compensate for the attenuation of the EXAFS amplitude at high k values, $\chi(k)$ is often multiplied by some power of k to give $k^n \chi(k)$. This is equivalent to applying a weighting scheme which goes as k^n (i.e., giving the high k region exponentially higher weights). This procedure is important in preventing the larger amplitude oscillations from dominating the smaller ones in the determination of interatomic distances which depends only on the frequency and not the amplitude. Weighting schemes with $n = 1$, 2, and 3 have been suggested by Teo and Lee (1979) for backscatterers with $Z > 57$, $36 < Z < 57$ and $Z > 36$, respectively. The factor k^3 effectively compensates for the $1/k$ dependence in the EXAFS equation and roughly cancels the $1/k^2$ behavior of the backscattering function $F(k)$ (viz., in the Born approximation at high k, $F(k)$ is roughly proportional to $1/k^2$.) The k^3 multiplication has the effect of weighting the EXAFS oscillations more uniformly over the data range of $k = 3 - 16 Å^{-1}$ as depicted in Fig. 6.2(b). The $k^3 \chi(k)$ may be approximated to a pseudo-charge density which is least sensitive to k_{min} and E_0. Also, the k^3 weighting assures that chemical effects on the EXAFS information, which are most significant at small k, are minimized. One variation of the k^3 multiplication factor that has been used (Ashley and Doniach 1975, Lee and Beni 1977) is to divide the experimental data by a theoretically calculated $F(k)$. In principle, this should exactly equalize the EXAFS oscillations if the Debye-Waller and inelastic loss factors are also correctly included in the division.

We note that weighting scheme may have a significant effect on the peak heights and peak positions in the Fourier transforms as well as the fitting parameters in the curve fittings. It is therefore important to use the same weighting scheme when comparing unknowns with the models. Parameters are transferable or comparable only if the data are analyzed in a similar fashion.

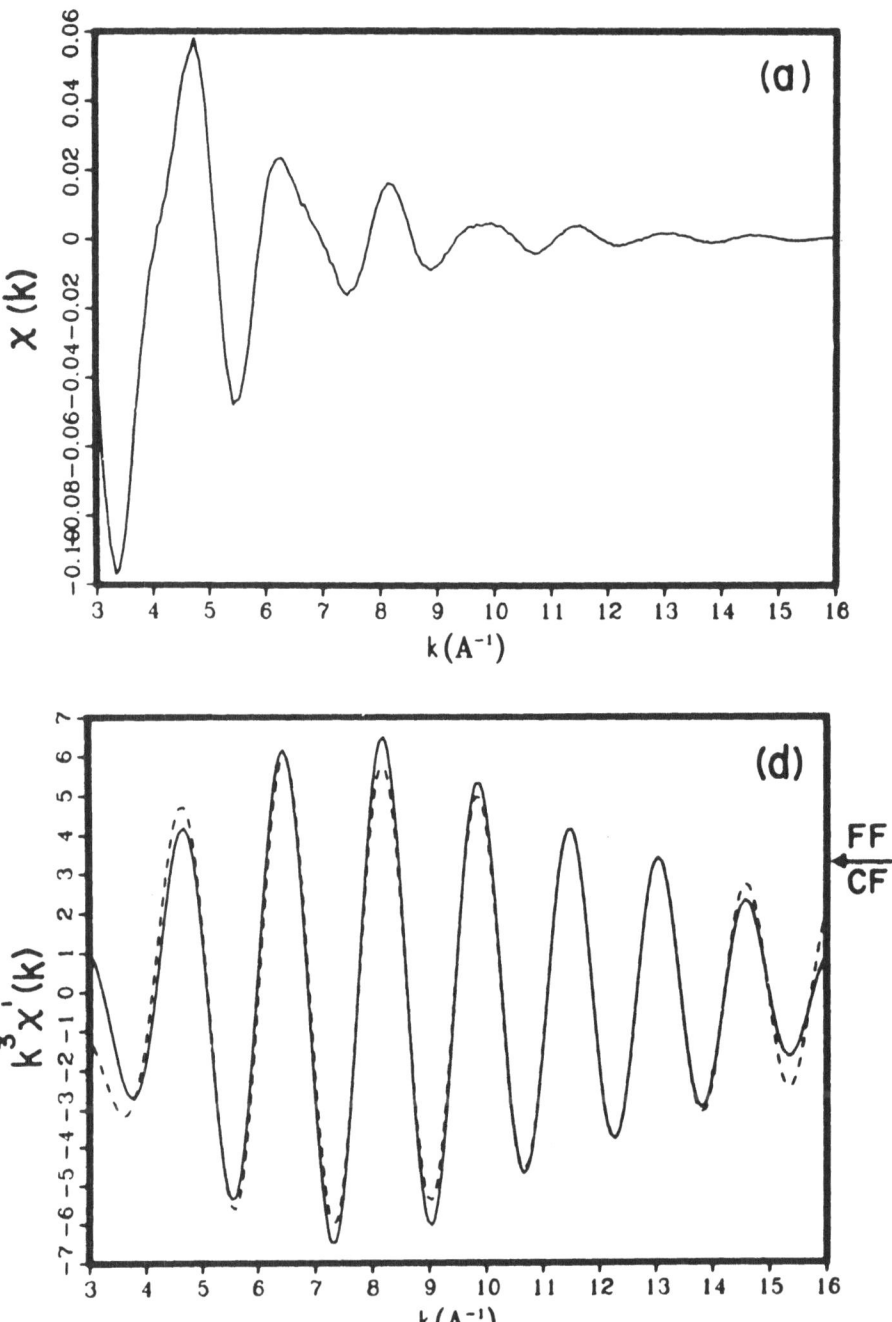

Fig. 6.2. Data reduction and data analysis in EXAFS spectroscopy: (a) EXAFS spectrum $\chi(k)$ vs k after background removal, normalization, and E to k conversion; (b) the solid curve is the weighted EXAFS spectrum $k^3\chi(k)$ vs k (after multiplying $\chi(k)$ by k^3). The dashed curve represents an attempt to fit the data with a two-distance model by the curve-fitting (CF) technique; (c) Fourier transformation (FT) of the weighted EXAFS spectrum in momentum (k) space into the radial distribution function $\rho_3(r')$ vs r' in distance space. r'

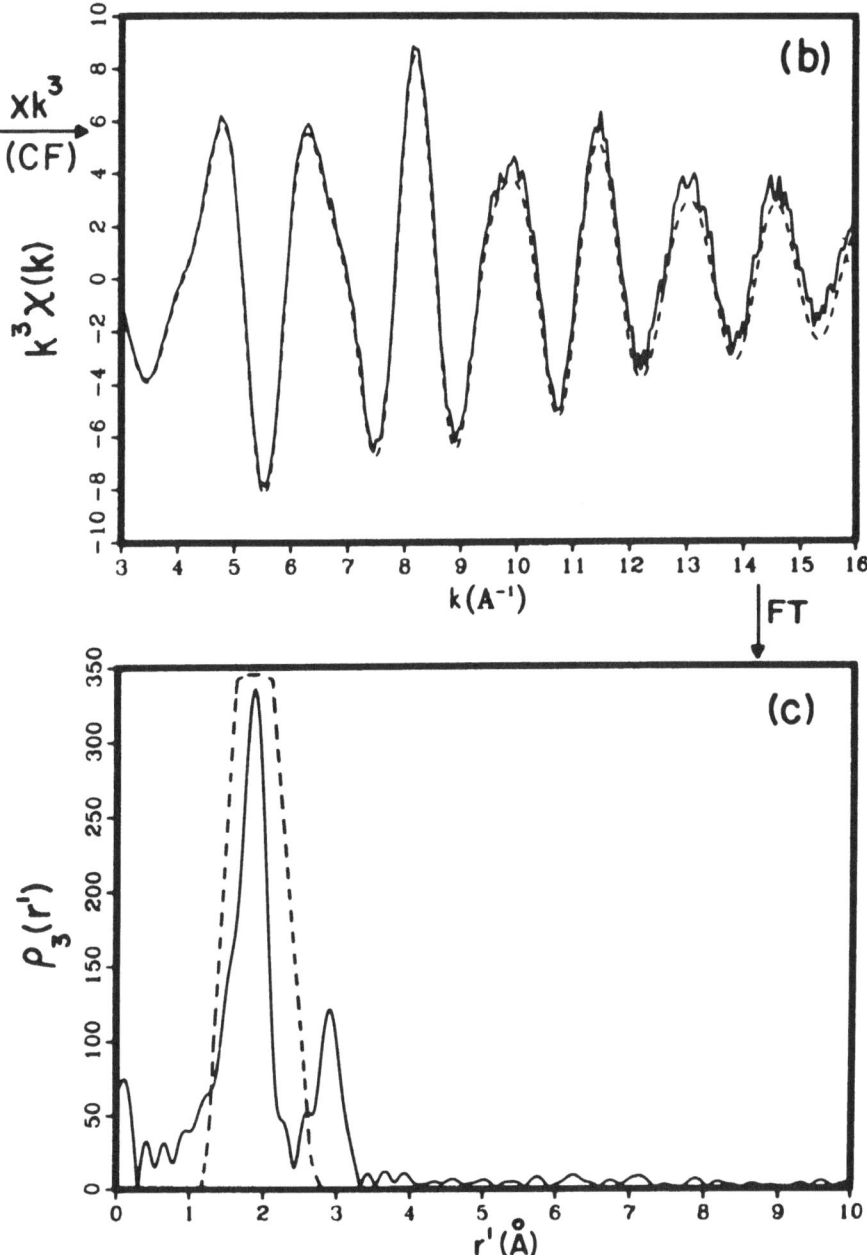

is related to the true distance r by a "phase shift" $\alpha = r - r'$. The dashed curve is the window function used to filter the major peak in Fourier-filtering (FF); (d) Fourier-filtered EXAFS spectrum $k^3\chi'(k)$ *vs* k (solid curve) of the major peak in (c) after backtransforming into k space. The dashed curve attempts to fit the filtered data with a single distance model. Likewise the minor peak in (c) can be Fourier filtered and backtransformed into k space and curve-fitted with a different single-distance model. (Reproduced from Teo, 1980).

6.1.6 Deglitching and Truncation

It should be mentioned that any discontinuities such as "glitches," spikes, or steps in the data will adversely affect the background removal process and hence the accuracy of the EXAFS data. These discontinuities should be removed either before or after background removal. This is most conveniently accomplished by a linear interpolation or other smoothing schemes through the questionable data region. Deglitching can also be performed in momentum k space (i.e., after background removal, conversion to k, and weighted) for sharp glitches covering only a few eV.

Furthermore, in order to avoid the edge effect and poor signal-to-noise at high energy (where the EXAFS oscillations are normally less than a few percent of the maximum oscillations), the data should be truncated at low and high k, respectively, before further data analysis (*vide infra*). The extent of truncation obviously depends on the quality of the data.

6.2 Fourier Transform (FT)

One of the earliest analyses of EXAFS (Sayers, Stern, Lytle, 1971) was based on the Fourier transform (FT) of the data expressed in momentum space. The absolute value of the transform was found to peak at distances shifted from the known values by several tenths of angstroms. By correcting for these shifts using systems with known distances, bond length information can be extracted (Stern, 1974; Stern, Sayers, Lytle, 1975; Ashley and Doniach, 1975; Lee and Pendry, 1975; Lee and Beni, 1977).

The Fourier transformation of $k^n\chi(k)$ in momentum (k) space over the finite k range k_{min} to k_{max} gives rise to a modified radial distribution function $\rho_n(r')$ in distance (r') space as shown in Figure 6.2(c).

$$\rho_n(r') = \frac{1}{(2\pi)^{1/2}} \int_{k_{min}}^{k_{max}} w(k)k^n\chi(k)e^{i2kr'}dk \qquad (6.7)$$

Each peak in $\rho_n(r')$ is shifted from the true distance r by $\alpha = r - r'$ where α amounts to ca $0.2 \sim 0.5\text{Å}$ depending upon, among others, the elements involved, the E_0 chosen, and the weighting scheme. α can be obtained from model compounds and transferred to the unknown systems to predict distances. If one assumes, to a first approximation, a linear phase function $\phi(k) = p_0 + p_1k$, the peaks are shifted by $\alpha = p_1/2$ since

$\chi(k) \propto \sin(2kr + \phi(k)) = \sin[p_0 + (p_1 + 2r)k]$. For closely related systems, the approximate number of neighboring atoms can be calculated by $N = \dfrac{Ar^2}{A_s r_s^2} N_s$, where N_s, A_s, r_s and N, A, r are the number of atoms, the Fourier transform peak magnitudes (or peak height if the two Debye-Waller factors are similar), and the interatomic distances in the standard (with subscripts s) and the unknown compounds, respectively.

Also included in the Fourier transform is a window function $w(k)$ which selects the k range to be transformed. $w(k)$ can be a square window if care is taken to choose the cutoffs to be where $\chi(k)$ is small or a smooth window such as the following Hanning function can be applied.

$$w(k) = \frac{1}{2}\left[1 - \cos 2\pi \left(\frac{k - k_{min}}{k_{max} - k_{min}}\right)\right] \qquad (6.8)$$

It is clear that $w(k) = 0$ at $k = k_{min}$ and k_{max}. This window is normally applied to the first and the last 5-10% of the data while keeping $w(k) = 1$ for the remaining data set. It is clear that this window smoothly sets the data to zero ($w(k) = 0$) at k_{min} and at k_{max}.

In general, this method works well for systems with well separated peaks. The weakness of it is that since the amplitude $F(k)$ and the phase $\phi(k)$ functions have characteristic k dependence (as we shall see in Chapter 7, $F(k)$ has peaks and valleys in k space while $\phi(k)$ is a nonlinear function of k), the Fourier transform peak magnitude A and the phase shift α in the distance space depend on E_0, the weighting of the data, the data range in k space, and the Debye-Waller factors, etc.

In fact, since heavy scatterers have amplitude envelopes ($\chi(k)$ in k space) which peak at higher k region than the light scatterers, it may be possible to differentiate scatterers with significantly different atomic numbers by calculating the ratio of the amplitude of a particular peak in the Fourier transforms ($\phi_n(r)$ in r space) with respect to a reference peak (such as the major coordination peak which involves only light scatterers) of known nature. Such a ratio should increase as n increases (say from 2 to 5) for a heavy scatterer as a result of the increasing weighting on the higher k region. On the other hand, the opposite (or little effect) may be true when a light scatterer is involved. Furthermore, owing to the nonlinearity of the phase shifts as a function of k (viz., the absolute value of the slope decreases as k increases), the position of a light scatterer peak in the Fourier transform will shift to (successively) larger distances as n increases. For heavy scatterers, this effect is less profound and therefore little or no shift will be observed.

Other problems of the Fourier transform technique include: (1) "side lobes" (due mainly to finite data length) which can interfere with the weaker peaks; (2) a "background" peak at or below a distance of ca 1Å due to residual background which may interfere with the short distances of interest (say at ca 1.5Å); (3) the position, magnitude, and shape of peaks are to some extent affected by the "boundary conditions" as such as E_0, weighting scheme, and data range; and (4) skewing of the transformed envelope (due to the nonlinear character of the phase and amplitude) which may affect the peak position.

It is obvious that the Fourier transform technique has the advantage of providing a simple physical picture of the local structure around the absorber. It must be emphasized, however, that while the magnitude of the Fourier transform does look like the radial distribution function $g(r)$, and the major peaks do correspond to shells of neighboring atoms, it is only a modified version of the true radial distribution function. The EXAFS equation (K edge) can be generalized for the case of a continuous radial distribution function to give

$$\chi(k) = - \int_0^\infty \frac{g(r)}{kr^2} \sin(2kr + \phi(k)) dr. \qquad (6.9)$$

Here the backscattering amplitude function, the inelastic electron mean free path, and Debye-Waller factors have been omitted for the sake of clarity and only one kind of atom-pair is considered. Hence, $\chi(k)$ is given by a modified Fourier transform of $g(r)$. If $\phi(k)$ is either known in advance or can be assumed to be closely approximated by a straight line, and if $\chi(k)$ is known for values of k from 0 to ∞, inversion of Eq. 6.9 to get $g(r)$ is possible. In real data, however, k covers only the range from k_{min}, about 3 or 4Å^{-1}, to k_{max}, usually no more than about 16Å^{-1}. Only an approximation of $g(r)$, i.e., one which has been effectively low-pass and high-pass filtered in k, can be calculated. The lack of high k information limits the resolution of the Fourier transform and later analysis, and the absence of low k information filters out all slowly varying components in $g(r)$ as a function of r (Lee, et al., 1981). As already mentioned in the previous chapter, this can produce very peculiar results in the EXAFS study of amorphous materials or other systems with a large amount of thermal disorder (Eisenberger and Brown, 1979, Hayes et al., 1978).

In particular, it is found that EXAFS can predict a *shrinkage* of interatomic distances with increasing temperature rather than the actual thermal expansion, along with a *decrease* in the number of nearest neighbors by as much as an order of magnitude. This is due to the fact that the shape of the radial distribution function changes as the temperature (or disorder) is increased. Since EXAFS sees only the more sharply peaked or rapidly varying parts of $g(r)$ and since thermal expansion is due to an increase in width and a distortion of $g(r)$ due to anharmonic terms in the interatomic potential, only the more ordered part of the distribution, which is the part at shorter distances, is observed in the EXAFS spectrum. The area under the $g(r)$ curve in this close-approach region decreases with increasing temperature, so the apparent coordination number determined by EXAFS also decreases.

6.3 Fourier Filtering (FF)

Fourier filtering involves Fourier transforming the $k^n\chi(k)$ data into the distance space, selecting the distance range of interest with some smooth window (dashed curve in Fig. 6.2(c)), and back transforming the data to k space (Fig. 6.2(d)). The resulting 'filtered' EXAFS spectrum $k^n\chi'(k)$ can then be fitted with simpler models (dashed curve in Fig. 6.2(d)) containing fewer number of distances. Generally speaking, for multi-term systems with significantly different distances ($\Delta r \geq 0.4$Å) in the Fourier transform, each individual shell can be isolated in this way and analyzed separately.

Though it is possible to use curve fitting techniques to extract information from the unfiltered data, it is, in practice, difficult to do so because of the noise in the data, the large number of variable parameters needed and their correlations, which can lead to false minima in the fit and misinterpretation of the results (Tullius et al. 1978). Fourier filtering has the advantage of simultaneous removal of the high frequency noise and the residual background as well as providing equally-spaced data points in k space. The latter is important since the raw data $\mu(E)$ with equally-spaced data points in energy space, after converting E to k, will give rise to $\chi(k)$ with increasing density of data points as k increases because of the $k \propto \sqrt{E}$ relation.

Fourier filtering is performed by first interpolating the non-uniformly spaced data onto a uniformly spaced k mesh and then extending the data by adding 0's to a range from $k = 0$ to about $k = 150$ Å^{-1}. The extended data is then Fourier transformed. The effect of the added zeros is equivalent to transforming only the original data and then applying an interpolation formula to the resulting transform. The dashed curve in Fig. 6.2(c) is a smooth filter window-function for isolating one coordination shell. The filter is shown applied to the magnitude of the transform, but in fact both the real and imaginary parts of the Fourier spectrum are filtered identically so that a phase error is minimized. The essential feature of a good filter window-function is that it does not produce severe distortions in the data, and many smoothly varying function satisfies this requirement. After filtering, the inverse transform is performed to give the solid curve in Fig. 6.2(d).

Fourier filtering, however, can cause some distortions (especially in the amplitude) at the boundaries of the spectrum. It is therefore beneficial to choose a sufficiently wide window and to use a data length shorter than the original one after filtering (viz., truncated at low and high k extremes). A common practice is to eliminate the first and the last wiggles in the data; and for subsequent data analysis, a data range of $k = 3$ to 13Å^{-1} is generally adequate. It should also be noted that the number of true degrees of freedom in the data is reduced; this must be taken into account in later analysis of the filtered data.

6.4 Curve Fitting (CF)

The curve fitting (CF) technique attempts to best fit the $k^n\chi(k)$ spectra in k space with some phenomenological EXAFS models (dashed curves in Fig. 6.2(b) and (d)). There are several variants of this method. The refinement procedure can be based on least-squares (commonly used), absolute-difference minimization, or other criteria. One could multiply $\chi(k)$ by some power of k to give $k^n\chi(k)$ before fitting so as to emphasize different k regions (higher n values imply more weights on the higher k regions) or simply apply a weighting scheme as a function of k. One could fit the entire EXAFS spectrum directly as is often done or, in cases where strong correlation exists between the amplitude and the phase, one could fit the experimental amplitude and phase functions separately as we shall discuss in a later section.

6.4.1 Parameterization

It is convenient to parameterize the amplitude and phase functions with simple analytical forms. The amplitude function $F(k)$ of scatterers can be parameterized with a simple sum of Lorentzians (Teo, et al, 1977)

$$F(k) = \sum_i \frac{A_i}{1 + B_i^2(k - C_i)^2} \qquad (6.10)$$

where A is the peak height, $2/B$ is the width and C is the peak position of the amplitude function in k space. For $Z \leq 36$, $36 \leq Z \leq 57$, and $57 \leq Z \leq 86$, one, two, and three Lorentzians are needed for the weighting schemes of $k^3\chi(k)$, $k^2\chi(k)$, and $k\chi(k)$, respectively. Other functional forms ranging from the simple two-parameter form (Cramer, *et al*, 1976)

$$F(k) = \frac{C}{k^\beta} \tag{6.11}$$

where $\beta \approx 2$, to the complicated ten-parameter form (Shulman, *et al*, 1975)

$$F(k) = \frac{a_0 + a_1 k + \ldots + a_5 k^5}{1 + a_6(k - a_{10}) + \ldots + a_9(k - a_{10})^4} \tag{6.12}$$

have also been used in the literature. The Lorentzian form is preferred in that it describes nicely the characteristic maxima of experimental amplitude curves with a minimum number of parameters. At high enough energy $(k \gg C)$, it reduces to the well-known Born approximation of scattering amplitude for fast electrons scattered elastically by a spherically symmetrical atom.

The phase functions can be parameterized by either a linear function such as Eq. 6.13a (Stern, Sayers, and Lytle, 1975), a quadratic function such as Eq. 6.13b (Citrin, *et al*, 1976), or a more complicated four-parameter form as Eq. 6.13c (Lee, *et al*, 1977)

$$\phi(k) = p_0 + p_1 k \tag{6.13a}$$

$$\phi(k) = p_0 + p_1 k + p_2 k^2 \tag{6.13b}$$

$$\phi(k) = p_0 + p_1 k + p_2 k^2 + a_3/k^3 \tag{6.13c}$$

While the central atom phase functions $\phi_a(k)$ can be fitted quite adequately with the quadratic form (Eq. 6.13b), the backscatterer phase functions $\phi_b(k)$ often require a functional form similar to (for $Z \leq 36$) or more complicated than (for $Z \geq 36$) Eq. 6.13c (Teo and Lee, 1979). In practice, since $\phi_a(k)$ has a stronger k dependence than $\phi_b(k)$, the combined phase $\phi(k)$ can often be parameterized more or less adequately with Eq. 6.13b or c.

By plotting the amplitude (A, B, C) or phase $(p_i, i = 0-3)$ parameters vs. the atomic number Z, Teo, *et al* (1977) and Lee, *et al* (1977) first noted that these parameters vary smoothly as a function of Z and, consequently, suggested that these parameters (or the theoretical functions) can be used for identification of *atom types* in unknown systems.

Parameterization of the amplitude and phase functions has the distinct advantage of reducing these functions into a few parameters. These parameters can first be obtained by fitting either the experimental EXAFS of structurally known model compounds or the theoretical curves and then transferred to the unknown systems for structural determination. Parameterization also provides some smoothing of the functions. In some cases, however, it can introduce error if the fits are not so good.

To alleviate problems due to errors caused by parameterization, it is possible to fit experimental curves with the amplitude and the phase functions (either determined experimentally or calculated theoretically) in numerical forms via interpolation using, say, the spline technique.

6.4.2 Phenomenological EXAFS Models

Given a set of amplitude $F_j(k)$ and phase $\phi_j(k)$ functions for each jth type of neighboring atoms, the experimental EXAFS spectrum $k^n\chi(k)$ can be fitted with an appropriate model such as

$$k^n\chi(k) \approx Y(k) = S \sum_j N_j F_j(k_j) k_j^{n-1} e^{-2\sigma_j^2 k_j^2} e^{-2r_j/\lambda_j(k_j)}$$

$$\times \frac{\sin[2k_j r_j + \phi_j(k_j)]}{r_j^2} \qquad (6.14)$$

where N_j, σ_j, r_j, λ_j and k_j denote the number of atoms, the Debye-Waller factor, the distance, the inelastic electron mean free path, and the photoelectron wave vector, respectively. The overall scale factor S can either be a constant or some function of k_j. Here a different E_0 is allowed for each type of neighboring atom by allowing $\Delta E_{0_j} = E_{0_j} - E_0$

$$k_j = \sqrt{k^2 - 0.2625(\Delta E_{0_j})} \qquad (6.15)$$

to least-squares refined. Here k is the experimental wave vector with E_0 chosen in the data reduction (E to k conversion) process and k_j is the refined wave vector with E_{0_j} which is consistent with the amplitude $F(k)$ and phase $\phi(k)$ functions (derived from either theory or model compounds) used.

6.4.3 Least-squares Refinements

In nonlinear least-squares refinements, the sum of the squares of the residuals:

$$\Delta = \sum_i [k^n\chi(k) - Y(k)]^2 \qquad (6.16)$$

is minimized iteratively with respect to relevant parameters such as ΔE_0, r_j, S, N_j, σ_j, λ_j, etc. Within each cycle, k_j and $Y(k)$ are calculated for each data point. The residual Δ is then calculated from Eq. 6.16 where i runs through all the data points. Many nonlinear least-squares methods have been used in EXAFS curve fitting. For example, an iterative scheme which combines the advantages of the gradient and the Taylor series methods has been devised by Marquardt (1963). At each step of the iteration a compromise is made between the parameter corrections predicted by the two methods. When the iterative process is in its early stage, the compromise criterion favors the gradient method, since the Taylor series method is likely to be unstable far from convergence. As the parameters get close to their best fit values, more use is made of the Taylor series method, and this avoids the very slow convergence of the gradient method in the neighborhood of the correct solution.

6.4.4 Correlations

It is obvious that curve fitting methods can provide higher resolution and more accurate results for systems with closely spaced interatomic distances. The problems of curve fitting methods include: (1) multi-dimensional parameter spaces which can be very time-consuming; and (2) correlations among various parameters.

In fact, each EXAFS wave contains two sets of highly correlated variables: $\{F(k), \sigma, \lambda, N\}$ and $\{\phi(k), E_0, r\}$. Significant correlations can occur both *within* and *between* these two set of variables as well as *between* different scattering terms. For example, increase in σ will cause an attenuation in the entire EXAFS spectrum though the effect increases with increasing k. In order to best fit the data, N or S values must be increased accordingly. Hence, if the experimental amplitude envelope is such that to give rise an erroneously large σ, the best fitted value of N will be to some extent artificially too high due to this correlation between σ and N even though in principle the *exponential* $e^{-2\sigma^2 k^2}$ and the *linear* N have different k dependence that they should not mix. Similarly, ΔE_0 and r can correlate to some extent even though they again have different k dependences. As a rule of thumb, increasing ΔE_0 by *ca* 3 eV will cause an increase in r by *ca* 0.01Å. These problems are more acute for the minor component(s) (in terms of the EXAFS contribution) of a multi-distance system since the least-squares refinement is dominated by the major component.

Correlation can also occur between these two sets of variables. One typical example is that between σ and the spread of distances Δr as described in the previous chapter. Here Δr contribute directly to the static component of σ. Fitting with two nearly equal distances of the same term often gives rise to unreasonable parameters. On the other hand, if the signal-to-noise ratio is good and the systematic errors in the data analysis are small, the curve fitting technique can be used to detect the presence of the higher order terms in Eq. 5.26 and hence extract more information than just the root-mean-square variation from the average distance.

One way to estimate the resolution by EXAFS is to note that $(\Delta r)(\Delta k) \sim 1$ where Δr is the spread of distances and $\Delta k \sim k_{max} - k_{min}$ is the data length. This implies that as Δk increases, Δr decreases. Hence, in order to resolve a distance spread of $\leq 0.1\text{Å}$ one must, in theory, have a data length of $\geq 10\text{Å}^{-1}$ in k space. The practical limit for resolving Δr with k range of 3-13Å^{-1} is ca 0.05Å.

Correlation can also occur between different terms if the atomic number Z and hence $F(k)$ and $\phi(k)$ are similar. Tradeoffs, or in some worst cases reversals, of the EXAFS contributions may be observed during the refinement. For these cases, additional information and/or assumptions may be needed to deduce meaningful parameters from the data. In this regard, model compounds of known structure will be helpful in that some of the parameters participating in the correlation can be fixed at the same values as those deduced from the model compounds. We shall see some of these techniques later in this chapter.

6.4.5 Errors

Since curve fittings are normally done with Fourier filtered data, it is important to consider the maximum number of parameters allowed in a multi-distance fit. The number of degrees of freedom can be estimated from the ratio of the width of the filter window to the Fourier transform resolution (note that the latter is inversely proportional to the length of the data set, $\Delta k \sim k_{max} - k_{min}$ (Brillouin, 1962)) as $N_{free} \approx \dfrac{2\Delta r_{filt}\Delta k}{\pi}$. For a filtering window width of about 1Å and a Δk of 12Å^{-1}, N_{free} is about 7. Increasing Δr_{filt} to 2Å will give N_{free} of 15. This consideration clearly points out the futility of trying to fit filtered data with EXAFS models containing more parameters than N_{free} and the need to use a wide enough filtering window.

The fitting errors for each parameter in curve fitting can be obtained by changing that particular parameter, x, while least-squares refining the others within the same terms, until the Σ^2 contribution from that particular term is doubled. All parameters associated with the other terms (except the overall scale factor) are held constant. A plot of Σ^2 *vs* x normally gives rise to a well defined minimum from which the fitting errors can be deduced. Reasonable errors of *ca* 0.01-0.03Å are commonly obtained by this method for first-shell distances. Other systematic errors including background removal and Fourier filtering may give rise to uncertainties of 1 and 10% in r and σ, respectively.

6.5 Parameter Correlation and the FABM Method

The major drawback of the curve-fitting method is parameter correlations and false minima. To alleviate, albeit partially, these problems, a fine adjustment technique based on models (FABM) has recently been developed by Teo, Antonio, Averill (1983). The technique involves first curve fitting the EXAFS spectra of the unknown and model compounds with the appropriate EXAFS equation using the *ab initio* backscattering amplitude and phase functions of Teo and Lee (1979). We shall refer to this as "best fit based on theory" (BFBT). The multidimensional parameter correlations often encountered in the curve fitting, which can affect the accuracy, are subsequently quantified to yield phase and amplitude *correlation curves* for each term in the EXAFS equation. These curves are characteristic of the type of compounds and hence carry chemical information not available by best fitting alone. With a good model, these curves can be used to improve the accuracy of the EXAFS structural determination. This "fine adjustment based on model" compounds technique will be referred to as FABM.

The advantage of the FABM technique lies with the detailed exploration of the multi-dimensional parameter correlation space. In particular, one can (1) determine error estimates for each parameter; (2) discover and discriminate against false (local) minima, if any, in the curve fitting; and most importantly, (3) differentiate a good model from a bad one. The prime criterion for a good model is that the chi-square minimum surface in the parameter correlation space must have *parameters* and *curvature* similar to those of the unknown. (Note that these criteria are also applicable to other curve fitting techniques.) While it is almost impossible to explore all the parameters, attention was focussed on the correlations of the most important ones (ΔE_β *vs.* Δr, and B *vs.* σ), which correspond to certain cross-sections of the multi-dimensional parameter space.

6.5.1 Fine Adjustment Based on Models

The fine adjustment of the distances involves the transfer of the characteristic ΔE_0^* for each type of backscatterer from the model to the unknown system. A series of fits are obtained by fixing the distance of interest at different values of up to $\pm 0.10\text{Å}$ away from the best fit value and allowing both scale factors (B_j) and the remaining parameters within the same term to vary during the least-squares refinement. All parameters associated with the other term, except the scale factor, are held constant at their best fit values. Thus, the four parameters refined are the Debye-Waller σ and the ΔE_0 of the particular term under consideration, as well as both scale factors. A plot of ΔE_0^* versus Δr produces the distance correlation curves shown schematically in Fig. 6.3 for model and unknown systems. *It is obvious that as the distance deviates from the best fit value, the goodness of fit deteriorates progressively.* The characteristic value ΔE_0^* for the model compound can be determined from the curve at the crystallographic value $\dot{r}_m$. The characteristic ΔE_0^* is then transferred to the unknown to determine the distance correction Δr_u. Here the subscripts m and u refer to the model and the unknown, respectively.

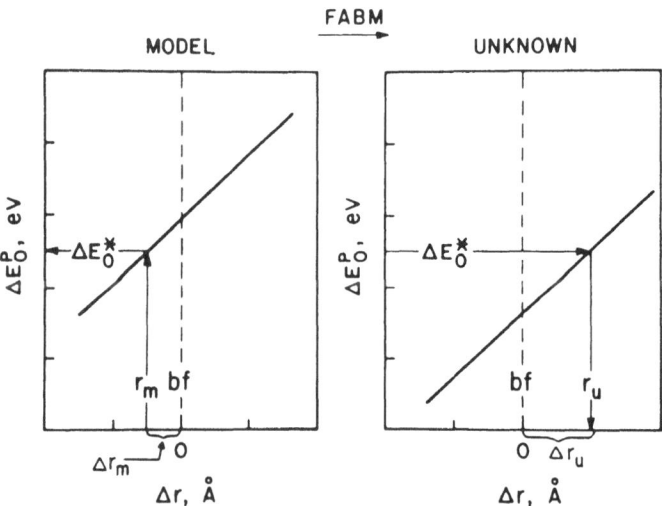

Fig. 6.3. Schematic description of the FABM distance adjustment. At the known crystallographic distance for the model (r_m) a characteristic ΔE_0^* is obtained which is then transferred to the unknown system to yield the distance adjustment to the BFBT $(\Delta r_u \approx 0)$ refined distance (r_u). (Reproduced from Teo, Antonio, Averill, 1983).

A similar procedure can be applied to the fine adjustment of coordination numbers. A series of fits are made by holding the Debye-Waller factor (σ) of one term at different values (on a specified interval) covering the range from zero to at least σ^* (including the best fit σ) while fixing the distances, Debye-Waller factors, and ΔE_0 values of the other term(s) at their best fit values. Four parameters are refined in the curve fitting, the distance r and the ΔE_0^β of the particular term as well as the scale factor B. A plot of B *versus* σ will then allow the determination of the B value at the characteristic σ^* value as depicted graphically in Fig. 6.4. From the model, $S^* = B_m/N_m$. Both σ^* and S^* are then transferred to the unknown to yield $N_u = B_u/S^*$ at σ^*. Once again, as σ deviates from the best fit value marked by $\times$, the goodness of fit deteriorates progressively.

It is claimed that, with the FABM technique, interatomic distances and coordination numbers can be determined to better than 0.5 and 10% and 1 and 20% accuracy for the major (*e.g.*, metal-ligand bonds) and the minor (*e.g.*, metal-metal bonds) terms, respectively.

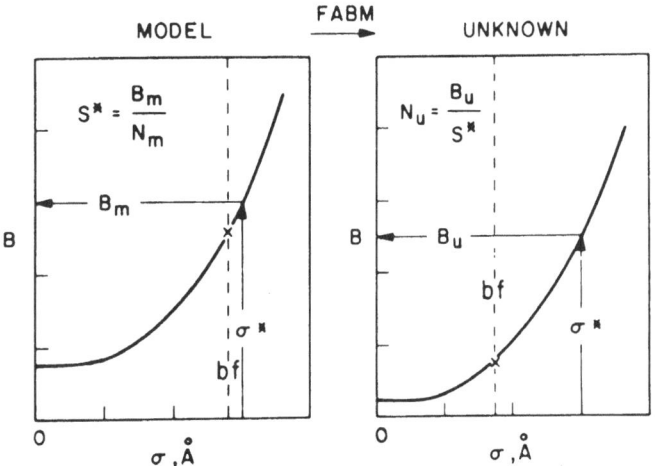

Fig. 6.4. Schematic description of the FABM coordination number determination. At the characteristic σ^* (determined from the model, subscript m), the scale factor B_m is obtained and the amplitude reduction factor S^* is calculated from the known coordination number N_m. The unknown (subscript u) scale factor B_u is then obtained at the value of the characteristic σ^*. The coordination number then follows from B_u and S^* as shown. The BFBT Debye-Waller factors are indicated by "$\times$." (Adapted from Teo, Antonio, Averill, 1983).

It should be emphasized that the FABM technique is model dependent. As with any other model-dependent data analysis technique, the results are accurate to the extent that the parameters and parameter correlations for a given model compound mimic those of the unknown system. However, the FABM method is distinct from other model dependent techniques in that only one parameter (ΔE_0^*), *and not the total phase function*, needs to be obtained from the model to calculate the distance in the unknown. Similarly, only two parameters (σ^*, S^*), *and not the entire amplitude function*, need to be obtained from the model to calculate the number of neighbors in the unknown. In the FABM approach, theoretical phase and amplitude functions, rather than functions extracted from models, are used in the curve fitting. In this respect, the method is less critically dependent on the models than other model-based techniques.

6.5.2 Critera for the Selection of Good Models

Since BFBT is model independent and FABM is model dependent, *the FABM distance corrections and coordination number determinations can improve upon the BFBT results if and only if good models are used.* Furthermore, since most EXAFS data analysis techniques are model dependent, it is desirable to find a way to distinguish a "good" model from a "bad" model. The parameter correlations obtained in the FABM method provide a convenient way for such a differentiation. The following criteria have been developed for choosing an appropriate compound as well as for applying the FABM method. It should be noted that Bunker, Stern, Blankenship, Parson (1982) have also reported criteria for a good model and indicated that a good model is critical to assure the accuracy of the structural parameters.

First, consider the ΔE_0^* vs. Δr plot. The criteria are: (1) the two slopes must be similar; and (2) the resulting correction in distance Δr for the unknown must be within $\pm 0.1 \text{Å}$ of the best fit value. For this range the ΔE_0^* vs. Δr plots are usually linear. The criterion of $\Delta r = \pm 0.1 \text{Å}$ is based on the experience that theory alone can predict distances to better than $\pm 0.06 \text{Å}$. In cases where the two lines are not parallel, Δr in excess of $\pm 0.1 \text{Å}$ are often observed. In this situation, the fine adjustment technique is not applicable (neither are most model-dependent data analysis techniques). In other words, for such cases the model is a bad one. As shown schematically in Fig. 6.5, model 1 is a good model for the unknown system because the two curves are nearly parallel, and the resulting distance correction Δr (r_{u1}) is within 0.1Å. Model 2, however, has a significantly different slope and Δr (r_{u2}) for the unknown is greater than 0.1Å. Here $\Delta r = 0$ corresponds to the best fit distances, and the fits at the crystallographic distances are indicated as r_{m1}, $\Delta E^*_{0_{m1}}$, and r_{m2}, $\Delta E^*_{0_{m2}}$ for models 1 and 2, respectively.

For the B *vs.* σ plot, we impose the following two criteria: (1) the best fit Debye-Waller factors for the model and the unknown system must be similar;

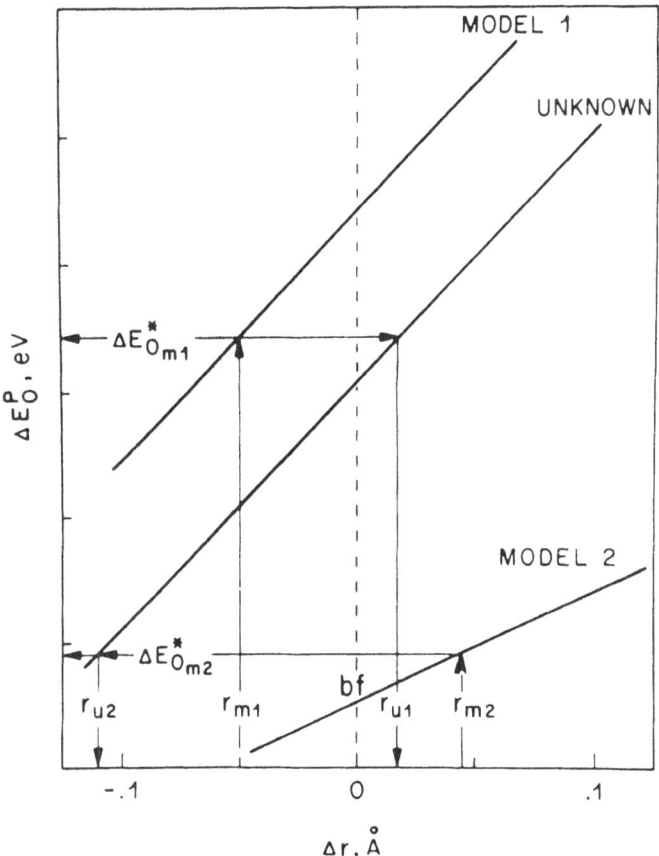

Fig. 6.5. Schematic showing different choices of a model for the distance adjustment in the FABM method. Model 1 is a good model for the unknown because the lines are nearly parallel and the resulting distance correction r_{u1}, obtained at the characteristic $\Delta E^*_{0_{m1}}$, is within 0.1Å of the BFBT distance for the unknown. Model 2 is a bad model for the unknown because the slopes are not similar and the distance correction r_{u2}, obtained at the characteristic $\Delta E^*_{0_{m2}}$, exceeds 0.1Å. (Reproduced from Teo, Antonio, Averill, 1983).

and (2) the ratio of the B values (B_u to B_m) for the two systems must be more or less independent of σ (*i.e.*, the ratio should be constant over all σ). In other words, the FABM coordination number, which is calculated from the value of B_u at σ^*, should be more or less independent of the choice of σ^*. As illustrated in Fig. 6.6, model 1 is good since it satisfies both criteria. Model 2 is inappropriate since it has a best fit σ (marked by 'x') significantly different from that of the unknown (although it satisfies the second criterion). Model 3 is also inadequate because it does not follow the second criterion (although it satisfies the first one). In fact, model 3 is schematically drawn such that the B values are displaced by a *constant difference*, Δ, from the corresponding values of the unknown. The S^* value obtained from model 3 is not strictly transferable to the unknown.

In addition, to achieve the highest accuracy with the FABM method, the unknown and model(s) must be measured and analyzed in a similar fashion.

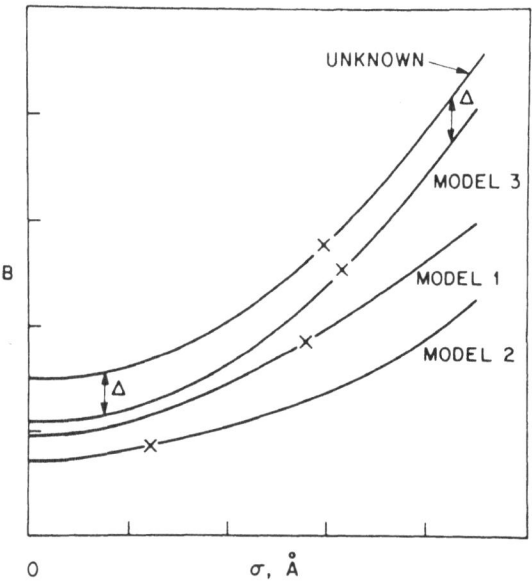

Fig. 6.6. Schematic showing the different choices of a model for coordination number determinations in the FABM method. Model 1 is a good model because the BFBT Debye-Waller factor (indicated by "x") is similar to that of the unknown, and also the ratio of the B values is constant over all σ. Model 2 is inappropriate since it has a BFBT σ significantly different from that of the unknown. Model 3 is also a bad model because the ratio of the B values (here displaced by a constant difference, Δ) is not independent of σ. (Adapted from Teo, Antonio, Averill, 1983).

The parameters (r, σ, and ΔE_0) of the model should be similar to that of the unknown. (This is particularly important if the fitting routine leads to more than one best fit minimum for the unknown and/or the model.)

It should be emphasized that it is risky to base the entire analysis on one point in the multi-dimensional parameter correlation space. For a good model compound, transfer of such parameters as ΔE_0 and σ will probably allow an accurate determination of interatomic distances and coordination numbers. However, unless and until the suitability of the model is ascertained via the parameter correlation method suggested by the FABM technique, or by other established criteria, the results remain *model dependent*.

6.6 The "Difference" Technique

To get at minor components in the EXAFS data analysis, a "difference" technique has been developed (Teo, *et al*, 1978; Cramer, *et al*, 1976). This method combines Fourier filtering, curve fitting, and a difference spectrum technique in separating different components or more importantly, in isolating minor component(s) from the major component(s) in the EXAFS. It involves first fitting the filtered data, in k space, with a model containing the dominant term(s) in the EXAFS. The latter is then subtracted from the data, resulting in a "difference" EXAFS spectrum which presumably contains the minor component and some residual background. This "difference" EXAFS spectrum is then Fourier transformed, and the minor component(s), if present, is Fourier filtered and backtransformed to give a filtered "difference" spectrum in k space. The latter can be fitted with a model which corresponds to the minor component(s). Fig. 6.7 illustrates the separation of the minor peak from the major peak in Fig. 6.2(c) by this technique. For example, the major peak may be due to metal-ligand bonds while the minor peak may be due to metal-metal distance(s). We first best fit the filtered EXAFS, $k^n\chi(k)$ (solid curve in Fig. 6.7(a)), with a model, $k^n\chi_L(k)$ (dashed curve in Fig. 6.7(a)) containing metal-ligand distance(s) only. The latter is then subtracted from the data to give the difference EXAFS (Fig. 6.7(b)):

$$k^n\Delta\chi(k) = k^n\chi(k) - k^n\chi_L(k) \tag{6.17}$$

Fourier transform (Fig. 6.7(c)) of $k^n\Delta\chi(k)$ reveals the metal-metal distance which can be isolated by Fourier filtering and fitted with a model, $k^n\chi_M(k)$ (dashed curve in Fig. 6.7(d)), containing metal-metal distance(s) only.

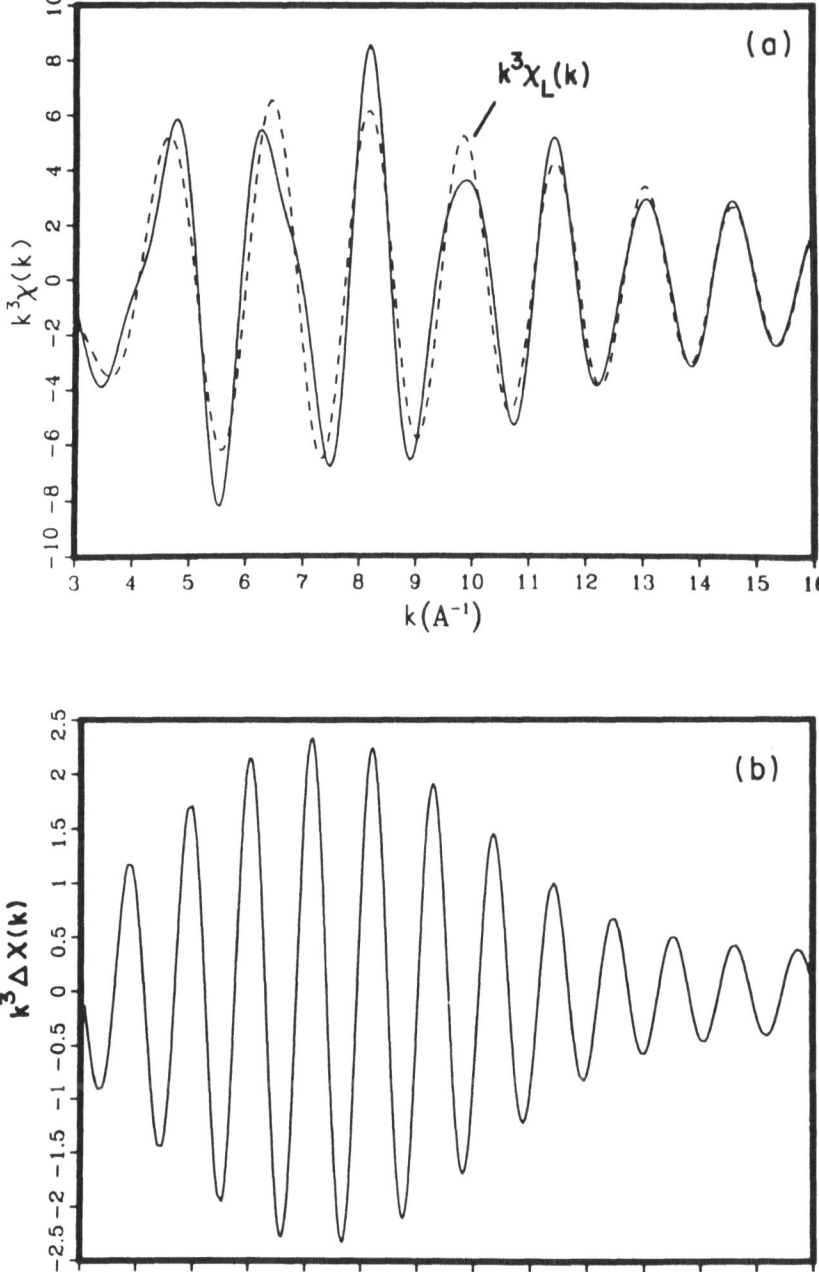

Fig. 6.7. The "Difference" Technique. The Fourier-filtered EXAFS $k^n\chi(k)$ spectrum (solid curve in (a)) is fitted with a model $k^n\chi_L(k)$ containing the major peak in the Fourier transform (dashed curve in (a)). The latter is then subtracted from the data to give the difference EXAFS $k^n\Delta\chi(k)$ (b). Fourier

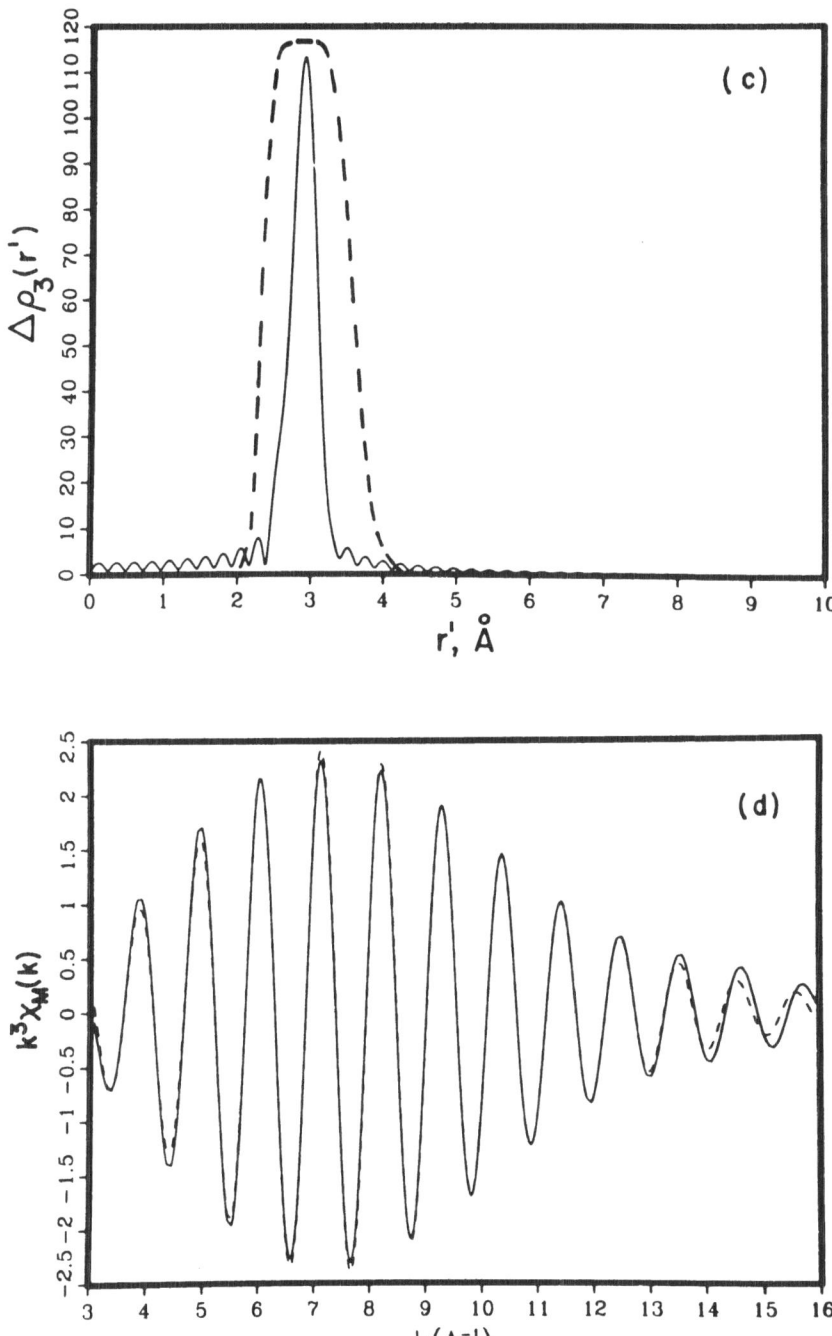

transform (c) of $k^n \Delta \chi(k)$ reveals the minor peak which can be isolated by Fourier filtering (filter window depicted as the dashed curve in (c)) and fitted with a model $k^n \chi_M(k)$ (dashed curve in (d)) containing the minor peak only.

The "difference" technique works well for systems in which the major and the minor components are "orthogonal" to each other. This "orthogonality" criterion is more or less satisfied for neighboring atoms with significantly different atomic numbers. That is, a low Z neighbor will have amplitude and phase functions as well as distances different from a high Z neighbor such that the two EXAFS contributions will peak in different regions in k space with different frequencies and phases. On the other hand, for systems in which the major and minor components are composed of neighboring atoms with similar atomic numbers, error can arise owing to the "trade off" or "overlap" of the two EXAFS contributions.

Similarly, the "difference" method is more applicable for systems where there is a large disparity in EXAFS contribution between the major and the minor components. In other words, if the major component(s) account for most (say, >80%) of the EXAFS signal, a good fit, and hence accurate structural parameters, can be obtained in the initial fitting. Since the EXAFS is dominated by the major component(s), correlation between the two components is minimized in the initial fitting and the minor component(s) can therefore be determined more accurately via this separation method than via direct curve fitting.

For systems in which either or both of the foregoing criteria are not satisfied, certain constraints may need to be imposed in the initial curve fitting in order to obtain meaningful structural information via this technique.

In addition to the two broad categories of data analysis techniques (Fourier transformation and curve fitting) discussed thus far, other methods are also available. Some of them are more specialized while others are general enough to merit consideration as alternatives in EXAFS data analysis as will be discussed in the following sections.

6.7 The Min-Max Method

The min-max method can be used to estimate interatomic distance in a single-shell system without elaborate computations. It is included here to illustrate that EXAFS is an excellent technique for the determination of distances.

Given the X-ray absorption spectrum $\mu(E)$ of a single-shell system, one can count the successive oscillations of the absorption coefficient in energy space, starting from the edge (disregarding, of course, the sharp absorption peaks at or near the edge). For typical interatomic distances of $\geq 2\text{Å}$, these oscillations have typical periodicity of $\geq 30\text{eV}$; the periodicity increases (i.e., the frequency decreases) as the photon energy increases. One can label the maxima as $m = 1, 5, 9, ...$ and the corresponding minima as $m = 3, 7, 11, ...$ as shown in Fig. 6.8(a) in energy space and Fig. 6.8(b) in k space. The next step is to convert the photon energy E into photoelectron wave vector k for each of these extrema by defining an E_0 near the edge (e.g., at the first maximum and using Eq. 6.6). One then plots m as a function of k values at which the extrema occur. For a single-shell system, one should obtain a linear curve as shown in Fig. 6.9 for Fe(SR)_4^- containing four equivalent Fe-S bonds.

To see how one can determine the interatomic distance r from such a curve, consider the argument of the sine term in the EXAFS expression which is responsible for the oscillations. The extrema should occur at

$$\frac{m\pi}{2} = 2kr + \phi(k) \tag{6.18}$$

since $\sin(m\pi/2) = 1$ for $m = 1, 5, 9, ...$ (maxima) and -1 for $m = 3, 7, 11, ...$ (minima). Assuming that the phase is linear in k:

$$\phi(k) = a + bk \tag{6.19}$$

one obtains

$$m = \frac{2}{\pi}(2r+b)k + \frac{2}{\pi}a \tag{6.20}$$

Hence a plot of m vs k should give rise to a linear curve with a

$$\text{slope} = \frac{2}{\pi}(2r+b) \tag{6.21}$$

and an intercept at $k=0$

$$m_{k=0} = \frac{2}{\pi}a \tag{6.22}$$

At $m=0$,

$$k_{m=0} = -\frac{a}{2r+b} \tag{6.23}$$

From Fig. 6.9, we find a slope of 2.38Å, $m_{k=0} = -5$, and $k_{m=0} = 2.1\text{Å}^{-1}$. Using Eqs. 6.21-23, we find $a = -7.85$ and $(2r+b) = 3.74\text{Å}$. It is apparent

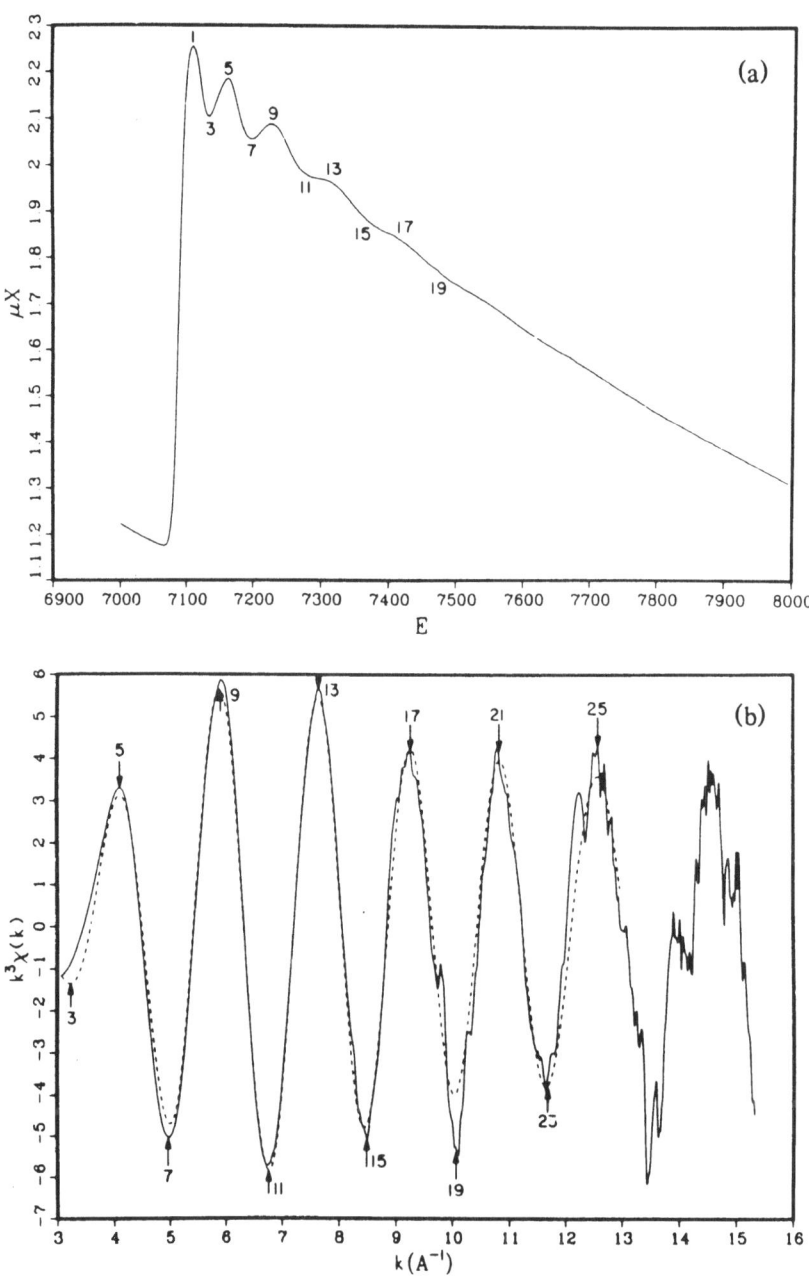

Fig. 6.8. The min-max method. It amounts to label the maxima as $m = 1, 5, 9, \ldots$ and the corresponding minima as $m = 3, 7, 11, \ldots$ as shown in (a) in energy (E) space and in (b) in momentum (k) space. For (a), it is necessary to convert the photon energy E into photoelectron wave vector k for each of these extrema by defining an E_0 near the edge. One then plots m as a

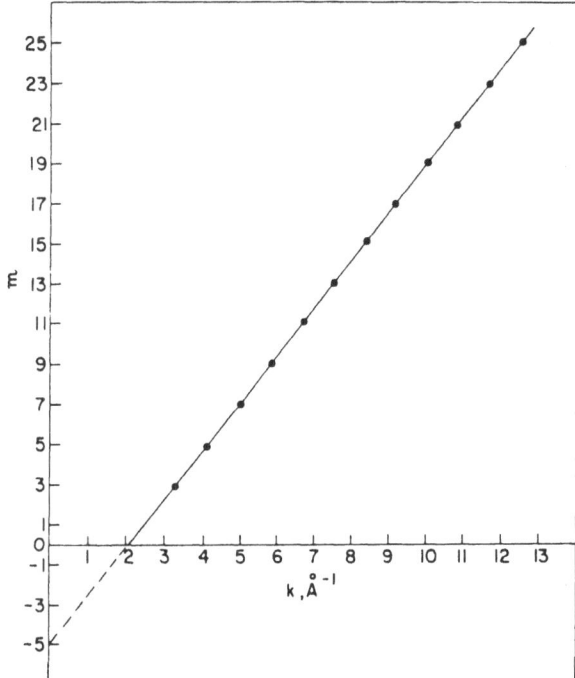

Fig. 6.9. The linear plot of m vs k for the data shown in Fig. 6.8 for $Fe(SR)_4^-$ containing four equivalent Fe-S bonds. A slope of 2.38Å, $m_{k-0} = -5$, and $k_{m-0} = 2.1$Å^{-1} can be obtained from which a distance of $r = 2.29$Å for the Fe-S bonds in $Fe(SR)_4^-$ can be determined which is in good agreement with the known value of 2.267(2)Å (Lane, $et\ al$, 1977).

function of k values at which the extrema occur. For a single-shell system, one should obtain a linear curve as shown in Fig. 6.9. Both (a) and (b) are for $Fe(SR)_4^-$ which contains four Fe-S bonds; (b) can be obtained from (a) using $E_0 = 7109$ eV. (Data taken from Shulman, Eisenberger, Teo, Kincaid, Brown, 1978.)

that in order to determine r, one needs to know the coefficient of the linear terms in the phase function, b. Conversely, the phase function $\phi(k) = a + bk$ can be determined from a model compound of known structure with the same atom pair or from theoretical calculations. If one fits the theoretical phase function of Teo and Lee (1979) for a Fe-S distance one obtains $\phi(k) \approx -7.78 - 0.844k$. Using $b = -0.844$Å, one obtains $r = 2.29$Å for the Fe-S bonds in $Fe(SR)_4^-$ which is in good agreement with the known value of $2.267(2)$Å (Lane, et al, 1977).

It should be noted that had one label the extrema as $m = 1 \pm 4n$, $3 \pm 4n$, $5 \pm 4n$, ... instead of $m = 1, 3, 5, ...$, the constant a will become $(a \pm 2n\pi)$ while b as well as r will remain unchanged since $\sin x = \sin(x \pm 2n\pi)$ where $n = 0, 1, 2, ...$

It is interesting to point out that for a single-term system, half a period of oscillation (one maximum and one minimum or any two well-defined and far-apart points in k space) is enough to determine the distance r if the phase function $\phi(k)$ is known.

6.8 Decomposition into Amplitude and Phase

Fourier filtering not only has the ability of separating the EXAFS into individual shells with a limited distance range, but also can be used to decompose the EXAFS wave into amplitude and phase functions which can then be analyzed separately.

For any EXAFS signal

$$\chi(k) = A(k)\sin \Phi(k) \tag{6.24}$$

where $A(k)$ and $\Phi(k) = 2kr + \phi(k)$ represents the amplitude and phase functions, respectively, we can write

$$\chi(k) = \frac{1}{2i}A(k)e^{i\Phi(k)} - \frac{1}{2i}A(k)e^{-i\Phi(k)} \tag{6.25}$$

If we now Fourier transform $\chi(k)$ into $\rho(r)$, the first term in Eq. 6.25 corresponds to positive r values while the second term to negative r values. If we now replace the negative r values of $\rho(r)$ by zeros and then backtransform $\rho(r)$ into k space, we obtain

$$Z(k) = \frac{1}{2i}A(k)e^{i\Phi(k)} \tag{6.26}$$

It is then straightforward to extract $A(k)$ and $\Phi(k)$ from the magnitude $z(k)$ and the argument $\alpha(k)$ of $Z(k) = z(k)e^{i\alpha(k)}$

$$A(k) = 2z(k) \tag{6.27}$$

$$\Phi(k) = \alpha(k) + \frac{\pi}{2}. \tag{6.28}$$

This method was originally developed by Kincaid and has been used widely in EXAFS data analysis. Fig. 6.10 shows the Fourier transform of the $k^3\chi(k)$ data of crystalline Ge. The dashed curves are the filtering windows. The Fourier filtered spectrum (first shell) is depicted in Fig. 6.11 where the dashed curve is the amplitude function, $A(k)$. The corresponding phase function, $\Phi(k)$, is shown in Fig. 6.12.

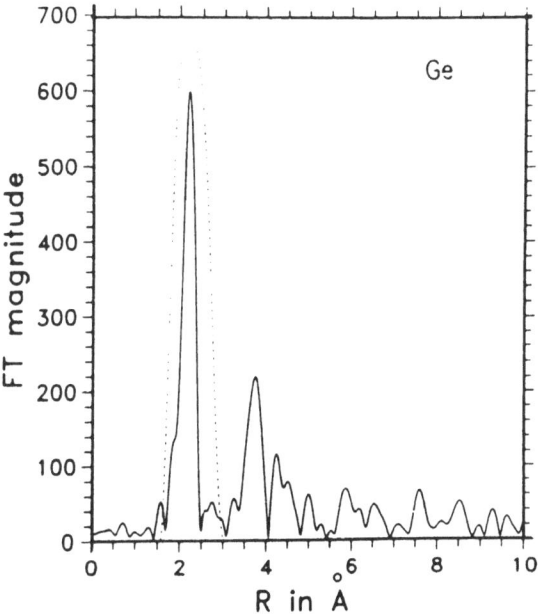

Fig. 6.10. Magnitude of the Fourier transform of the $k^3\chi(k)$ spectrum of crystalline Ge. The transform was obtained using the fast Fourier transform algorithm with added 0's to produce a smoothly interpolated curve. The dashed line is a smooth window function used to filter out the first-neighbor shell EXAFS from the high frequency noise and outer shells. (Reproduced from Lee, *et al*, 1981).

148

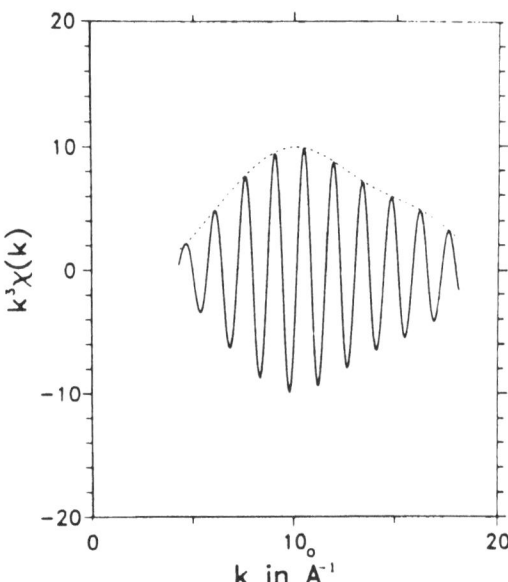

Fig. 6.11. Inverse transform of Fourier transformed data in Fig. 6.10 after multiplication by the window function. Note that all higher frequency noise and higher-order shells have been removed. Dashed curve is the amplitude function $A(k)$ derived using the Fourier transform. (Reproduced from Lee, *et al.*, 1981).

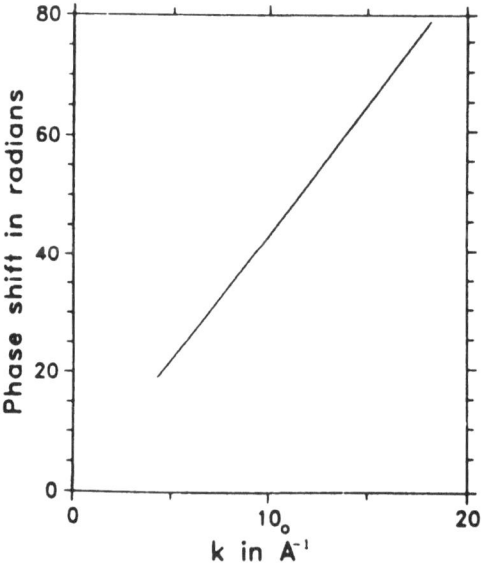

Fig. 6.12. Total phase $\Phi(k) = 2kr + \phi(k)$ of the filtered spectrum (first shell) shown in Fig. 6.11. This curve looks like a straight line because $2kr$ is much larger than $\phi(k)$. (Reproduced from Lee, *et al*, 1981).

6.8.1 Phase Information

The phase component $\Phi = 2kr + \phi(k)$ for a single-distance system (cf. Fig. 6.12) obtained via the Fourier filtering method can be used to extract the interatomic distance r provided that $\phi(k)$ is known from a model compound or from theory (phase transferability).

With a carefully chosen model compound (designated by the subscript m) of known and similar structure, the difference in the phase functions is simply

$$\Phi(k') - \Phi_m(k) = 2k'r - 2kr_m + \phi'(k') - \phi(k) . \qquad (6.29)$$

A plot of $\Phi(k') - \Phi_m(k)$ as a function of k therefore produces a roughly linear curve which normally does not pass through the origin due to the nonzero difference in phase $\phi'(k') - \phi(k)$ caused by the difference in E_0. By adjusting $\Delta E_0 = E_0 - E_0'$ for the unknown such that $\phi'(k') - \phi(k) = 2(k - k')r$, Eq. 6.29 becomes

$$\Phi(k) - \Phi_m(k) = 2k(r - r_m) \qquad (6.30)$$

which passes through the origin (i.e. zero intercept). In practice a plot of the difference $\Phi(k') - \Phi_m(k)$ vs k is made and a linear curve (dashed line) least-squares fit to it as shown in Fig. 6.13(a). The E_0 for the unknown system is then least-squares refined until the fitted linear curve, when extrapolated, passes through the origin (cf. Fig. 6.13(b)), giving rise to a slope of $2(r - r_m)$ from which the unknown distance r can readily be calculated.

A closely related method which calls for varying E_0 until the "slope" $[\Phi(k) - \Phi_m(k)]/k$ remains constant over a large range of k has also been devised by Martens et al. (1978a). The $1/k$ factor weighs the low k data more heavily but Martens et al. compensate for this by cutting off low k information.

6.8.2 Amplitude Information (The Ratio Method)

The amplitude function (cf. Fig. 6.11, dashed curve) produced by Fourier filtering contains information about the number N of nearest neighbors, the degree of disorder (Debye-Waller factor σ), the effective photoelectron mean free path $\lambda(k)$, and other inelastic and/or intrinsic loss factors. It is possible in principle to extract these quantities assuming chemical transferability of amplitudes.

The procedure of amplitude transferability is quite straightforward (Stern et al. 1975). Using the simplest expression for $\chi(k)$, the amplitude function for a

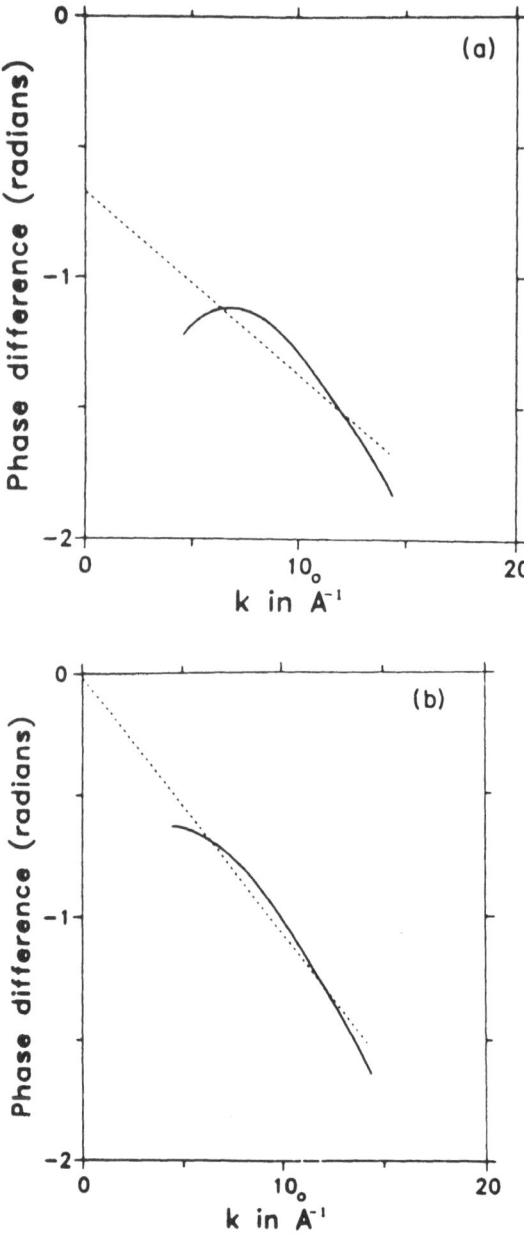

150

Fig. 6.13. (a) A plot of the difference in the total phase functions of the unknown, $\Phi(k)$, and the model, $\Phi_m(k)$, as a function of k. A linear curve (dashed line) is least-squares fit to it. Extrapolation of the fitted line generally does not pass through the origin, indicating apparent violation of the assumption of phase-shift transferability. (b) Here ΔE_0 of the unknown system has been least-squares refined, and the phase-difference fitted line now passes through the origin. (Adapted from Lee, *et al*, 1981).

single shell of N identical atoms is given by

$$A(k) = \frac{N}{kr^2} F(k) e^{-2k^2\sigma^2} e^{-2r/\lambda(k)} . \qquad (6.31)$$

where $F(k)$ is the scattering amplitude. Denoting the model system by the subscript m, the logarithm of the amplitude ratio gives

$$\ln\left[\frac{A(k)}{A_m(k)}\right] = \ln\left[\frac{N}{N_m}\frac{r_m^2}{r^2}\right] + 2k^2\left[\sigma_m^2 - \sigma^2\right] + 2\left[\frac{r_m}{\lambda_m(k)} - \frac{r}{\lambda(k)}\right] . \qquad (6.32)$$

If it is assumed that $\lambda(k) \approx \lambda_m(k)$ and $(r_m - r) \ll \lambda(k)$, the third term can be neglected. Then, a plot of $\ln[A(k)/A_m(k)]$ vs k^2 gives rise to a linear curve with an intercept of $\ln(N/N_m \times r_m^2/r^2)$ and a slope of $2(\sigma_m^2 - \sigma^2)$, assuming other factors being equal or similar for the two systems. In practice, $A(k)$ and $A_m(k)$ are obtained via Fourier filtering. A plot of $\ln[A(k)/A_m(k)]$ vs k^2 is then made and a linear curve is least-squares fitted to it. The possible difference in E_0 is not as important here since the amplitude $(A(k))$ functions are less sensitive to E_0 than the phase $(\phi(k))$ functions. Knowing σ_m, N_m, r_m from the model compound as well as r from the phase analysis, it is then possible to determine the unknown σ and N from the amplitudes.

6.9 The Beat-node Method

For systems with a large disparity ($\geq 0.1 \text{Å}$) in distances, a beat-node technique, originally developed by Martens, et al. (1977), may also be used.

Consider a system with one scatterer type but two sets, N_1 and N_2, of distances R_1 and R_2, respectively. The sum of two sine waves at different frequencies produces beating in the amplitude with a period of $2k(R_1 - R_2)$. If the data extend to k_{max} such that $k_{max} \geq \pi/[2(R_1 - R_2)]$, the minimum in the amplitude (beat-node) as well as the associated kink in the phase shift function can be measured. (Since k_{max} is typically limited to about 15Å^{-1}, the distance resolution is $ca.$ 0.1Å.) The EXAFS can be written as

$$\chi(k) = \frac{1}{k} \tilde{A}(k) \sin[2k\tilde{R} + \tilde{\phi}(k)] \qquad (6.33)$$

where

$$\tilde{A}(k) = A_1(k)[1 + C(k)^2 + 2C(k)\cos(2k\Delta R)]^{1/2} \qquad (6.34)$$

$$\tilde{\phi}(k) = \phi(k) + \arctan\left[\frac{1 - C(k)}{1 + C(k)} \tan k\Delta R\right] \qquad (6.35)$$

$$A_i(k) = \frac{N_i}{R_i^2} F(k) e^{-2\sigma_i^2 k^2} e^{-2R_i/\lambda} \qquad (i = 1, 2) \tag{6.36}$$

$$C(k) = \frac{A_2(k)}{A_1(k)} = \frac{N_2}{N_1} \frac{R_1^2}{R_2^2} e^{-2(\sigma_2^2 - \sigma_1^2)k^2} e^{-2\Delta R/\lambda} \tag{6.37}$$

$$\bar{R} = (R_1 + R_2)/2 \tag{6.38}$$

$$\Delta R = R_1 - R_2 \tag{6.39}$$

Here $C(k)$ can be varied as a single parameter or fitted as a function

$$C(k) = ce^{-2sk^2} \tag{6.40}$$

where $c = (N_2/N_1)(R_1^2/R_2^2)e^{-\Delta R/\lambda}$ and $s = \sigma_2^2 - \sigma_1^2$. The EXAFS now involves modified amplitude $\bar{A}(k)$ as well as phase $\bar{\phi}(k)$ functions which are related to the first (often the dominant) term denoted by the subscript 1. Knowing $F(k)$ and $\phi(k)$ from theory or model compounds, it is possible to deduce the average distance $\bar{R}$ and the bond length difference ΔR (as well as the sign of it) along with information concerning the number of bonds and Debye Waller factors. Using this method, Martens et al. (1977) were able to resolve the second and third shell distances in CuO which differ by 0.19Å.

6.9.1 Derivations of Eq. 6.33-40

We shall derive Eq. 6.33-40 as follows. The two-term EXAFS can be written and combined to give

$$\chi(k) = \frac{1}{k}\{A_1(k)\sin[2kR_1+\phi(k)]+A_2(k)\sin[2kR_2+\phi(k)]\} \tag{6.41a}$$

$$= \frac{1}{k}A_1(k)\text{Im}\{e^{i(2kR_1+\phi(k))} + C(k)e^{i(2kR_2+\phi(k))}\} \tag{6.41b}$$

$$= \frac{1}{k}A_1(k)\text{Im}\{e^{i(2k\bar{R}+\phi(k))}[e^{ik(\Delta R)} + C(k)e^{-ik(\Delta R)}]\} \tag{6.41c}$$

where $C(k)$, $\bar{R}$, and ΔR are given by Eq. 6.37, 38, and 39, respectively. If we define

$$Z(k) = e^{ik(\Delta R)} + C(k)e^{-ik(\Delta R)} = z(k)e^{i\alpha(k)} \tag{6.42}$$

then by simple manipulation of complex variables, we can write

$$z(k) = [(\text{Re } Z)^2 + (\text{Im } Z)^2]^{1/2} \tag{6.43}$$

$$= \{[(1+C(k))\cos(k\Delta R)]^2 + [1-C(k))\sin(k\Delta R)]^2\}^{1/2}$$

$$= [1 + C(k)^2 + 2C(k)\cos(2k\Delta R)]^{1/2}$$

$$\alpha(k) = \text{arc tan}\left[\frac{\text{Im } Z}{\text{Re } Z}\right] \tag{6.44}$$

$$= \text{arc tan}\left[\frac{1-C(k)}{1} + C(k)\,\tan(k\Delta R)\right]$$

Substituting Eq. 6.43 and 6.44 into Eq. 6.41c:

$$\chi(k) = \frac{1}{k}\,A_1(k)z(k)\text{Im}[e^{i(2k\bar{R}+\phi(k)+\alpha(k))}] \tag{6.45}$$

$$= \frac{1}{k}\tilde{A}(k)\sin(2k\bar{R}+\tilde{\phi}(k))$$

where

$$\tilde{A}(k) = A_1(k)z(k) \tag{6.46}$$

$$= A_1(k)[1 + C(k)^2 + 2C(k)\cos(2k\Delta R)]^{1/2}$$

$$\tilde{\phi}(k) = \phi(k) + \alpha(k) \tag{6.47}$$

$$= \phi(k) + \text{arc tan}\left[\frac{1-C(k)}{1+C(k)}\,\tan(k\Delta R)\right].$$

6.10 The Lee and Beni Method

A Fourier transform method which makes use of theoretically calculated amplitude and phase functions has been developed by Lee and Beni (1977). This technique calls for a Fourier transform of the data by first removing the phase and amplitude functions as follows:

$$\rho(r') = \int_{k_1}^{k_2}\left[\frac{k\chi(k)}{F(k)}e^{-i\phi(k)}e^{2\sigma^2k^2}\right]e^{i2kr'}dk \tag{6.48}$$

In Eq. 6.48, the experimental EXAFS $\chi(k)$ is divided by the theoretical backscattering amplitude function $F(k)$ and multiplied by the theoretical phase factor $e^{-i\phi(k)}$ as well as multiplied by k and by the Debye-Waller factor $e^{2\sigma^2 k^2}$ before Fourier transformation over the data range k_1 to k_2. These prefactors essentially cancel the amplitude and phase factors in the experimental $\chi(k)$ and we will be transforming basically a sine function. Since $\sin x = (e^{ix} - e^{-ix})/2i$, we can write

$$\rho(r') = \int_{k_1}^{k_2} \left[\frac{e^{-2r/\lambda}}{r^2} \left(\frac{e^{i2kr} - e^{-i2kr}}{2i} \right) \right] e^{i2kr'} dk \tag{6.49}$$

$$= \frac{e^{-2r/\lambda}}{r^2} \int_{k_1}^{k_2} \left[\frac{e^{i2k(r'+r)} - e^{i2k(r'-r)}}{2i} \right] dk$$

$$= \frac{e^{-2r/\lambda}}{4r^2} \left[\frac{e^{i2k_2(r'-r)} - e^{i2k_1(r'-r)}}{r' - r} - \frac{e^{i2k_2(r'+r)} - e^{i2k_1(r'+r)}}{r' + r} \right]$$

Ignoring the slowly varying second term and consider only the first term which is a delta function:

$$\rho(r') \approx \frac{e^{-2r/\lambda}}{4r^2} \left[\frac{e^{i2k_2(r'-r)} - e^{i2k_1(r'-r)}}{r' - r} \right] \tag{6.50}$$

The imaginary, real, and magnitude of $\rho(r')$ are, respectively,

$$\operatorname{Im} \rho(r') \approx \frac{e^{-2r/\lambda}}{4r^2} \left[\frac{\sin 2k_2(r'-r) - \sin 2k_1(r'-r)}{r' - r} \right] \tag{6.51}$$

$$= \frac{e^{-2r/\lambda}}{2r^2} \left[\frac{\cos[(k_2+k_1)(r'-r)]\sin[(k_2-k_1)(r'-r)]}{r' - r} \right]$$

$$\operatorname{Re}\rho(r') \approx \frac{e^{-2r/\lambda}}{4r^2} \left[\frac{\cos 2k_2(r'-r) - \cos 2k_1(r'-r)}{r'-r} \right] \tag{6.52}$$

$$= -\frac{e^{-2r/\lambda}}{2r^2} \left[\frac{\sin[(k_2+k_1)(r'-r)]\sin[(k_2-k_1)(r'-r)]}{r'=x} \right]$$

$$|\rho(r')| = \frac{e^{-2r/\lambda}}{2r^2} \left[\frac{\sin[(k_2-k_1)(r'-r)]}{r'-r} \right] \tag{6.53}$$

It is apparent that both Im $\rho(r')$ and $|\rho(r')|$ peak, while Re $\rho(r')$ vanishes, at $r' = r$ in the Fourier transform. That is, the peak of the imaginary part of the Fourier transform $\rho(r')$ should coincide with the peak of its absolute magnitude if the calculated phase shift is consistent with the experimental data. In fact, Lee and Beni (1977) proposed the use of this criterion in the distance determination by varying the energy threshold E_0 until the peaks (within the same shell) in Im $\rho(r')$ and $|\rho(r')|$ match. One example is shown in Fig. 6.14 for Br_2 where it can be seen that Im $\rho(r')$ (solid curve) is fairly symmetric about its peak at 2.30 Å which is close to the peak at 2.26Å in $|\rho(r')|$ (dashed curve). The E_0 was chosen at 13 eV above the bound state. By varying E_0 until the peaks of Im $\rho(r')$ and $|\rho(r')|$ coincide, a distance of 2.281Å is obtained which agrees very well with the known value of 2.284Å. As a result, the E_0 was found to be 7 eV above the bound state. It should be cautioned that different

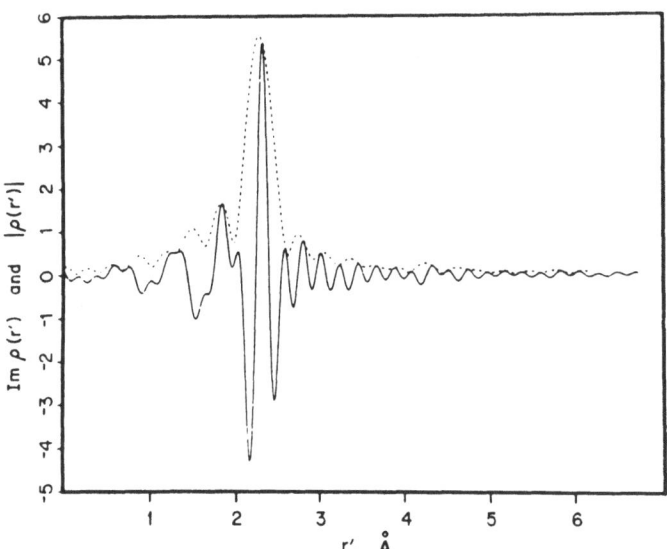

Fig. 6.14. Fourier transformation of the EXAFS $k\chi(k)$ data for Br_2 with E_0 chosen at 13 eV above the bound state. Note that the imaginary part of the transform Im $\rho(r')$ (solid curve) is fairly symmetric about its peak at 2.30Å which is close to the peak at 2.26Å in the absolute magnitude $|\rho(r')|$ (dashed curve). By varying E_0 until the peaks of Im $\rho(r')$ and $|\rho(r')|$ coincide, a distance of 2.281Å is obtained which agrees very well with the known value of 2.284Å. Consequently, E_0 was found to be 7 eV above the bound state. (Adapted from Lee and Beni, 1977.)

shells may require different E_0 values and the peak matching procedure must be performed separately for each individual shell.

6.11 The r Space Method

A Fourier transform technique which has an explicit expression for the atomic pair correlation function has been developed by Hayes and coworkers (1975, 1976, 1978). Curve fitting is done in the Fourier (or distance) space. Consider the K-shell EXAFS of atom species A (denoted by the subscript a). Assume for simplicity that the nearest neighbors consist only of atom species B (denoted by the subscript b) distributed closely about a mean nearest neighbor distance r. The EXAFS due to these nearest neighbors alone can be represented approximately by

$$\chi(k) = \frac{(2\pi)^{\frac{1}{2}}}{kr^2} 2 \operatorname{Re}[P_{ab}(k) \, \Lambda_{ab}(k,r)], \tag{6.54}$$

where

$$P_{ab}(k) = (2\pi)^{-\frac{1}{2}} \int_0^\infty e^{i2kr'} p_{ab}(r') dr' \tag{6.55}$$

and

$$\Lambda_{ab}(k,r) = -2i\pi^2 f(\pi,k) e^{i2\delta_a} e^{-2r/\lambda(k)} . \tag{6.56}$$

In these expressions, the EXAFS $\chi(k)$ has been divided into two distinct contributions: structural information in P and the complicated momentum dependences in Λ. $P_{ab}(k)$ is the Fourier transform of the first peak in the radial distribution of atom, $p_{ab}(r)$, defined so that $\int p_{ab}(r) dr$ equals the number of nearest neighbor atoms. In Λ_{ab} are included those factors which express the complex interactions of the final state electron with the excited central atom (δ_a), the backscattering atoms $f(\pi,k)$, and the intervening media (λ).

The usual expression for the EXAFS is obtained for the case where the peak in $p(r)$ is a Gaussian of half-width σ representing a single shell of N atoms at r:

$$\chi(k) = \frac{N}{kr^2} |\Lambda(k,r)| \exp[-2\sigma^2 k^2] \, 2 \cos\{2kr + \text{phase } [\Lambda(k,r)]\} . \tag{6.57}$$

The advantage of this method is that it casts the peak shape (in the distance space) in an explicit form which is convenient for peak shape analysis for systems with non-Gaussian or asymmetric peaks.

6.12 The Phase Linearization Method

Stearns (1982) found that there exists an energy threshold, E_c, that gives rise to a linear phase function $\phi(k) = a - bk$ and a distance shift α in the Fourier transform which is independent of the weighting factor n. A method was proposed to determine E_c and to use the resulting linear phase function in distance determination. More recently, Stearns and Stearns (1983) proposed the use of the "linearized phase" as a test of phase transferability in EXAFS data analysis. The method circumvened the need for E_0 adjustment. The rationale for the method is that as $k \rightarrow \infty$, the photoelectron must behave like a free electron; and so the phase must become linear in this limit. At $E_0 = E_c$, each pair of absorber-scatterer atoms is characterized by the two unique parameters a and b in the linear phase function which can be qualitatively interpreted, in a square-well picture, as the depth of the well and the scattering size of the atom. The method, however, is quite time-consuming. Furthermore, the phase functions of heavy scatterers (e.g. Pt) are known to have complicated nonlinear k dependence (cf. Chapter 7) which cannot be linearized. As a result, the method is applicable only for relatively light scatterers with atomic number $Z \leq 50$.

6.13 The Regularization Algorithm

Babanov, Vasin, Ageev, and Ershov (1981,1983) proposed a new interpretation of EXAFS spectra in real space using a regularization algorithm of "ill-defined" problems. These authors claimed that the method has less instability in deriving the radial distribution function than the commonly used Fourier transform methods. Ageev, et al (1983) also compared the radial distribution function of $Fe_{80}B_{20}$ obtained via this method to that obtained via diffraction studies.

6.14 Other More Specialized Methods

Other more specialized methods are also available. Examples include maximum entropy spectra estimation (Labhardt and Yuen, 1979), moment expansion (discussed in Chapter 5), the probabilistic approach using the cumulant expansion (Bunker, G., 1983; Rehr, 1983; Bouldin and Stern, 1982), etc. The readers are referred to the original literature for discussions on the strengths and weaknesses of these methods as well as specific applications.

Chapter 7

Theoretical Amplitude and Phase Functions

7.1 Introduction

It is clear from previous chapters that structural determinations by EXAFS spectroscopy rely heavily on our knowledge concerning the amplitude and phase functions. In other words, both the Fourier transform and the curve fitting techniques commonly used in EXAFS data analysis require a detailed knowledge of the amplitude $F(k)$ and phase $\phi(k)$ functions for the determination of chemical information such as coordination number N, Debye-Waller factor σ, and interatomic distance r through the assumptions of amplitude and phase transferabilities.

Furthermore, in the basic EXAFS formulation (cf. Chapters 2 and 4), we note that each EXAFS wave contains two sets of highly correlated variables: $\{F(k), \sigma, \lambda, N, S\}$ and $\{\phi(k), E_0, r\}$. Significant correlations can occur both *within* and *between* these two sets of variables as well as *between* different scattering terms. In order to determine N and σ, $F(k)$ must be known reasonably well; similarly, in order to determine r, $\phi(k)$ must be known accurately. In fact, the accuracy of EXAFS results is intimately related to the accuracy of these functions. This is particularly true for multiatom and/or multidistance systems where the accuracy is critically dependent upon the resolution of the EXAFS into individual waves.

Experimentally the amplitude and phase functions can be obtained from the EXAFS spectra of model compounds as described in Chapter 6. However, it is often possible to obtain only the product (vide infra) $F_b(k)e^{-2\sigma^2k^2}e^{-2r/\lambda}$ (assuming N is known) for each type of backscatterer B and the combined phase $\phi_{ab}(k)$ for each pair of atoms AB (assuming r is known) from experimental data, where a and b refer to the absorber A and the backscatterer B, respectively. The extraction of amplitude function $F_b(k)$ alone requires knowledge of the Debye-Waller factor σ (from a separate study of vibration frequencies or from temperature-dependent measurements) and the electron mean free path λ, whereas the separation of the total phase shift ϕ_{ab} into individual phases ϕ_a (due to the absorber A) and ϕ_b (due to the backscatterer B) can only be achieved by measuring the phase shifts of various combinations of pairs of atoms and arbitrarily defining ϕ_a or ϕ_b for one atom. For example, if one arbitrarily defines ϕ_a of atom A (absorber), one can deduce ϕ_b of atom B (scatterer) from the experimental phase shift ϕ_{ab} for that atomic pair A-B. From ϕ_b one can then determine the central atom phase $\phi_{a'}$ of any atom A′ by measuring $\phi_{a'b}$ which is the total phase shift of atom pair A′B where A′ and B denote the (new) absorber and the (old) scatterer, respectively. Similarly, from ϕ_a it is possible to deduce the scatterer phase $\phi_{b'}$ of any atom B′ by measuring $\phi_{ab'}$ for the atom pair AB′ with A and B′ being the (old) absorber and the (new) scatterer, respectively. All individual phase functions constructed in this manner are "relative" to the arbitrarily defined ϕ_a of absorber A.

To avoid the tedious task of searching, measuring, and analyzing model compounds, it is clearly desirable to calculate the amplitude $F_b(k)$ and the individual phase shifts $\phi_a(k)$ and $\phi_b(k)$ from first principle. With an accurate method, this not only represents a major saving in time and effort, but also greatly reduces the danger of introducing experimental errors from the analyses of model compounds. In particular, separation of the total phase $\phi(k)$ into individual contributions, $\phi_a(k)$ due to the absorber and $\phi_b(k)$ due to the backscatterer, allows a simple combination of phase shifts to be taken for any pair of atoms.

In this chapter, a brief discussion will be given for various techniques of theoretical calculations followed by extensive calculations performed by Teo and Lee (1979) via an electron-atom scattering theory originally developed by Lee and Beni (1977). The characteristic shape (k dependence) of theoretical amplitude and phase functions, tabulated by Teo and Lee (1979) for the majority of the elements in the Periodic Table, along with their Z (atomic number) dependence, allows identification of atom types in unknown systems (Teo, *et al*, 1977; Lee, *et al*, 1977) in addition to the determination of structural parameters, thereby greatly enhances the chemical content of EXAFS spectroscopy.

7.2 Theoretical Methods

The calculation of EXAFS amplitude and phase functions is basically an electron-atom scattering problem if one assumes that for the kinetic energy of 60-1000 eV, the ejected photoelectron is mainly (and increasingly) scattered by the core electrons of the neighboring atom(s). Thus, the distribution and the binding energies of the valence electrons, either on the absorber or on the scatterers, will not affect these functions to a substantial extent. Hence, to a first approximation, chemical or bonding effects related to the valence electrons can be ignored and the problem reduces to the calculation of the scattering of an electron with varying kinetic energy (60-1000 eV) by an atom (neighbors).

There are basically two types of theoretical methods for calculating these functions. The first is the Hartree-Fock (HF) method in which the Hartree-Fock equation for the atom plus the external electron is solved by a self-consistent field iterative procedure. The atom is described by, and held frozen at the tabulated atomic wave functions. The external electron is allowed to exchange with the atomic wave functions but not to polarize the atom. This method is reasonable for the high energy regime where the fast moving photoelectron does not have time to significantly polarize the atom. The second is the Hartree-Fock-Slater (HFS) method in which the atom is replaced by an electron gas of varying density $\rho(r)$ calculated from the tabulated atomic wave functions. The exchange between the external electron and this electron gas is replaced by the local potential

$$V(r) = -e^2 \left[\frac{3}{\pi} \rho(r) \right]^{1/3} \tag{7.1}$$

This latter equation is in fact basically Slater's local $X\alpha$ approximation (Slater, *et al*, 1951, 1969, 1971; Gaspar, 1954; Kohn and Sham, 1965) to the nonlocal exchange-correlation term. This "local density function" approximation has been tested for low energy electrons in band structure calculations and is believed to include both the exchange and the correlation effects. Hence, it is a reasonable theory for the low energy region where the electrons on the atom response adequately to the photoelectron.

The HF method has been used by Lee and Pendry (1975), Kincaid and Eisenberger (1976), and Pettifer and McMillan (1977) whereas the HFS method has been used by Ashley and Doniach (1975) and by Lagarde (1976). Neither of these two methods provides satisfactory agreement with experiments throughout the entire EXAFS energy spectrum (60-1000 eV).

Clearly a compromise is needed which will take into account the exchange and correlation effects adequately only at low energies (where these effects are important) but not at high energies (where these effects should be switched off gradually). In 1977, Lee and Beni developed a scheme which is capable of interpolating between the low energy regime, where the HFS method is most successful, and the high energy region, where the HF approach is more reasonable. Basically the theory involves the construction of an effective complex scattering potential that adequately accounts for the exchange and correlation effects caused by the electrons in the atom using a modified Thomas-Fermi approach which amounts to replacing the atom by an electron gas with spatially varying density and calculating the self-energy using the plasmon pole approximation. Specifically, the method amounts to first calculating the spatial dependence of the slowly varying electron density $\rho(r)$ and the Fermi energy $E_F(r)$ from the tabulated Hartree-Fock atomic wave functions. If E is the kinetic energy of the photoelectron, then the local Fermi energy of the electron-atom complex can be obtained by using the Thomas-Fermi description of an atom:

$$E_{\text{loc}}(r) = E + E_F(r) \qquad (7.2)$$

The exchange and correlation potential can then be approximated by the self-energy Σ_{homo} of the homogeneous electron gas of the local density $\rho(r)$ as a function of the radial distance r and the kinetic energy E of the incoming electron:

$$U_{xc}(r,E) = \Sigma_{\text{homo}}(E_{\text{loc}}(r), \rho(r)) . \qquad (7.3)$$

This complex potential is added to the electrostatic potential and the Schrodinger equation for such a system is solved iteratively to yield a set of complex phase shifts, $\delta_l(k)$ where $l = 0, 1, 2, \cdots$ is the orbital angular momentum.

It should be emphasized here that the *complex effective potential* implies that inelastic effects are included. In other words, inelastic scattering due to the excited central atom and the backscattering atoms is included in the calculation, but not that due to the intervening media (*cf. Section 5.2.2*).

If one now approximates the spherical wave of the outgoing photoelectron by a plane wave, the scattering of the electron by an atom at any arbitrary scattering angle β (defined as the angle of diffraction of the electron trajectory) is given by

$$f(\beta,k) = F(\beta,k)e^{i\theta(k)} \tag{7.4}$$

$$f(\beta,k) = \sum_{l=0}^{\infty} \left[\frac{2l+1}{k}\right] e^{i\delta_l(k)} \sin \delta_l(k) P_l (\cos \beta) \tag{7.5}$$

where $P_l(\cos \beta)$ is the Legendre polynomial and $\delta_l(k) = \delta_l^R(k) + i\, \delta_l^I(k)$ with $\delta_l^R(k)$ and $\delta_l^I(k)$ being the real and imaginary parts of $\delta_l(k)$. $F(\beta,k)$ and $\theta(\beta,k)$ are the scattering amplitude and phase functions, respectively. (Strictly speaking, $f(\beta,k)$ is also weakly dependent upon the distance r from the absorber which is ignored in the plane wave approximation.) Equation 7.4 will be used in the following chapter to describe multiple scattering effects and bond angle determinations.

For single-scattering theory of EXAFS with only backscattering, $\beta = \pi$, Eq. 7.4 and 7.5 reduce to

$$f(\pi,k) = F(k)e^{i\theta(k)} \tag{7.6}$$

$$f(\pi,k) = \sum_{l=0}^{\infty} \left[\frac{2l+1}{2ik}\right] (e^{2i\delta_l(k)} - 1)(-1)^l \tag{7.7}$$

Hence $F(k)$ and $\theta(k)$ are the *backscattering* amplitude and phase functions, respectively, and are readily obtainable from the following set of equations

$$f^R(\pi,k) = \sum_{l=0}^{\infty} \left[\frac{2l+1}{2k}\right] (\sin 2\delta_l(k)) \tag{7.8}$$

$$f^I(\pi,k) = \sum_{l=0}^{\infty} \left[\frac{2l+1}{2ik}\right] (\cos 2\delta_l(k) - 1)(-1)^l \tag{7.9}$$

$$F(k) = \left[f^R(\pi,k) + f^I(\pi,k)\right]^{1/2} \tag{7.10}$$

$$\theta(k) = \tan^{-1} \frac{f^I(\pi,k)}{f^R(\pi,k)} \tag{7.11}$$

7.3 Theoretical Amplitude and Phase Functions

Using Lee and Beni (1977)'s electron-atom scattering theory described in the previous section, the absorber (central atom) phase shifts $\phi_a^l(k) = 2\delta_l'(k)$ and the backscattering amplitude $F_b(k) = F(k)$ and phase $\phi_b(k) = \theta(k)$ functions have been calculated by Teo and Lee (1979). We note here that the photoelectron experiences the central atom phase shift $\delta_l'(k)$ twice but the scatterer phase shift $\theta(k)$ only once.

The results for nearly half of the elements in the periodic table with atomic number $Z < 86$ are tabulated in Tables I-VIII in Appendix V.

Tables I-III were calculated using Clementi-Roetti (1974) wave functions. The central atom phase shifts $\phi_a^l(k)$ were obtained using the relaxed $Z + 1$ ion approximation (Lee and Beni, 1977) which amounts to using the $Z + 1$ atomic wave function with one outer (valence) electron missing (see Chapter 4). The results using Herman-Skillman (1963) wave functions are tabulated in Tables IV-VIII. For the central atom phase shifts the relaxed Z ion with one core electron (1s or 2s) missing was used (cf. Chapter 4). In all backscattering amplitude (Tables I and IV) and phase (Tables II and V) function calculations, the wave functions were truncated at 1.5 times the covalent radius and a uniform charge density was added to preserve charge neutrality within this radius. For all central atom phase shift (Tables III, VI-VIII) calculations, the Coulomb field was cut off at twice the covalent radius. All calculations were performed on neutral atoms except for a few alkali and alkali-earth elements, which were treated as cations. The difference between various oxidation states, though small but significant, can be compensated by changing the threshold energy (vide infra).

The atomic ground state electronic configuration was used for most elements. For example, the valence shell configurations for groups 4A and 7A are ns^2np^2 and ns^2np^5, respectively. The valence electronic configuration for the three transition metal series were $3d^{Z-20}4s^2$, $4d^{Z-37}5s^1$, and $5d^{Z-70}6s^2$. Again, the difference between various electronic configurations can largely be compensated by E_0 variation. It should also be pointed out that the Herman-Skillman wave functions are inadequate for heavy atoms (beyond the rare earths, for instance) because of relativistic corrections. However, the use of relativistic wave functions for tungsten showed that the difference is sufficiently small to justify the use of the more readily available nonrelativistic Herman-Skillman wave functions for all elements.

A comparison of Tables I-III (Clementi-Roetti wave functions) with Tables IV-VIII (Herman-Skillman wave functions) revealed that the two sets of results agree quite well with the exception of central atom (absorber) phase shifts. A comparison of Table III with the corresponding Table VII showed that the $(Z + 1)$ approximation in the former case results in more positive phase shifts. The difference (ca. 0.7-0.2 rad), however, decreases with increasing k values and therefore can be substantially removed by changing the energy threshold. For the sake of consistency, the results derived from Herman-Skillman wave functions should be used whenever possible.

Throughout this book, the ab initio theoretical EXAFS functions in the range of $k \simeq 4\text{-}15 \text{ Å}^{-1}$ are tabulated. The truncation at low k value $(k \simeq 4 \text{ Å}^{-1})$ is due to the fact that the theory is less reliable for $k \leq 4 \text{ Å}^{-1}$ as a result of inadequate treatment of valence electrons, particularly for light atoms with $Z \leq 9$ where the energy of the valence electrons is a substantial portion of that of the core electrons. Furthermore, other physical phenomena such as multiple scattering may become important at low k values. At high k values $(k \geq 15 \text{ Å}^{-1})$, the EXAFS signal is generally attenuated substantially by Debye-Waller factor.

It should also be mentioned that the "experimental" units of angstroms for the amplitude function $F(k)$, radian for the phase functions $\phi(k)$ and Å^{-1} for the electron wave vector k are used here and in Teo and Lee (1979)'s paper. The functions $F(k)$, $\phi_b(k)$, and $\phi_a^l(k)$ are equivalent to the functions $A(k)$, $\theta(k)$, and $2\delta'_l(k)$ (all in atomic units) used in Lee and Beni (1977)'s paper.

7.4 Properties of Amplitude and Phase Functions

If we have an absorbing atom A and a backscattering atom B, the backscattering amplitude is given by $F_b(k)$ and the phase function by $\phi_{ab}(k)$ where

$$\phi_{ab}(k) = \phi_a^l(k) + \phi_b(k) - \pi, \quad l = 1 \text{ for } K \text{ and } L_I \text{ edges} \qquad (7.12a)$$

$$\phi_{ab}(k) = \phi_a^l(k) + \phi_b(k), \quad l = 2,0 \text{ for } L_{II,III} \text{ edges} \qquad (7.12b)$$

ϕ_a^l is the phase shift of the central atom A and ϕ_b is the phase shift of the neighboring atom B.

The phase functions are listed and plotted as an increasing function of atomic number Z merely for clarity and for the purpose of facilitating interpolation of the phase shifts of intermediate Z elements. In practice, any phase functions can be modified by $\pm 2n\pi$ where $n = 0,1,2,3, \cdots$, since $\sin (\phi(k) \pm 2n\pi) = \sin (\phi(k))$. With these remarks, we shall now discuss the backscattering amplitude $(F(k))$, the backscattering phase $(\phi_b(k))$, and the central atom phase shifts $(\phi_a(k))$ with the aid of Figures 7.1-9.

7.4.1 Amplitude.

Figure 7.1a depicts the scattering amplitudes for two main groups 7A and 4A, respectively, whereas Figure 7.1b shows the amplitude functions of the transition metals Fe, Ru, and Os. It is apparent that as the atomic number Z increases the scattering amplitude at high k values generally increases. More importantly, there are peaks and valleys in the amplitude functions which move to higher k values as Z increases. These amplitude peaks correspond to the resonances of the electron-atom scattering process and can be associated with the $\pi/2$ crossings of phase shifts with different l values. Figures 7.2a-d show how these peaks and valleys progress within each series. For light atoms with $Z \leq 10$, the amplitude function peaks at the low k region (≤ 3 Å^{-1}) such that only a monotonically decreasing function is observed. This structureless tail is due to the fact that, when the electron exceeds the binding energy of the deepest shell, the electron is sampling mostly the nuclear potential. For elements with $10 \leq Z \leq 30$, we observe an amplitude envelope whose peak height decreases while its peak position advances in k as Z increases. For elements with $30 \leq Z \leq 54$, we find one peak and one valley in the amplitude function. A second peak starts to come in from the low k region. Both peaks increase in amplitude and advance in peak position as Z increases. For elements with $57 \leq Z \leq 71$, both peaks advance in peak position (in k) while diminish somewhat in peak height. Finally, for elements with $72 \leq Z \leq 82$, a third peak starts to come in from the low k region. The three peaks and the two valleys move to higher k values as Z increases.

The positions of the peaks P_i and valleys V_i of the amplitude functions follow roughly linear relationships:

$$P_1 = 0.204(Z + 8) \qquad (7.13a)$$

$$V_1 = 0.136Z \qquad (7.13b)$$

$$P_2 = 0.136(Z - 21) \qquad (7.13c)$$

$$V_2 = 0.136(Z - 39) \qquad (7.13d)$$

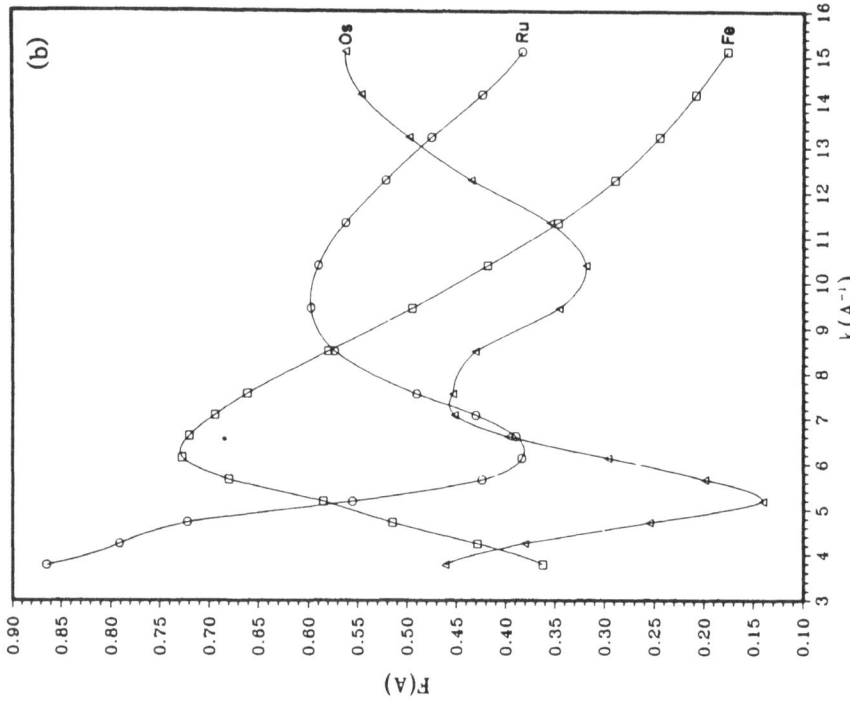

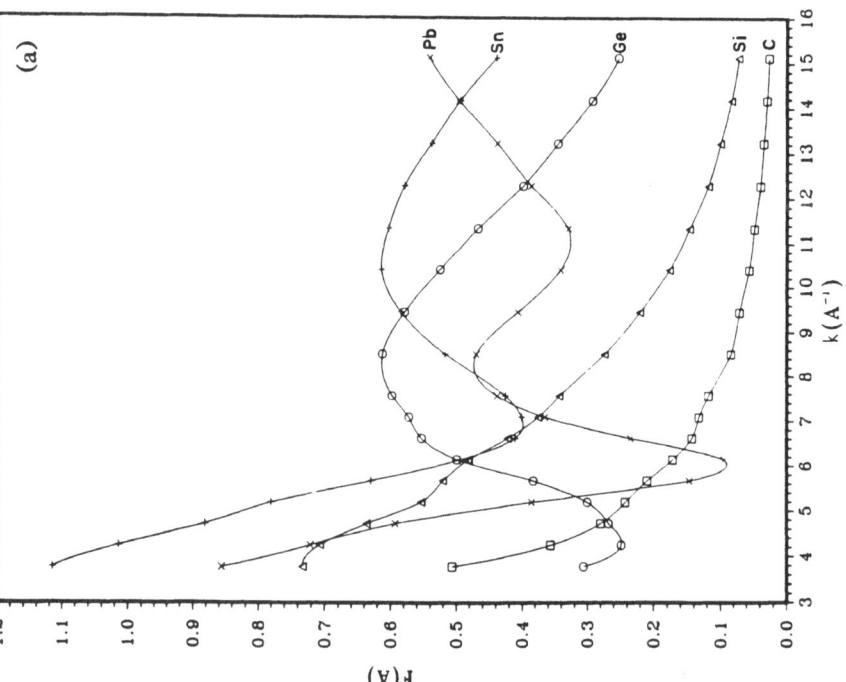

Figure 7.1. Backscattering amplitude functions for (a) group 4A elements; (b) transition metals Fe, Ru, and Os. (Reproduced from Teo and Lee, 1979).

The simple linear trends of these peaks and valleys allow chemical identification of unknown elements and differentiation of elements with sufficiently different Z values, as well as interpolation or extrapolation for other elements.

Some of the plots exhibit a considerable amount of scatter, especially at low k and for the heavier elements like the rare earths. These scatters are systematic and are due to the truncation of the atomic wave function at the muffin tin radius r_{mt}. Such a truncation introduces a discontinuity in the exchange and correlation potential at the muffin tin radius which introduces oscillations in the amplitude and phase functions (see below). It should be perfectly legitimate to smooth out such scatters before comparison is made with experiments. In practice, the scatter is small enough not to make too much difference.

7.4.2 Scatterer Phase.

Figures 7.3a,b depict the corresponding backscattering phase shifts for the two groups while Figures 7.4a-d show the scatterer phase variation for the four series of elements.

At high enough k values, $\phi_b(k)$ decreases almost linearly with increasing k, whereas at low k values it exhibits complicated patterns which are related to the amplitude function $F(k)$. That is, above a certain energy, the phase shift decreases with increasing electron energy, whereas below such energy the phase shift depends heavily on the "resonances" interactions between the photoelectron and the various electronic shells of the scattering atom. The plateau (slow varying regions) and the inflection points (fast varying regions) in the scatterer phase correspond to the peaks and valleys, respectively, in the scattering amplitude. These are illustrated in Figures 7.3a,b for the two groups of elements where the P_i arrows designate the plateau and the V_i arrows represent the inflection points in $\phi_b(k)$.

It is also interesting to note that the scatterer phase shift varies systematically with the atomic number Z. Within each shell, the phase shift increases linearly with Z (as a result of the increasingly positive potential) with breaks (changing slope) at $Z \simeq 21$ and 57 which correspond to the starts of the d and f shells, respectively.

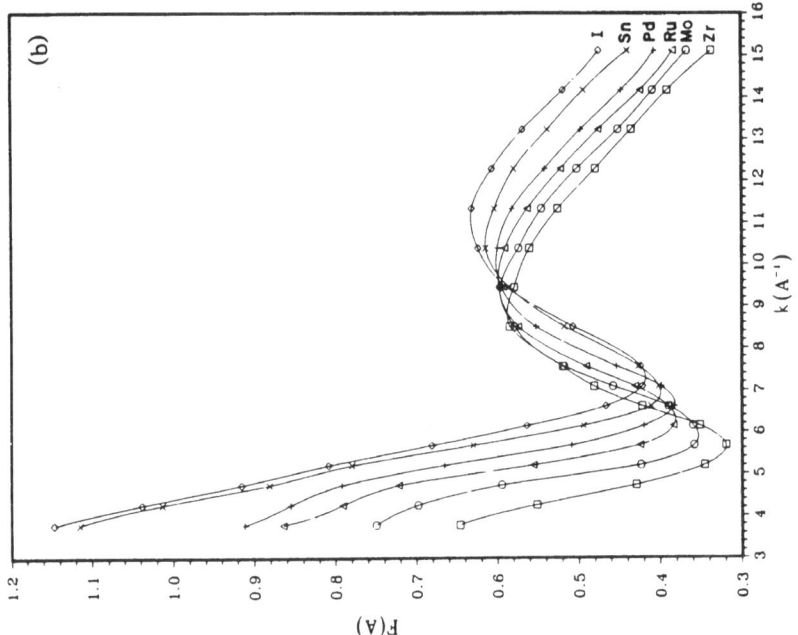

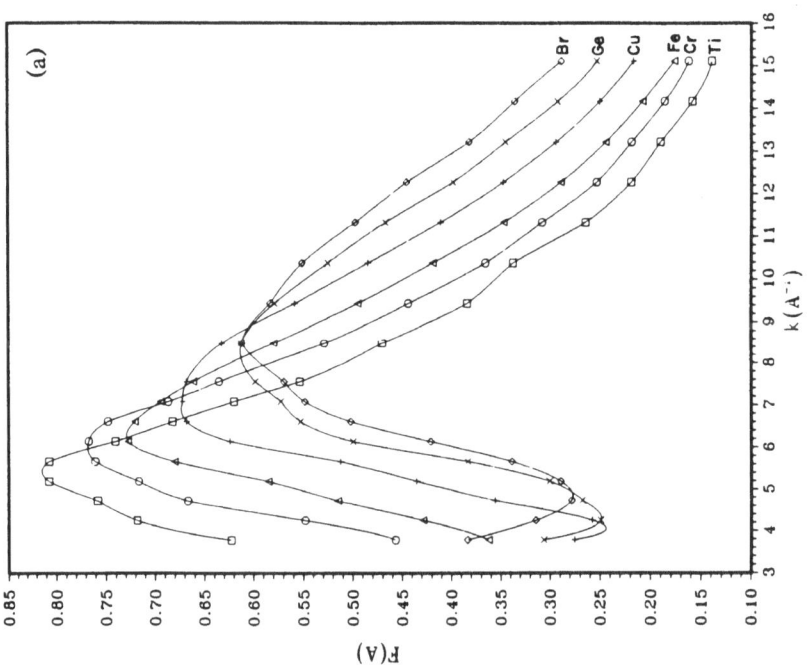

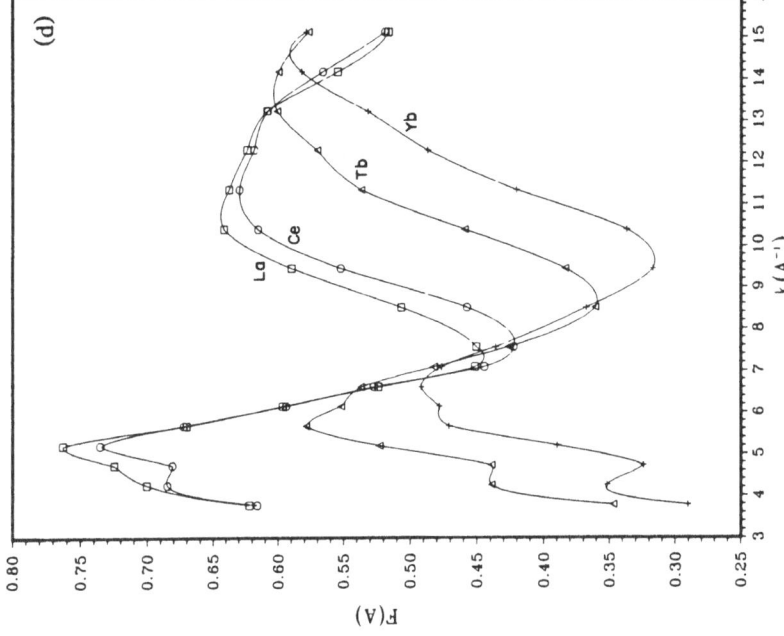

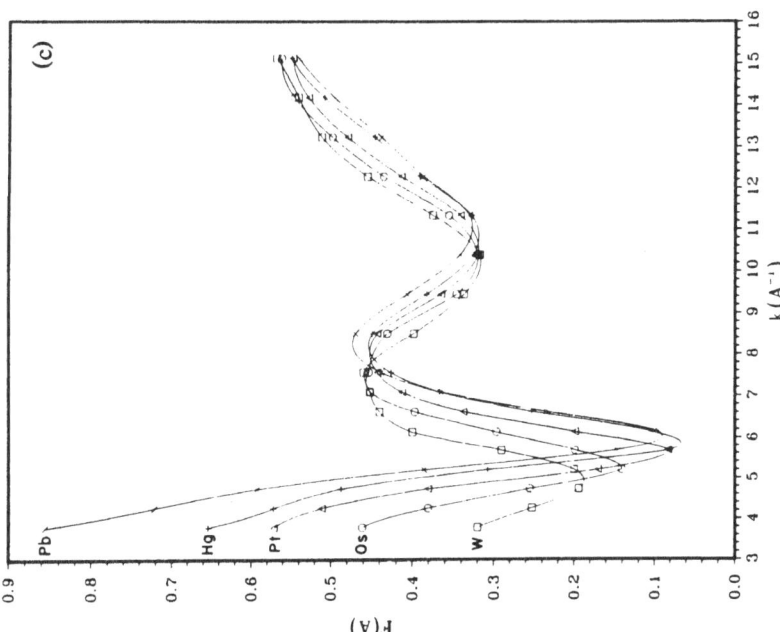

Figure 7.2. Backscattering amplitude functions for some representative elements in (a) first transition series and beyond; (b) second transition series and beyond; (c) third transition series and beyond; (d) lanthanides. (Reproduced from Teo and Lee, 1979).

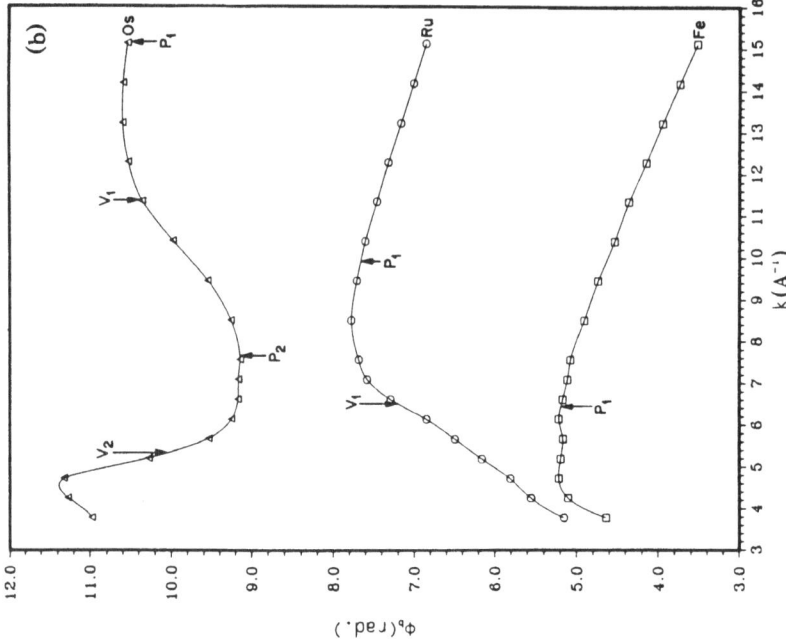

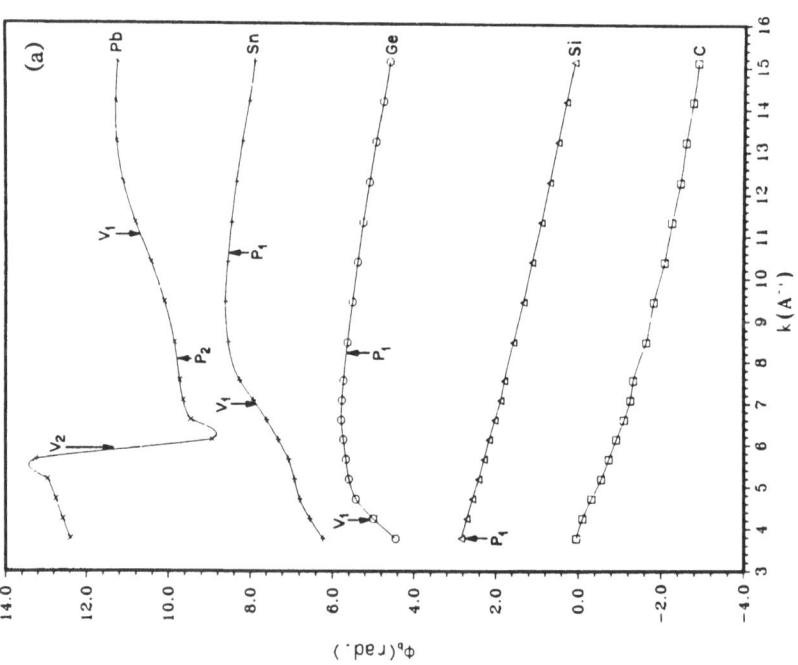

Figure 7.3. Backscattering phase functions for (a) group 4A elements; (b) transition metals Fe, Ru, and Os. The arrows P_i and V_i designate the "plateau" and the "inflection" points in $\phi_b(k)$ which correspond to the "peaks." (Reproduced from Teo and Lee, 1979).

7.4.3 Central Atom Phase Shift.

Central atom (or absorber) phase shift is a much *simpler* but *stronger* function of k. It generally decreases with increasing photoelectron energy (and hence k) as shown in Figures 7.5a,b for the two groups and Figures 7.6a-d for the three series of elements.

Again the central atom phase shift $\phi_a(k)$ varies systematically with increasing Z. In particular the potential is more attractive for increasing Z and we expect that the phase shift should increase. Indeed, if one plots the central atom phase shifts for a fixed k value as a function of Z, one gets a linear curve with, again, breaks at $Z \simeq 21$ and 57 which correspond to the injections of d and f electrons, respectively, into the electronic structure. The slope decreases at these "break points" which reflects the fact that the phase shift increases at slower rates for heavy elements than for light atoms.

7.4.4 Effect of Electronic Configuration.

The small but significant effects of valence shell electronic configuration on the amplitude as well as the scatterer and central atom phase functions are illustrated in Figures 7.7a-c for Pd. The configurations used are $4d^8 5s^2$, $4d^9 5s^1$, and $4d^{10} 5s^0$. It can be seen that the amplitude functions (Figures 7.7a) are little affected by changes in electronic configuration. The small variations at low k values are not unexpected because in this region the photoelectron energy is comparable to the valence shell binding energies. On the other hand, both the scatterer and the absorber phase shifts exhibit interesting systematic variations with electronic configuration. First, the scatterer phase (Figures 7.7b) increases with increasing population of the s orbital (or equivalently depopulation of the d orbitals). The difference between various configurations, however, diminishes as k increases and can largely be compensated for by changing E_0.

The effect of valence shell electronic configuration on central atom phase function follows the same trend. Figure 7.7c depicts the $\phi_a^0(k)$, $\phi_a^1(k)$, and $\phi_a^2(k)$ functions for three different electronic configurations ($4d^{10-n} 5s^n$ where $n = 0, 1, 2$) of Pd. In all cases, the phase shifts increase with increasing population of s orbital (or increasing depopulation of d orbitals). The difference, however, diminishes with increasing k (note that the near parallel appearance of the phase functions is an optical illusion). Again, such a difference can largely be compensated for by E_0 variation.

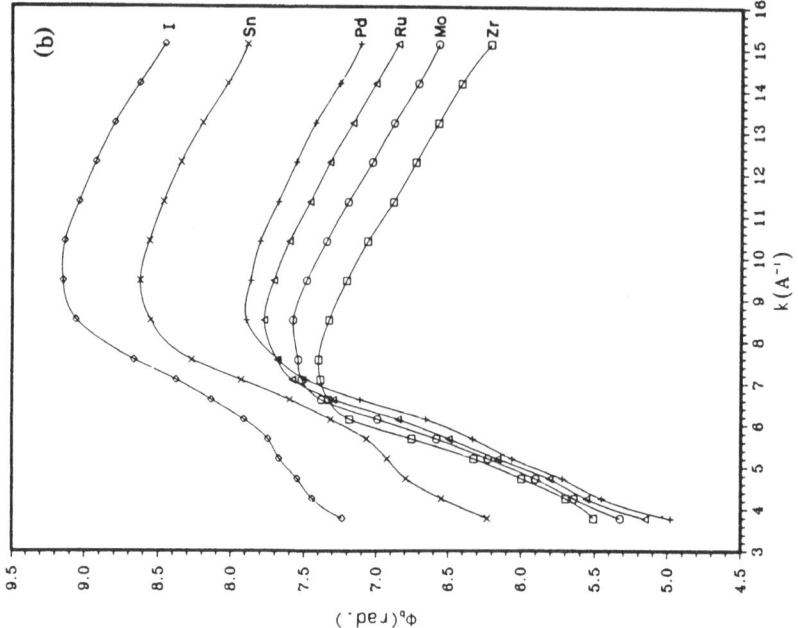

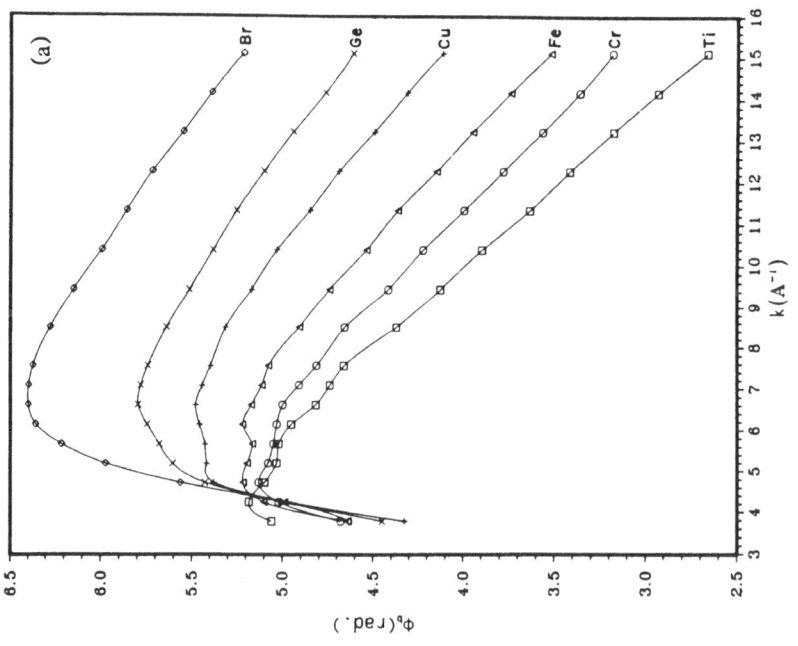

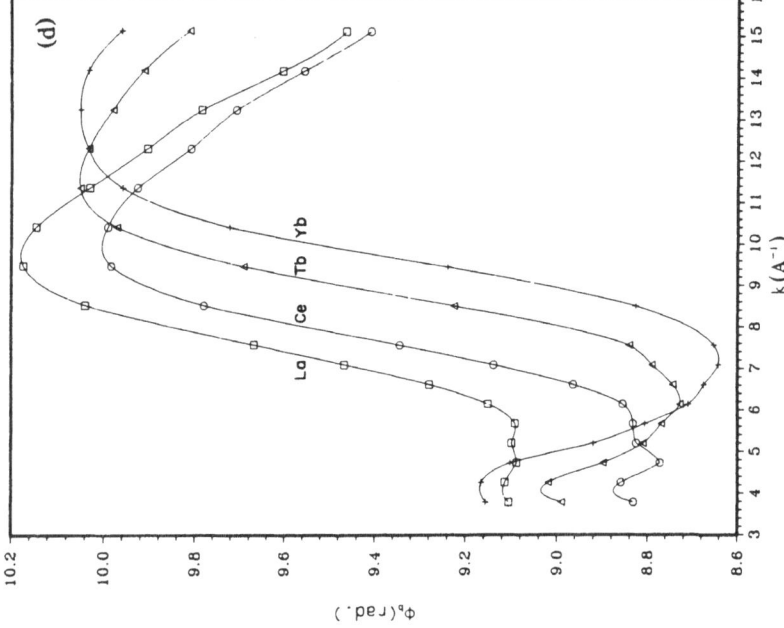

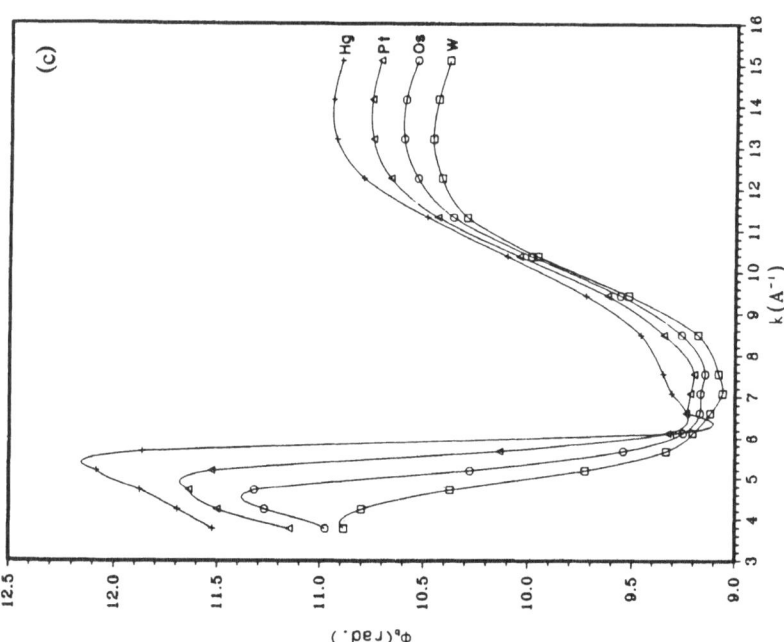

Figure 7.4. Backscattering phase functions for some representative elements in (a) first transition series and beyond; (b) second transition series and beyond; (c) third transition series and beyond; (d) lanthanides. (Reproduced from Teo and Lee, 1979).

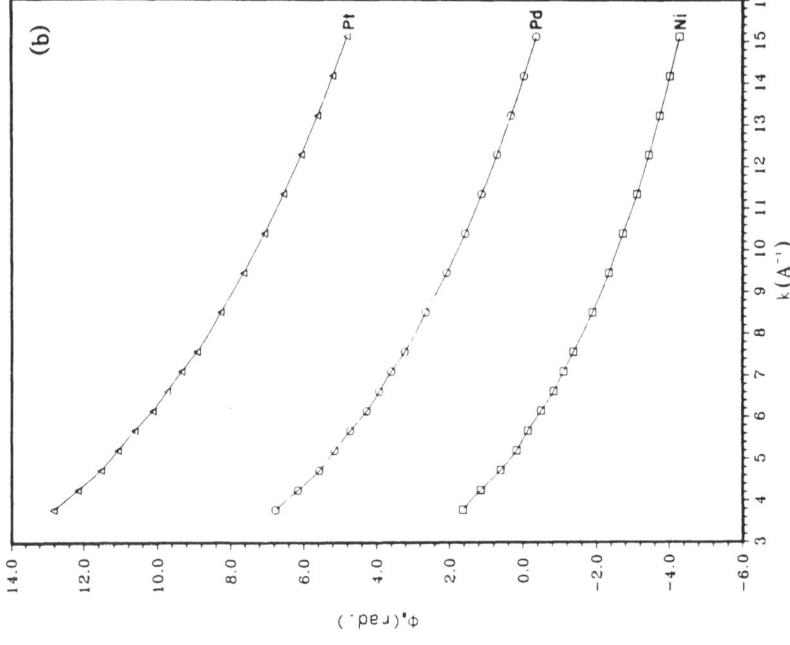

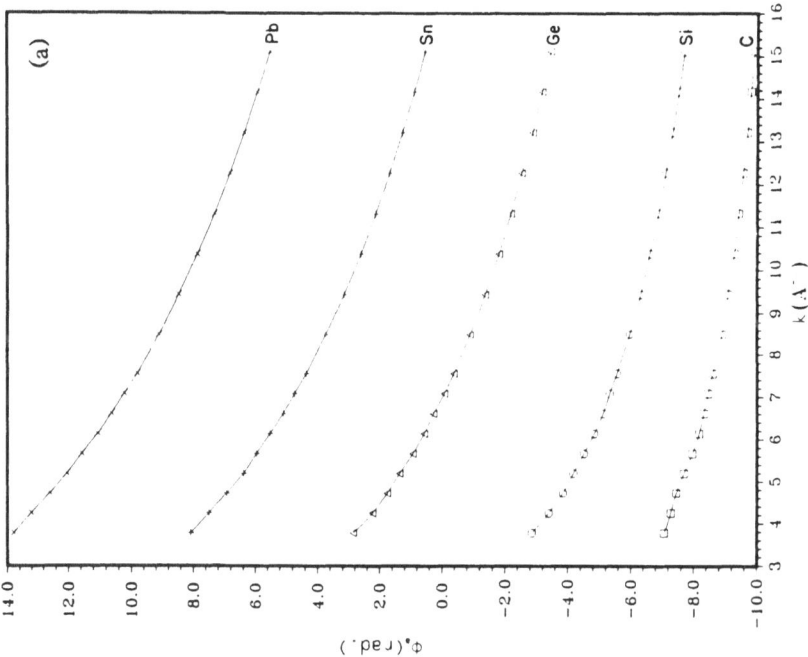

Figure 7.5. Central atom (absorber) phase functions ϕ_a^l for (a) group 4A elements; (b) transition metals Ni, Pd, and Pt. (Reproduced from Teo and Lee, 1979).

7.4.5 Charge Effect.

The effect of atomic charge on central atom phase shift is shown in Figure 7.8. Here we plot ϕ_a^0, ϕ_a^1, and ϕ_a^2 as a function of k for Ca and Ca^{2+}. It is obvious that the dication not only has a more positive phase shift but also a larger slope. It is also readily apparent that the effect of atomic charge on phase shifts is significantly larger than that of electronic configuration. This is not unexpected since the atomic charge exerts a significant effect on the central atom potential which is experienced by both the outgoing and the incoming photoelectrons. The difference, again, can partially be compensated for by E_0 variation.

7.4.6 Comparison of $\phi_a^l(l = 0, 1, 2)$ Functions.

The ϕ_a^l phase functions listed in Tables VI, VII, and VIII, where $l = 0, 1, 2$, are central atom (absorber) phase shifts for the transitions $p \rightarrow s$ ($L_{II,III}$ edges), $s \rightarrow p$ (K or L_I edge), and $p \rightarrow d$ ($L_{II,III}$ edges), respectively. Two examples are shown in Figure 7.7c for palladium and in Figure 7.8 for calcium. In both cases, there are large differences between the ϕ_a^l phase functions. The order $\phi_a^2 > \phi_a^1 > \phi_a^0$ as well as the divergence at large k values may not be physically meaningful since subtracting 2π from ϕ_a^2 and adding 2π to ϕ_a^0 yield an inverted order $\phi_a^2 < \phi_a^1 < \phi_a^0$ which converges at high k.

As discussed in Chapter 4, ϕ_a^1 functions are needed for K or L_I edges and ϕ_a^2 functions are used for $L_{II,III}$ edges since in the latter case, the $p \rightarrow d$ transition dominates. Ekardt and Tran Thoai (1983) argued that ϕ_a^2 functions can also be used for $M_{II,III}$ edges since, as discussed by Fano and Cooper (1968), for energies not too near the excitation edge the transition from a discrete (n,l)-level to a continuum $(n',l+1)$-state is favored by a factor of ~ 10 over the transition to the $(n',l-1)$-state. Consequently, for both $L_{II,III}$ and $M_{II,III}$ edges, the transitions from the p state to the d states are much more important than the transitions to the s states and hence the latter can be neglected.

The difference in absorber phase shifts for exciting 1s (K edge) vs. 2s (L_I edge) was found to be insignificant (Teo, unpublished results).

7.4.7 Relativistic Effect.

Relativistic wave functions have been used as input to calculate the backscattering amplitude and phase for tungsten. The results are compared

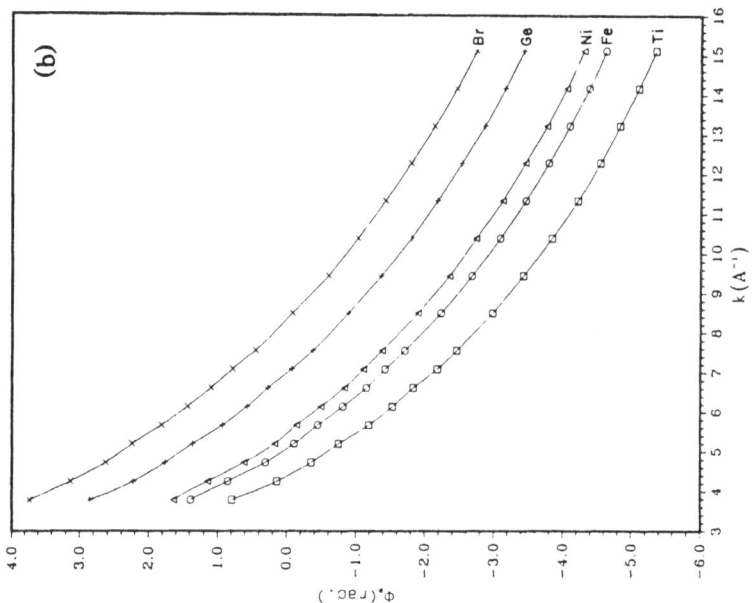

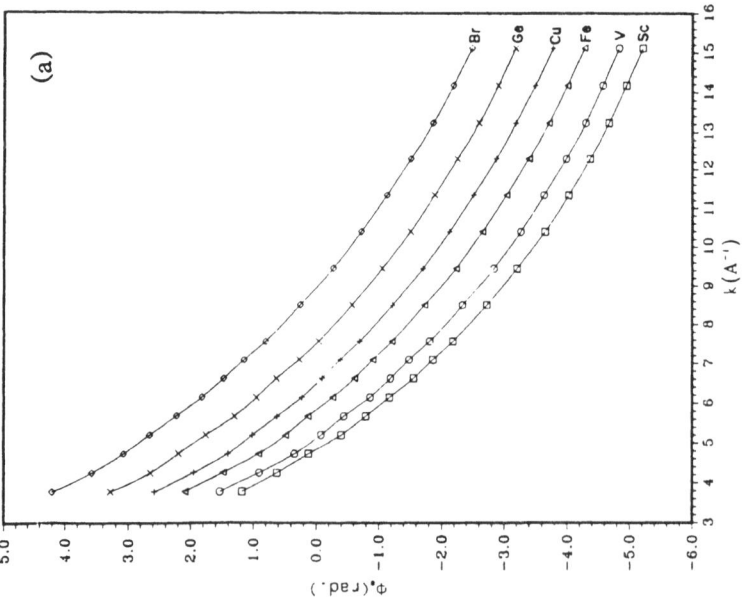

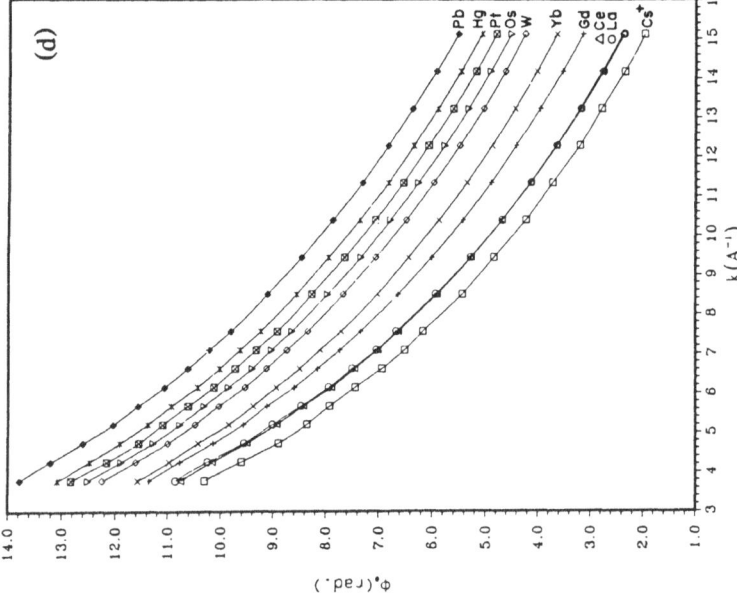

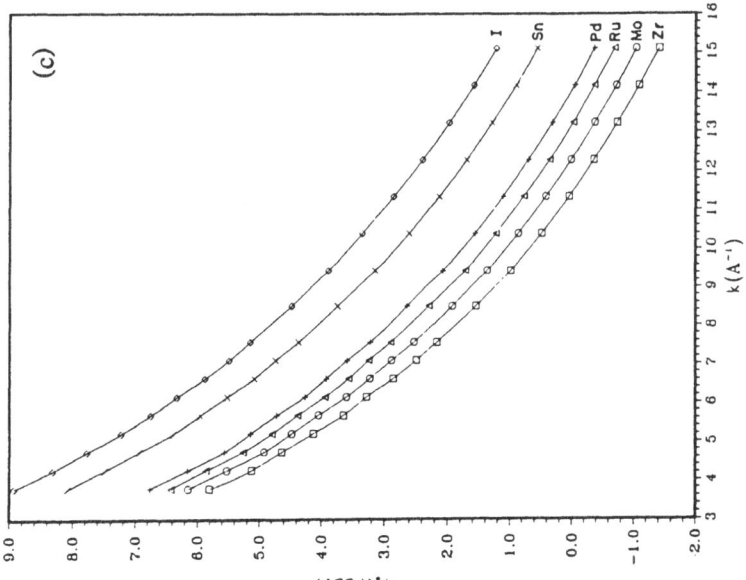

Figure 7.6. Central atom (absorber) phase functions ϕ_a^l for some representative elements in (a) first transition series and beyond (Clementi-Roetti wave functions); (b) first transition series and beyond (Herman-Skillman wave functions); (c) second transition series and beyond; (d) third transition series, lanthanides, etc. (Reproduced from Teo and Lee, 1979).

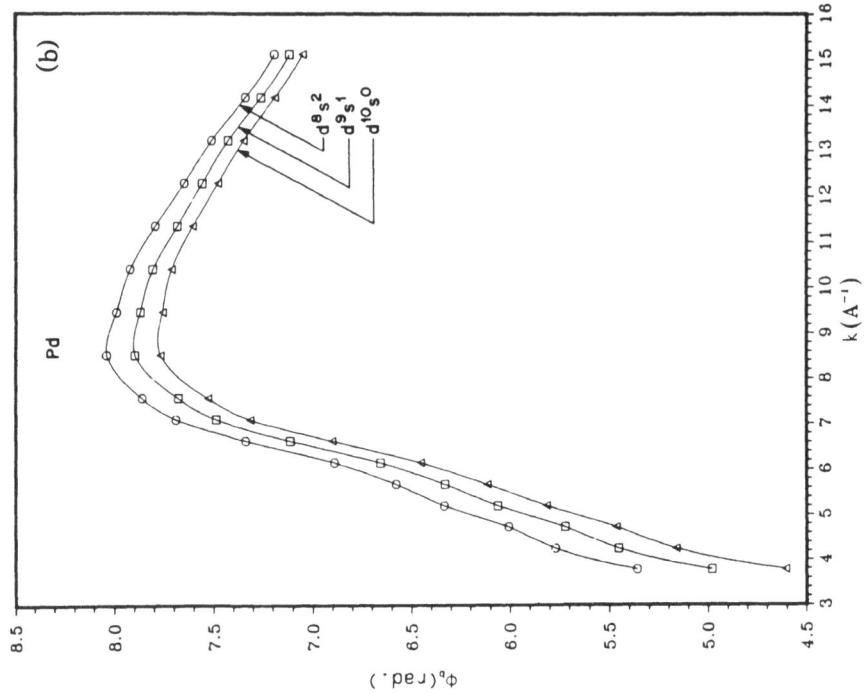

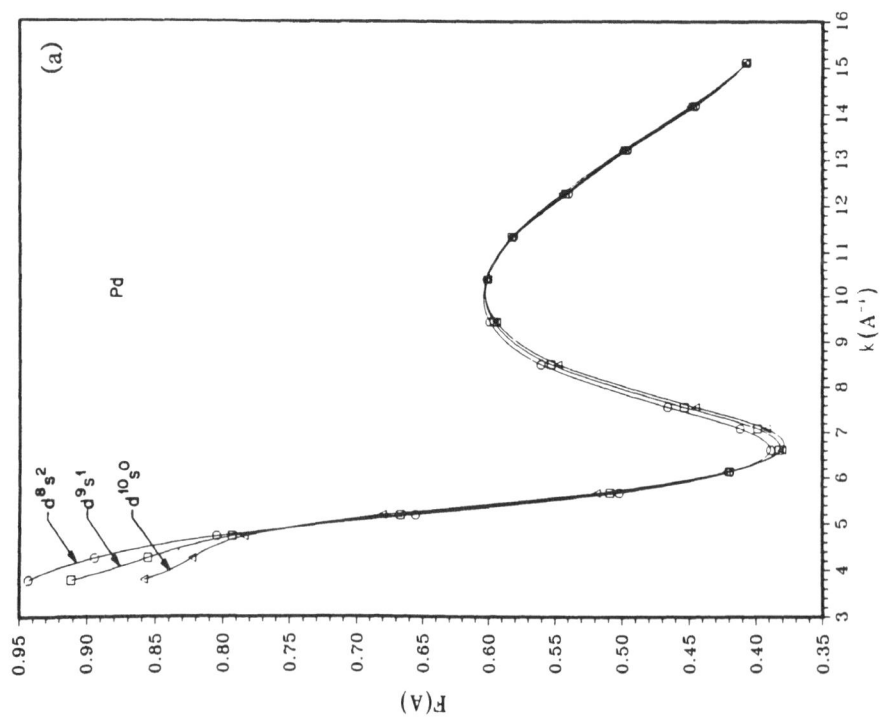

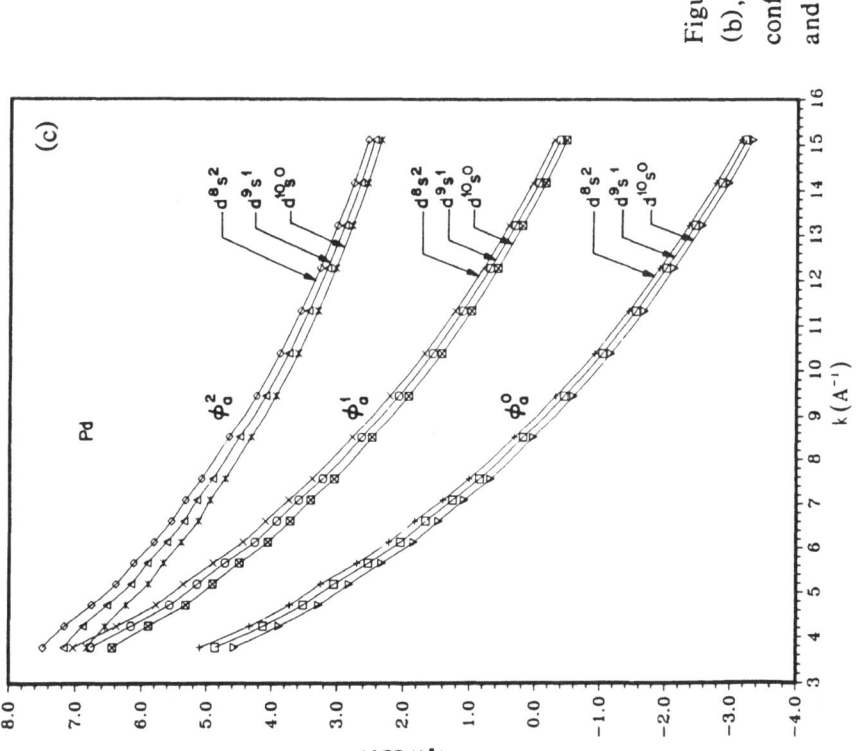

Figure 7.7. Comparisons of the amplitude (a), backscattering phase (b), and central atom phase (c) functions for Pd with electronic configurations $4d^8 5s^2$, $4d^9 5s^1$, and $4d^{10} 5s^0$. (Reproduced from Teo and Lee, 1979).

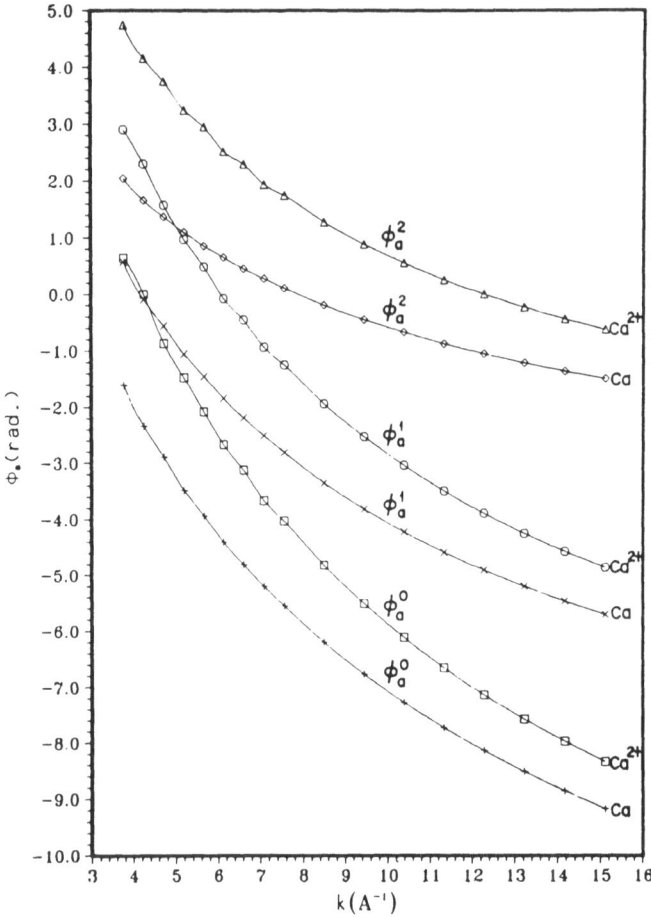

Figure 7.8. The effect of atomic charge on the central atom phase shifts as exemplified by Ca and Ca^{2+}. (Reproduced from Teo and Lee, 1979).

with that from Herman-Skillman wave functions in Figure 7.9. We see that the amplitudes are in substantial agreement, especially beyond $k = 6\ \text{Å}^{-1}$. The phase shows a systematic deviation which is progressively smaller for higher k. This kind of deviation is of the same order as that due to configuration differences and can be compensated for by changing E_0. Strictly speaking, for the heavier elements one should treat the electron scattering problem relativistically. However, it has been argued (Teo and Lee, 1979) that the relativistic corrections such as spin-orbit terms are small compared with the Hartree and exchange potential and that it can again be compensated for by E_0 variation.

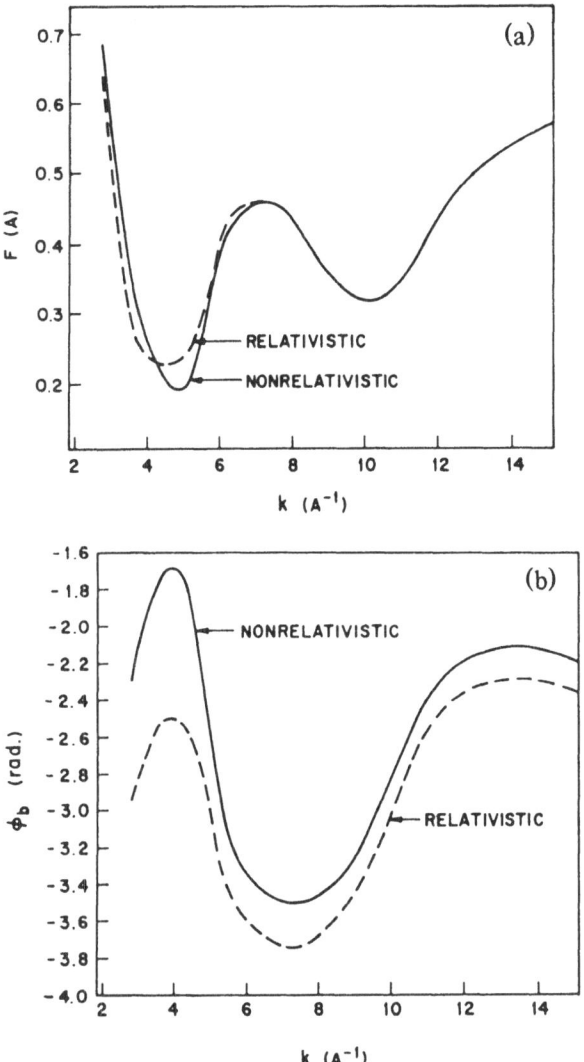

Figure 7.9. Relativistic effects on the amplitude (a) and scatterer phase (b) functions as exemplified by a calculation on W using relativistic (dashed curves) vs. nonrelativistic (solid curves) wave functions. (Reproduced from Teo and Lee, 1979).

7.5 Comparison of Theory and Experiment

Theoretical amplitude and phase functions described in this chapter have been widely used in EXAFS data analysis (see, e.g., review by Teo, 1980). Generally speaking, accuracy of 0.5-1% are to be expected for the distances and accuracy of 10-30% for the coordination numbers. These theoretical functions have also been compared extensively with experimental results. We shall discuss two such studies here.

Rabe, Tolkiehn, Werner (1981) compared the k and Z dependence of the backscattering amplitude $F(k) = |f(\pi, k)|$ functions calculated by Teo and Lee (1979) with their experimental results. As expected, they found that, in contrast to the good agreement between theory and experiment in terms of the functional shape (k dependence) and the variation of the peaks and valleys as a function of atomic number (Z dependence), theoretical amplitude is larger by almost a factor of two. A major portion of this difference is due to inelastic losses such as the shake up/shake off processes at the absorbing atom and the inelastic scatterings as the photoelectron travels to and from the neighboring atoms as discussed in Chapter 5. Another source of amplitude reduction, especially at low k regions, is the breakdown of the plane wave approximation which was discussed in Chapter 4.

A recent study by Bunker and Stern (1983) investigated the physical processes that affect the phase-shift factor in EXAFS. Comparison was made between the theoretical phase shifts of Teo and Lee (1979) and the experimental data on the tetrahedrally coordinated series of CuBr, ZnSe, GaAs, and Ge, using the zero of energy E_0 as a variable. They found that the most reliable criterion for determining E_0 was to set the intercept of the phase difference between theory and experiment to zero at $k = 0$. The results of this analysis verified that the phase shift is not strictly transferable, but much of the lack of transferability can be accounted for by varying E_0 in accordance with the above criterion. Different values of E_0 are required for each shell of neighboring atoms. The largest cause of the lack of transferability is the *breakdown of the small-atom or plane-wave approximation*, which, fortunately, can be accounted for by a shift in E_0 of -13 and -8 eV for the first and second shells of the samples, respectively. As discussed above, the breakdown of the small-atom approximation also produces large effects on the EXAFS amplitude at low k and this *cannot* be corrected simply by a change in E_0. These authors also found that the ionicity of bonding can be accounted for by a shift in E_0 of about 5 eV per electron-charge transfer.

Chapter 8

Multiple Scattering and
Bond Angle Determination

While EXAFS spectroscopy can provide structural information about the local environment of the absorbing atoms in terms of radial distribution functions (distances), no direct method of determining angular information is available; except, perhaps, for elaborate measurements on single crystals utilizing polarized X-rays. Furthermore, the very same advantageous characteristics of EXAFS (short-range, single-scattering) are also its serious limitations: distance determinations out to only *ca* 4Å. The situation, however, changes dramatically when atoms (including the X-ray absorbing atom and its neighbors) are arranged in a linear or nearly collinear fashion. In such cases, EXAFS contributions from neighboring atoms as far as 8Å can be observed. For these systems, both the amplitude and the phase of the EXAFS of a more distant neighbor are significantly affected by the intervening atom(s). In particular, the amplitude is greatly enhanced and is therefore commonly called "focusing" effect. The short-range single-scattering theory of EXAFS fails in these situations and one must take into account multiple scattering processes involving the intervening atoms.

The multiple-scattering effect in EXAFS was first observed when theoretical calculations of EXAFS were compared with measurements on copper metal (Lee and Pendry, 1975; Ashley and Doniach, 1975). The observed amplitude of the scattered wave for the fourth shell of copper was larger than the amplitude calculated from the single-scattering theory, and the observed phase shift was also off by approximately π from the calculated phase shift. These discrepancies were explained as an effect of first-shell atoms that intervene directly in the absorber-to-scatterer path to the fourth-shell atoms in the face-centered cubic lattice. Rather than occluding the EXAFS from the fourth-shell atoms as might be expected, the intervening atoms actually accentuate the EXAFS due to the shadowed atoms by enhanced forward scattering of both the outgoing and the backscattered photoelectron waves. These multiple-scattering events also cause additional phase shifts.

Recently, a new EXAFS formulation which takes into account the effect of multiple scattering has been developed (Teo, 1981). Theoretical scattering amplitude and phase functions have also been calculated for various scattering angles. Combination of the new multiple scattering formalism and the new theoretical functions facilitate the understanding of the "focusing" effect as well as the assessment of the relative importance of various multiple scattering pathways as the scattering angle varies. It also provides strong evidence that multiple scattering processes involving three atoms (including the absorber and two neighboring atoms) are important in determining the EXAFS of the distant shells, especially at large bond angles and low k values. Based on these, a new method for interatomic angle determinations by EXAFS has been devised which is applicable to systems with bond angles greater than ca 100°. The accuracy for such angle determinations is better than 6% for low Z ($Z \leqslant 10$) and 3% for high Z scatterers. In most cases, it amounts to an accuracy of ca 5°, which is comparable to the scattered range of crystallographically independent bond angles often observed in diffraction studies. The method requires no single crystal measurements and is applicable to wide varieties of materials (polycrystalline or amorphous solids, liquids and solutions, gases, surfaces, polymers, etc.).

A more formal derivation of the multiple scattering theory was also given by Boland, Crane, and Baldeschwieler (1982), and by Rennert and Vasvári (1983).

In this chapter, a description of the multiple scattering theory and bond angle determination will be given for various systems containing three (ABC), four (AB_1B_2C), or multiple ($AB_1...B_nC$) atoms at different geometries.

8.1 Scattering Amplitude and Phase

Let us first define a scattering angle β and atom B (and similarly γ at atom C) which is related to the bond angle A-B-C (α) (cf. Fig. 8.1) by

$$\beta = 180° - \alpha \qquad (8.1)$$

For any arbitrary angle β, the scattering of an electron by an atom B can be described by $f(\beta,k)$. If one approximates the spherical wave of the outgoing photoelectron by a plane wave (small-atom approximation):

$$f(\beta,k) = F(\beta,k)e^{i\theta(\beta,k)} \qquad (8.2)$$

where $F(\beta,k)$ and $\theta(\beta,k)$ are the scattering amplitude and phase functions, respectively. For *backscattering* where $\beta = 180° = \pi$ radians, Eq. 8.2 becomes Eq. 7.6 with $F(\pi,k) \equiv F(k)$ and $\theta(\pi,k) = \theta(k)$ in the previous chapter.

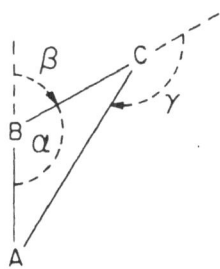

Fig. 8.1. Schematic representation of a three-atom ABC system where A is the X-ray absorbing atom (central atom), B is the nearest neighbor, and C is the next nearest neighbor. Here α is the A-B-C bond angle and β and γ are scattering angles at atom B and C, respectively.

For the formulation of the multiple scattering and the establishment of the bond angle determination method, these scattering amplitude and phase functions must be known accurately. In principle, these functions can be determined from studying the EXAFS of a large number of compounds with varying bond angles (one per each type of intervening atoms). In practice, however, it is extremely difficult to find model compounds with basically identical structural framework differing only in bond angles. Even if two or three such bond angles are known (e.g. M-N-O angles ranging from 120° to 180° have been found for metal nitrosyl compounds), it is still not enough to establish accurately $F(\beta,k)$ or $\theta(\beta,k)$ as a function β. It is therefore necessary to resort to theoretical calculations for these functions.

To calculate the scattering amplitude and phase of the photoelectron by an atom at any arbitrary angle β, we make use of the following equation

$$f(\beta,k) = \frac{1}{k} \sum_{l=0}^{\infty} (2l+1) e^{i\delta_l(k)} \sin \delta_l(k) P_l(\cos \beta) \qquad (8.3)$$

where δ_l is the phase shift with angular momentum l and $P_l(\cos \beta)$ is the Legendre polynomial.

Substituting $\delta_l(k) = \delta_l^R + i\,\delta_l^I$ (where δ_l^R and δ_l^I are real and imaginary parts of $\delta_l(k)$) in Eq. 8.3, we have

$$f(\beta,k) = \sum_{l=0}^{\infty} \left[\frac{2l+1}{2k} \right] [(\sin 2\delta_l^R) e^{-2\delta_l^I} + i(1 - \cos 2\delta_l^R\, e^{-2\delta_l^I})] P_l(\cos \beta) \qquad (8.4)$$

$$f^R(\beta,k) = \sum_{l=0}^{\infty} \left[\frac{2l+1}{2k} \right] (\sin 2\delta_l^R) e^{-2\delta_l^I} P_l(\cos \beta) \qquad (8.4a)$$

$$f^I(\beta,k) = \sum_{l=0}^{\infty} \left[\frac{2l+1}{2k} \right] (1 - \cos 2\delta_l^R e^{-2\delta_l^I}) P_l(\cos \beta) \qquad (8.4b)$$

Combining Eq. 8.2 and 8.4, we obtain

$$F(\beta,k) = [(f^R(\beta,k))^2 + (f^I(\beta,k))^2]^{1/2} \qquad (8.5)$$

$$\theta(\beta,k) = \tan^{-1} \frac{f^I(\beta,k)}{f^R(\beta,k)} \qquad (8.6)$$

It should be mentioned here that the forward scattering ($\beta \ll \pi$) amplitudes are more sensitive to chemical effects (especially at low k values), and hence more difficult to calculate, than the backscattering ($\beta = \pi$) amplitudes. Qualitatively speaking, the reason is that the backscattering process "samples" the *core* electrons of the neighboring atom whereas the forward scattering process "sees" both the *core* and the *valence* electrons of the atom.

8.1.1 $F(\beta, k)$ and $\theta(\beta, k)$

It is apparent from Eq. 8.4-8.6 that if we know the phase shifts $\delta_l(k)$, we can calculate $F(\beta, k)$ and $\theta(\beta, k)$. In fact, as an intermediate step in the calculations (Teo and Lee, 1979) of backscattering amplitude $(F(k) \equiv F(\pi, k))$ and phase $(\phi_b(k) \equiv \theta(\pi, k))$ functions described in the previous chapter, the $\delta_l(k)$ functions have already been calculated. It is therefore straightforward to calculate $F(\beta, k)$ and $\theta(\beta, k)$ from these complex phase shifts. Some of the results using Herman-Skillman wave functions are tabulated in Appendix VI. Other details of theory and calculations have been reported in the literature (Lee and Beni, 1977; Teo and Lee, 1979; Teo, 1981).

Fig. 8.2(a),(b) depict the amplitude functions $F(\beta, k)$ for β ranging from $0°$ to $70°$ for copper (cf. Appendix VII). It is apparent that the amplitude has its maximum at $\beta - 0°$ and attenuates rapidly as β increases. The high k region, however, drops off much faster than the low k region. At $\beta \geq 30°$, $F(\beta, k)$ is generally quite small (≤ 1) with some fine structure which changes as β varies. The complexity of the fine structure also seem to increase with increasing atomic number Z.

Fig. 8.3(a),(b) show plots of $\theta(\beta, k)$ for β ranging from $0°$ to $70°$ for copper (cf. Appendix VII). As β increases, the scatterer phase increases, first slowly at low β then at a faster rate. Again, at high β values ($\beta \geq 30°$), complex structures tend to develop which is related to the sampling of the core levels of the scattering atoms and hence is Z (atomic number) dependent.

In Fig. 8.4, we show the dependence of scattering amplitude $F(\beta, k)$ on β for a few representative k values for copper (cf. Appendix VII). For all k values, the amplitude attenuates rapidly from its maximum value at $\beta - 0°$ to $F \leq 1$ at $\beta \geq 30°$. At low k values ($k \leq 9\text{Å}^{-1}$), the amplitude exhibits both maxima and minima. For example, at $k - 3.78\text{Å}^{-1}$, $F(\beta, k)$ reaches its minima at $\beta \approx 36°$, $80°$, and $141°$ and its maxima at $\beta \approx 0°$, $52°$, $108°$, $180°$. The maxima (minima) correspond to β values where multiple scattering effect is most (least) important and vice versa. The number of these extrema increases with increasing atomic number Z, corresponding to the sampling of the increasing number of electronic shells in the scattering atom. It is also obvious from Figure 8.4 that the positions of these extrema (except the maxima at $\beta - 0°$ and $180°$) also change somewhat (generally to lower β values) as k increases. At sufficiently large k values ($k \geq 9 \text{ Å}^{-1}$) on the other hand, these extrema vanishes, resulting in monotonically decreasing amplitude functions.

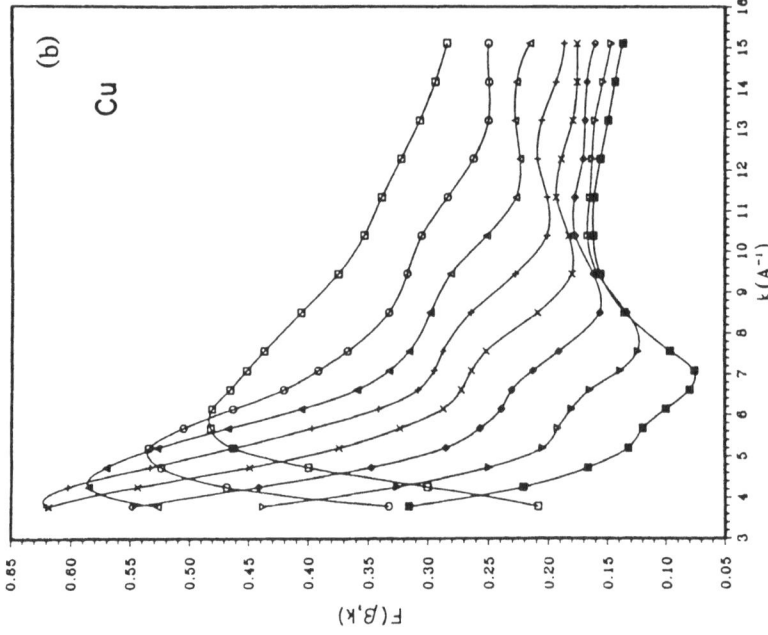

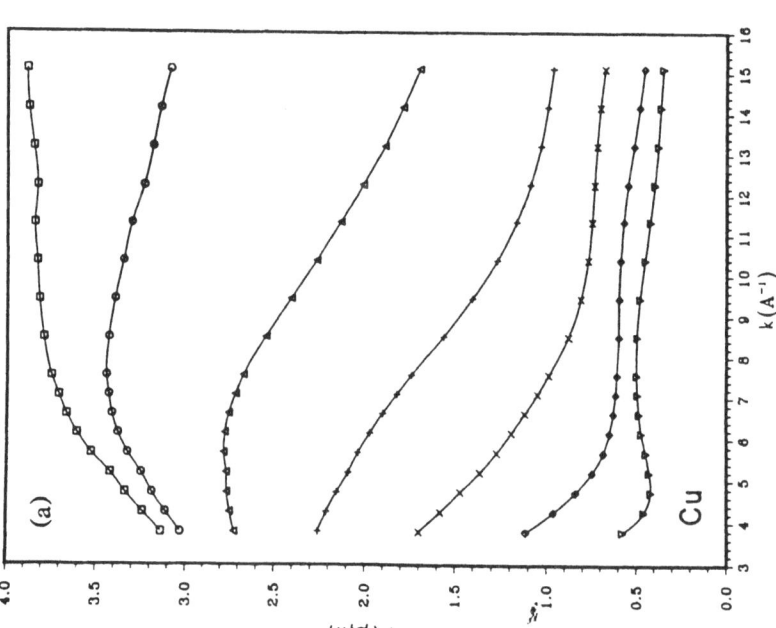

Fig. 8.2. Scattering amplitude $F(\beta,k)$ in Å vs photoelectron wave vector k in Å^{-1} for copper as a function of scattering angle β where (a) $\beta = 0°$ ($\square$), 5° ($\bigcirc$), 10° ($\triangle$), 15° (+), 20° ($\times$), 25° ($\diamondsuit$), 30° ($\triangledown$), 45° ($\triangle$), 50° (+), 55° ($\times$), 60° ($\diamondsuit$), 65° ($\triangledown$), 70° ($\boxtimes$). and (b) $\beta = 35°$ ($\square$), 40° ($\bigcirc$), 45° ($\triangle$),

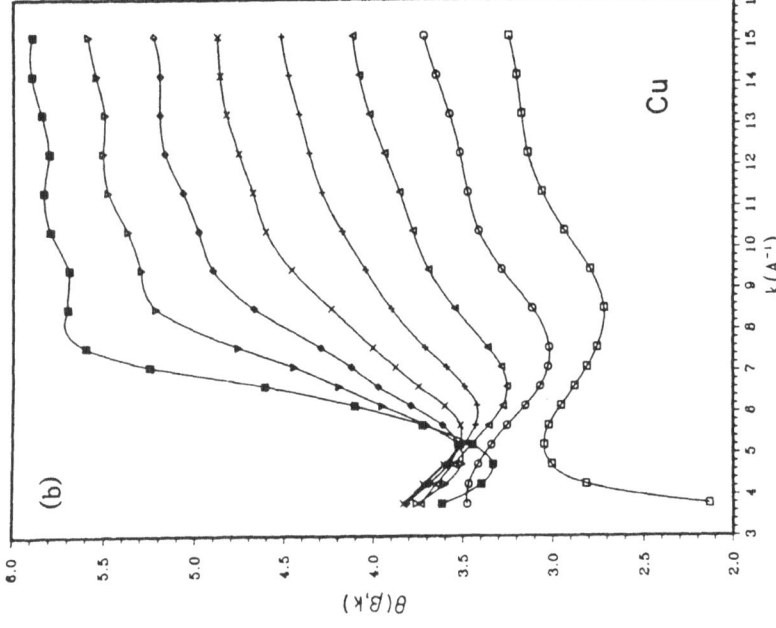

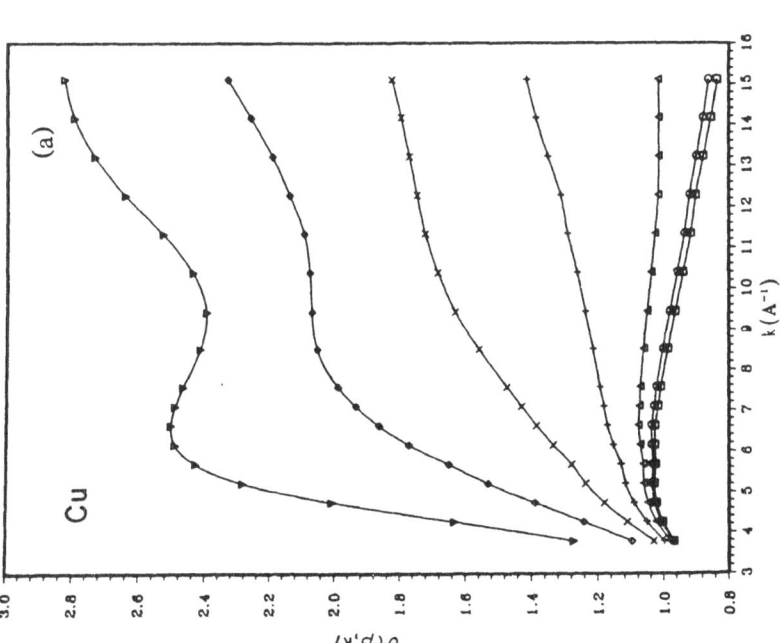

Fig. 8.3. Scattering phase $\theta(\beta,k)$ in radian *vs* photoelectron wave vector k in Å^{-1} for copper as a function of scattering angle β where (a) $\beta = 0°$ (□), 5° (○), 10° (△), 15° (+), 20° (×), 25° (◇), 30° (▽) and (b) $\beta = 35°$ (□), 40° (○), 45° (△), 50° (+), 55° (×), 60° (◇), 65° (▽), 70° (⊠).

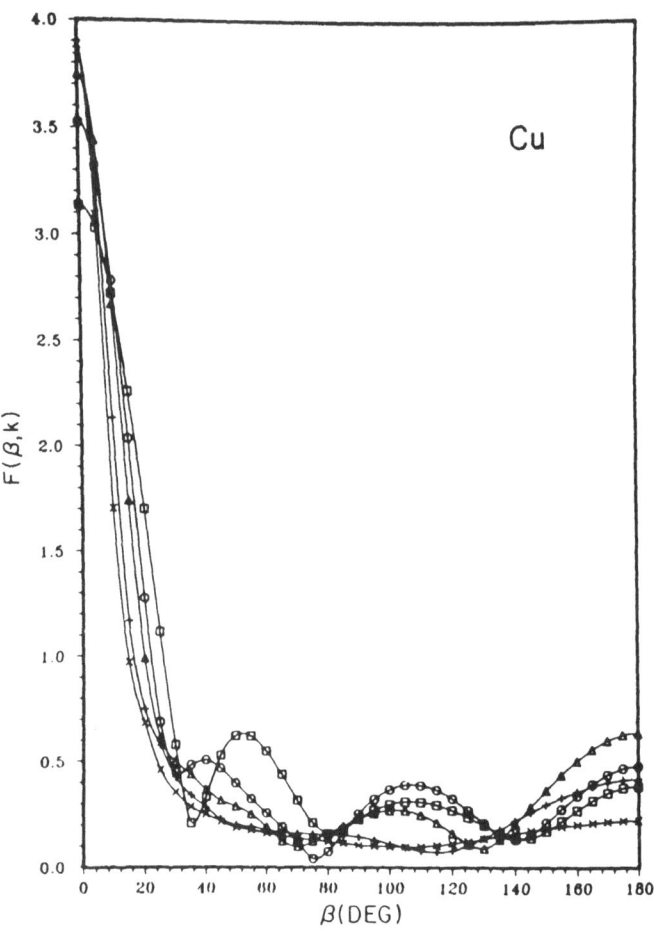

Fig. 8.4. Scattering amplitude $F(\beta,k)$ in Å vs scattering angle β in degree for copper at different k values where $k = 3.7795(\square)$, $5.6692(\bigcirc)$, $7.5589(\triangle)$, $11.3384(+)$, $15.1178(\times)Å^{-1}$. (Reproduced from Teo, 1983).

In Fig. 8.5 we show the corresponding plots of scattering phase $\theta(\beta,k)$ vs β (*cf.* Appendix VII). Again, at low k values, the phases exhibit plateau and inflection points which correspond to the maxima and minima in the $F(\beta,k)$ vs β plots in Fig. 8.4. At high k values, $\theta(\beta,k)$ is a simpler function of β.

In short, *the characteristic features of $F(\beta,k)$ and $\theta(\beta,k)$ as functions of β, k, and Z* form the theoretical basis for bond angle determination by EXAFS (to be discussed later in this chapter).

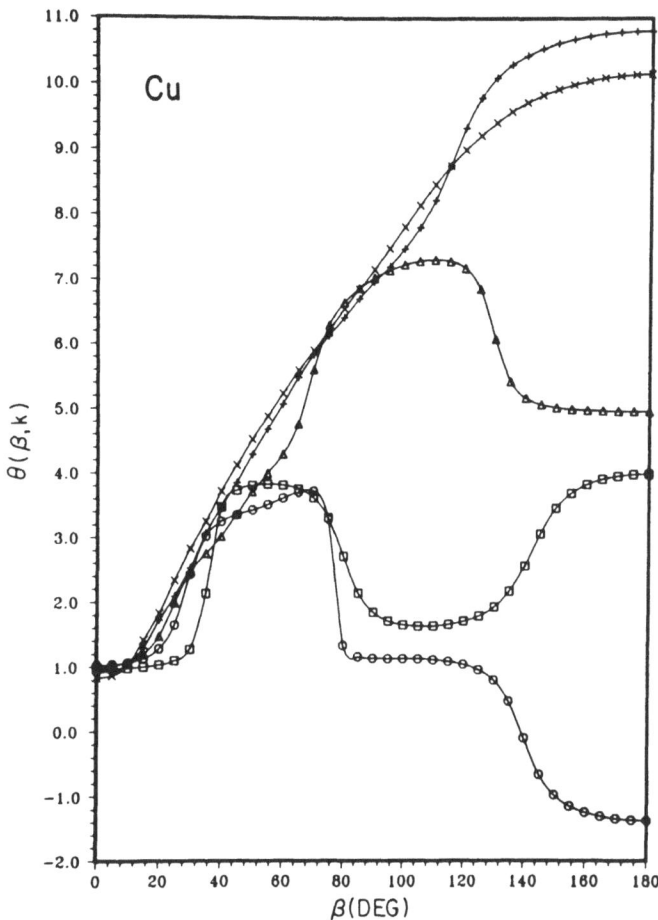

Fig. 8.5. Scattering phase $\theta(\beta, k)$ in radian *vs* scattering angle β in degree for copper at different k values where $k = 3.7795(\square)$, $5.6692(\bigcirc)$, $7.5589(\triangle)$, $11.3384(+)$, $15.1178(\times)\text{Å}^{-1}$.

Finally, to demonstrate the accuracy of the approximation made in Section 8.2.1.1, we show in Fig. 8.6 and 8.7 the amplitude and phase functions for oxygen scattering at angles $\beta = 0°$, $30°$, $60°$, $90°$, $120°$, and $180°$. It is readily apparent that for $\beta \geq 120°$, the error introduced by the approximation is minimal, especially at large k values ($k \geq 5\text{Å}^{-1}$).

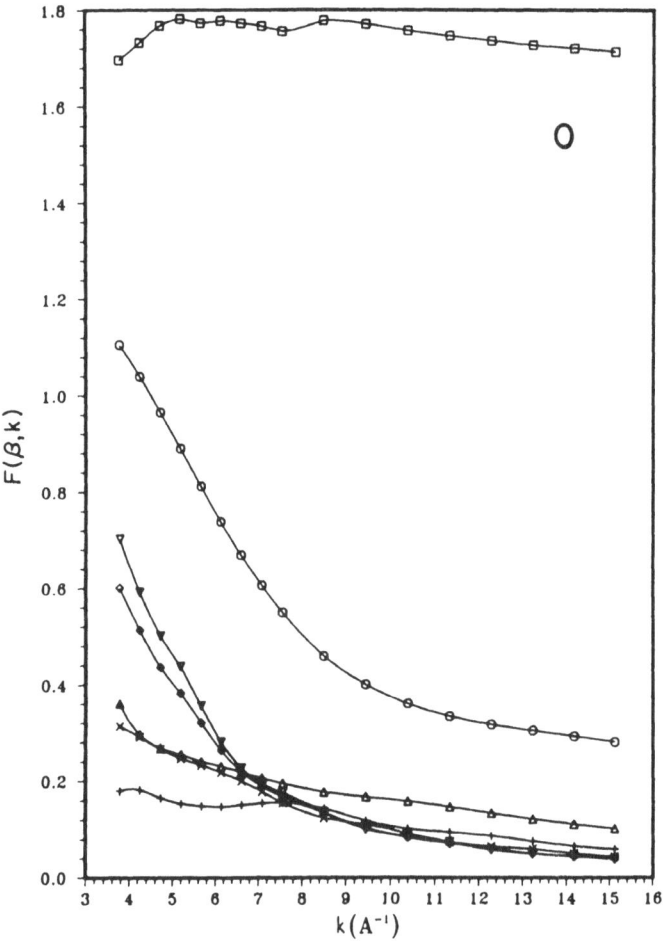

Fig. 8.6. Scattering amplitude $F(\beta,k)$ in Å vs photoelectron wave vector k in Å^{-1} for oxygen at different scattering angles β where $\beta = 0°$(□), $30°$(O), $60°$(△), $90°$(+), $120°$(×), $150°$(◇), $180°$(▽). (Reproduced from Teo, 1981).

8.2 Multiple Scattering

The single-electron single-scattering theory of EXAFS makes use of the fact that in most cases, multiple scattering is not important. (Strictly speaking, the single-electron single-scattering theory of EXAFS already includes one particular multiple scattering correction: $viz.$, the backscattering process involving the central atom which gives rise to the $2kr$ phase factor. Here, multiple scatterings refer to processes involving atoms other than the central atom.) This assumption is generally valid if one considers that multiple

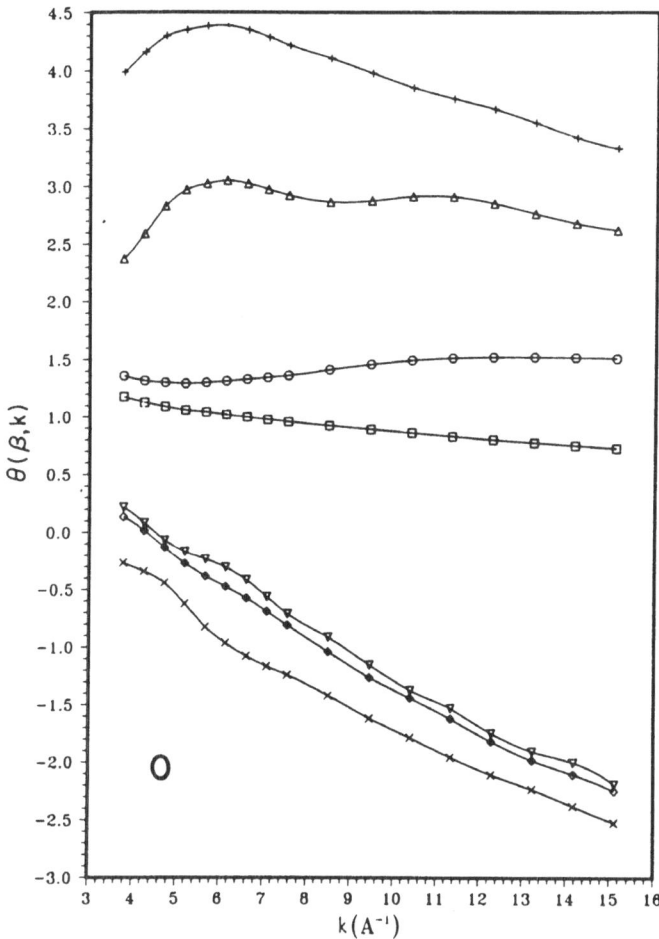

Fig. 8.7. Scattering phase $\theta(\beta, k)$ in radian *vs* photoelectron wave vector k in Å^{-1} for oxygen at different scattering angles β where $\beta = 0°(\square)$, $30°(\bigcirc)$, $60°(\triangle)$, $90°(+)$, $120°(\times)$, $150°(\diamondsuit)$, $180°(\nabla)$. (Reproduced from Teo, 1981).

scattering processes can be accounted for by adding all scattering paths that originate and terminate at the central atom (absorber). Each of these processes then behaves like sin $(2kr_{eff})$ where $2r_{eff}$ is the total scattering path length which is much larger than that of the backscattering(s) from the nearest neighbors. Thus, multiple scatterings will give rise to rapidly oscillatory waves in k space which tend to cancel out. The amplitude of these waves are also significantly attenuated by the large scattering path lengths, making it relatively unimportant in comparison with the direct backscattering.

On the other hand, multiple scattering in EXAFS can become important when atoms are arranged in an approximately collinear array. In such cases, the outgoing photoelectron is strongly *forward-scattered* by the intervening atom, resulting in a significant amplitude enhancement. In fact, both the amplitude and the phase are modified by the intervening atom(s) for bond angles ranging from 180° to ~75°. The effect, however, drops off very rapidly for bond angles smaller than *ca* 150°. For these systems, it is necessary to take into account multiple scatterings involving the intervening atom(s).

8.2.1 ABC Systems

Consider a three-atom array A-B-C (Fig. 8.1) where A is the central atom (absorber), B is the nearest neighbor (the intervening atom), and C is the next nearest neighbor. For such a system, the EXAFS of the absorber A comprises two contributions, one from the *backscattering* of B and the other from the *backscattering* of C via the intermediary atom B. We shall designate these two contributions as AB and ABC, respectively. These two contributions can generally be separated by Fourier filtering. The former can be described quite adequately by the *backscattering* from the atom B with the single-electron single-scattering theory. The latter, if it is affected by multiple scattering involving the intervening atom B, must be treated with a generalized formulation which takes into account three pathways shown schematically in Fig. 8.8. The first pathway (I) is the "single scattering" from A to C and back (*viz.* $A \rightarrow C \rightarrow A$). The second pathway (II) is the "double scattering" via atom B around the triangle in either directions (*viz.* $A \rightarrow B \rightarrow C \rightarrow A$ or $A \rightarrow C \rightarrow B \rightarrow A$). This term should therefore be counted twice. The third pathway (III) is the "triple scattering" via the intervening atom B, involving the pathway $A \rightarrow B \rightarrow C \rightarrow B \rightarrow A$. The EXAFS for an *unoriented* sample and/or using unpolarized source, corresponding roughly to the A-C distance, is then the sum of these three terms:

$$\chi^{ABC}(k) = \frac{(-1)^l}{k} \operatorname{Im} e^{i2\delta_l'(k)} \left\{ \frac{f_C(\pi,k)}{r_{AC}^2} e^{-2r_{AC}/\lambda} e^{i2kr_{AC}} e^{-2\sigma_{AC}^2 k^2} \right.$$

$$+ \frac{2\cos(\gamma-\alpha)f_B(\beta,k)f_C(\gamma,k)}{r_{AB}\, r_{BC}\, r_{AC}} e^{-(r_{AB}+r_{BC}+r_{AC})/\lambda} e^{ik(r_{AB}+r_{BC}+r_{AC})} e^{-(\sigma_{AB}^2+\sigma_{BC}^2+\sigma_{AC}^2)k^2}$$

$$+ \frac{f_B(\beta,k)^2 f_C(\pi,k)}{r_{AB}^2 r_{BC}^2} e^{-2(r_{AB}+r_{BC})/\lambda} e^{i2k(r_{AB}+r_{BC})} e^{-2(\sigma_{AB}^2+\sigma_{BC}^2)k^2} \left. \right\} \qquad (8.7)$$

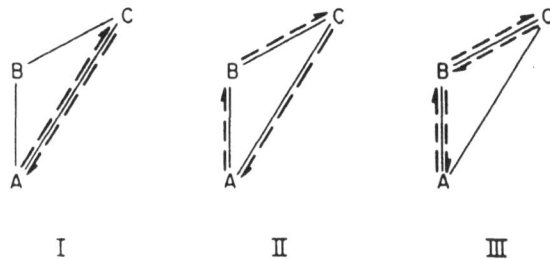

Fig. 8.8. Schematic representation of three scattering pathways for a three-atom ABC system. Each of these pathways originates and terminates at the absorbing atom A. Pathway I is the direct backscattering from atom A to atom C and back. Pathway II is the multiple scattering via atom B around the triangle in either direction (only one is shown) and pathway III is the multiple scattering via atom B in both outgoing and incoming trips. (Reproduced from Teo, 1981).

Here we assume that the Debye-Waller factors can be approximated by $e^{-2\sigma_{AC}^2 k^2}$, $e^{-(\sigma_{AB}^2 + \sigma_{BC}^2 + \sigma_{AC}^2)k^2}$, and $e^{-2(\sigma_{AB}^2 + \sigma_{BC}^2)k^2}$ for the three pathways I, II and III, respectively. For a more theoretical (normal modes) treatment of the effect of thermal vibrations, see Boland and Baldeschwieler (1984a,b). $\cos(\gamma - \alpha)$ is a geometrical factor where α is the A-B-C angle and $(\gamma - \alpha)$ is the B-A-C angle. If we now make the plane wave approximations $f_B(\beta, k) \simeq F_B(\beta, k)e^{i\theta_B(\beta, k)}$ and $f_C(\gamma, k) \simeq F_C(\gamma, k)e^{i\theta_C(\gamma, k)}$ where $F_B(\beta, k)$, $F_C(\gamma, k)$ and $\theta_B(\beta, k)$, $\theta_C(\gamma, k)$ are the scattering amplitude and phase functions for scattering angles β, γ at atoms B, C, respectively, Eq. 8.7 can be rewritten as

$$
\chi^{ABC}(k) \simeq \frac{(-1)^l}{k} \left\{ \frac{F_C(\pi, k)}{r_{AC}^2} e^{-2r_{AC}/\lambda} e^{-2\sigma_{AC}^2 k^2} \sin[2kr_{AC} + 2\delta_l'(k) + \theta_C(\pi, k)] \right.
$$

$$
+ \frac{2\cos(\gamma - \alpha) F_B(\beta, k) F_C(\gamma, k)}{r_{AB} r_{BC} r_{AC}} e^{-(r_{AB} + r_{BC} + r_{AC})/\lambda} e^{-(\sigma_{AB}^2 + \sigma_{BC}^2 + \sigma_{AC}^2)k^2}
$$

$$
\times \sin[k(r_{AB} + r_{BC} + r_{AC}) + 2\delta_l'(k) + \theta_B(\beta, k) + \theta_C(\gamma, k)]
$$

$$
+ \frac{F_B(\beta, k)^2 F_C(\pi, k)}{r_{AB}^2 r_{BC}^2} e^{-2(r_{AB} + r_{BC})/\lambda} e^{-2(\sigma_{AB}^2 + \sigma_{BC}^2)k^2}
$$

$$
\left. \times \sin[2k(r_{AB} + r_{BC}) + 2\delta_l'(k) + 2\theta_B(\beta, k) + \theta_C(\pi, k)] \right\} \tag{8.8}
$$

8.2.1.1 Approximations

It is apparent that these three terms correspond to the three pathways I, II, and III shown in Fig. 8.8 with effective distances r_{AC}, $(r_{AB} + r_{BC} + r_{AC})/2$, and $(r_{AB} + r_{BC})$, respectively. For a given pathway which can be detected and separated via Fourier filtering, it can be analyzed with the appropriate term in 8.8. However, for small β, the three pathways are generally not separable since $r_{AC} \lesssim r_{AB} + r_{BC}$. In such cases, one can assume $\gamma \approx \pi$ such that $\cos(\gamma - \alpha) \approx 1$ and

$$f_C(\gamma, k) \approx f_C(\pi, k) \tag{8.9}$$

Eq. (8.7) then becomes

$$\chi^{ABC}(k) = \frac{(-1)^l}{k} e^{-2\sigma_{AC}^2 k^2} \operatorname{Im} e^{i2\delta'_l(k)} \frac{f_C(\pi, k)}{r_{AC}^2} e^{-2r_{AC}/\lambda} e^{i2kr_{AC}}$$

$$\times \left\{ 1 + \frac{r_{AC}}{r_{AB}r_{BC}} f_B(\beta, k) e^{-(r_{AB} + r_{BC} - r_{AC})/\lambda} e^{ik(r_{AB} + r_{BC} - r_{AC})} \right\}^2 \tag{8.10}$$

Here we also assume that the root-mean-square displacements of the A-B and B-C bonds are uncorrelated and that $\sigma_{AC}^2 \approx \sigma_{AB}^2 + \sigma_{BC}^2$.

If we now make the plane wave approximations:

$$f_B(\beta, k) = F_B(\beta, k) e^{i\theta_B(\beta, k)} \tag{8.11a}$$

$$f_C(\pi, k) = F_C(k) e^{i\theta_C(k)} \tag{8.11b}$$

where $F_B(\beta, k)$ and $\theta_B(\beta, k)$ are the scattering amplitude and phase functions for the scattering angle β at atom B and define

$$\tilde{r} = \frac{r_{AC}}{r_{AB}r_{BC}} \tag{8.12}$$

$$\Delta r = r_{AB} + r_{BC} - r_{AC} \tag{8.13}$$

$$\bar{\theta} = \theta_B(\beta,k) + k(\Delta r) \qquad (8.14)$$

$$\Omega_B(\beta,k)e^{i\omega_B(\beta,k)} = \{1 + \bar{r}f_B(\beta,k)e^{-\Delta r/\lambda}e^{ik(\Delta r)}\}^2 \qquad (8.15)$$

we have

$$\chi^{ABC}(k) = \frac{(-1)^l}{k}e^{-2\sigma_{AC}^2k^2}\,e^{-2r_{AC}/\lambda}\,\frac{F_C(k)}{r_{AC}^2}\,\Omega_B(\beta,k)$$

$$\times \operatorname{Im} e^{i\left[2\delta_i'(k)+\theta_c(k)+\omega_B(\beta,k)+2kr_{AC}\right]} \qquad (8.16)$$

or equivalently,

$$\chi^{ABC}(k) = \frac{1}{kr_{AC}^2}\,\Omega_B(\beta,k)F_C(k)e^{-2\sigma_{AC}^2k^2}e^{-2r_{AC}/\lambda}$$

$$\times \sin(2kr_{AC} + \phi_{AC}(k) + \omega_B(\beta,k)) \qquad (8.17)$$

where

$$\phi_{AC}(k) = \phi_A^l(k) + \phi_C(k) - l\pi \qquad (8.18)$$

It is apparent from Eq. 8.17 that *the effect of multiple scattering via the intervening atom B is to multiply the amplitude $F_C(k)$ by $\Omega_B(\beta,k)$ and to add $\omega_B(\beta,k)$ to the phase $\phi_{AC}(k)$.* That is, if one substitutes the modified amplitude $F_C(k)\Omega_B(\beta,k)$ for $F_C(k)$ and the corrected phase $\phi_C(k) + \omega_B(\beta,k)$ for $\phi_C(k)$, the EXAFS data can be analyzed in the usual way. Clearly, both $\Omega_B(\beta,k)$ and $\omega_B(\beta,k)$ are functions of the scattering angle β. Conversely, as we will see later, Eq. 8.16 or 8.17 can be used to analyze EXAFS data to determine the scattering angle β.

Eq. 8.15 can be rewritten as

$$\Omega_B(\beta,k) = 1 + 2\,\bar{r}F_B(\beta,k)e^{-\Delta r/\lambda}\cos\bar{\theta} + \left[\bar{r}F_B(\beta,k)e^{-\Delta r/\lambda}\right]^2 \qquad (8.19)$$

$$\omega_B(\beta,k) = 2\tan^{-1}\frac{\bar{r}F_B(\beta,k)e^{-\Delta r/\lambda}\sin\bar{\theta}}{1 + \bar{r}F_B(\beta,k)e^{-\Delta r/\lambda}\cos\bar{\theta}} \qquad (8.20)$$

In deriving Eq. 8.16-20 from Eq. 8.7 or Eq. 8.8, we assume that the geometrical factor $\cos(\gamma-\alpha) \approx 1$. This is generally speaking a good approximation for two reasons. First, at low scattering angles $\beta \rightarrow 0°$ (or large bond angles $\alpha \rightarrow 180°$) where multiple scattering effects are most important, $\cos(\gamma-\alpha) \rightarrow 1$. Even at $\beta \approx 60°$ (or $\alpha \approx 120°$) where the multiple scattering effect is virtually disappearing, for example, $\cos(\gamma-\alpha) \approx \cos 30° = 0.87$ (assuming $r_{AB} \approx r_{BC}$) is still very close to 1. Second, as we shall see later, the double scattering pathway II is generally the least important one (by orders of magnitude) since at low scattering angles, the triple scattering pathway III dominates whereas at high scattering angles, only the single scattering pathway I remains. Hence overestimating pathway II by as much as 10% will not cause any problem. The advantage of making these assumptions is that it greatly simplies the EXAFS formulation by combining the three scattering pathways. In fact, it allow us to recast the multiple scattering formula in a form similar to the single scattering theory but with *modified* amplitude and phase functions which depend on the bond angle. Furthermore, the simplified multiple-scattering formalism (Eq. 8.16 or 8.17) reduces the number of parameters involved in curve fitting, thereby greatly facilitating bond-angle determination as we shall discuss in the next section.

It should also be noted that the approximation implicit in Eq. 8.8 becomes exact at $\beta = 0°$ (i.e. linear case). However, at this extreme, one should in theory consider only the multiple scattering pathway III in Fig. 8.8. As we shall see later that in actual fact, when the system approaches linearity ($\beta < 30°$) pathway III becomes much more important than pathways I or II (i.e., III >> II > I).

We shall now illustrate the effect of the amplitude and phase modification factors, $\Omega(\beta,k)$ and $\omega(\beta,k)$, as a function of the scattering angle β using oxygen as an example. For clarity, we set $e^{-\Delta r/\lambda} = 1$ in Eq. 8.19 and 8.20. We further assume $r_{AB} = 1.95\text{Å}$ and $r_{BC} = 1.28\text{Å}$ and calculate r_{AC} from r_{AB}, r_{BC}, and β (*cf.* Eq. 8.12-14).

Fig. 8.9 and 8.10 show the results for $\Omega(\beta,k)$ and $\omega(\beta,k)$ respectively, for $\beta = 0\text{-}70°$. It is immediately obvious that multiple scattering can lead to not only amplitude enhancement ($\Omega > 1$), but also amplitude reduction ($\Omega < 1$). At $\beta \approx 0°$, $\Omega(\beta,k)$ has the maximum magnitude ($\Omega(\beta,k)\approx9$) and is generally a flat function of k. As β increases, $\Omega(\beta,k)$ attenuates rapidly, especially at high k region. At $\beta \approx 30°$, for example, $\Omega(\beta,k)$ drops to ~2.5

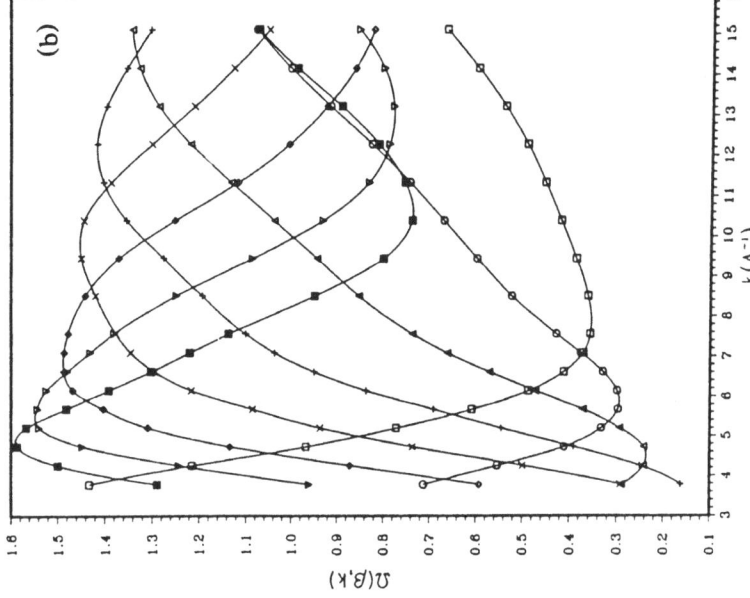

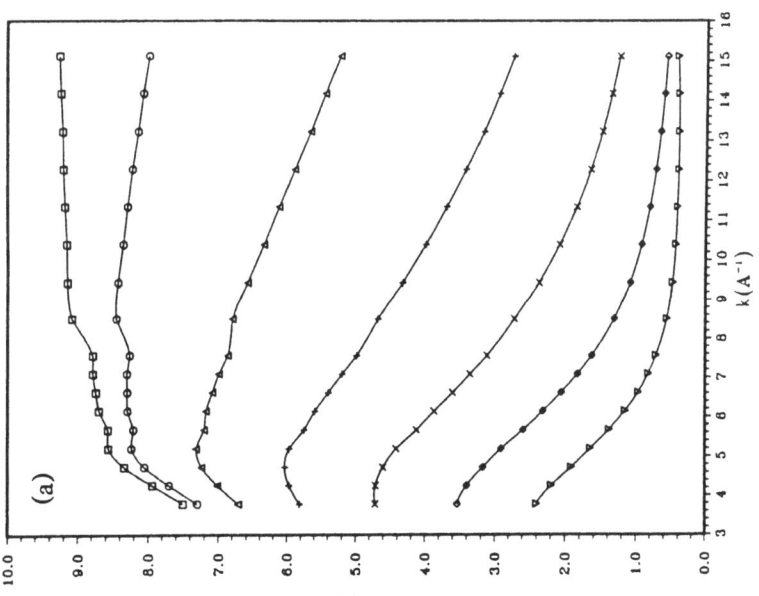

Fig. 8.9. Amplitude modification factor $\Omega(\beta, k)$ vs photoelectron wave vector k in $Å^{-1}$ for oxygen with $r_{AB} = 1.95Å$, $r_{BC} = 1.28Å$ as a function of scattering angle β where $\beta = 0°$ ($\square$), $5°$ ($\bigcirc$), $10°$ ($\triangle$), $15°$ ($+$), $20°$ ($\times$), $25°$ ($\diamond$), $30°$ ($\triangledown$) for (a) and $\beta = 35°$ ($\square$), $40°$ ($\bigcirc$), $45°$ ($\triangle$), $50°$ ($+$), $55°$ ($\times$), $60°$ ($\diamond$), $65°$ ($\triangledown$), $70°$ ($\boxtimes$) for (b). (Reproduced from Teo, 1981).

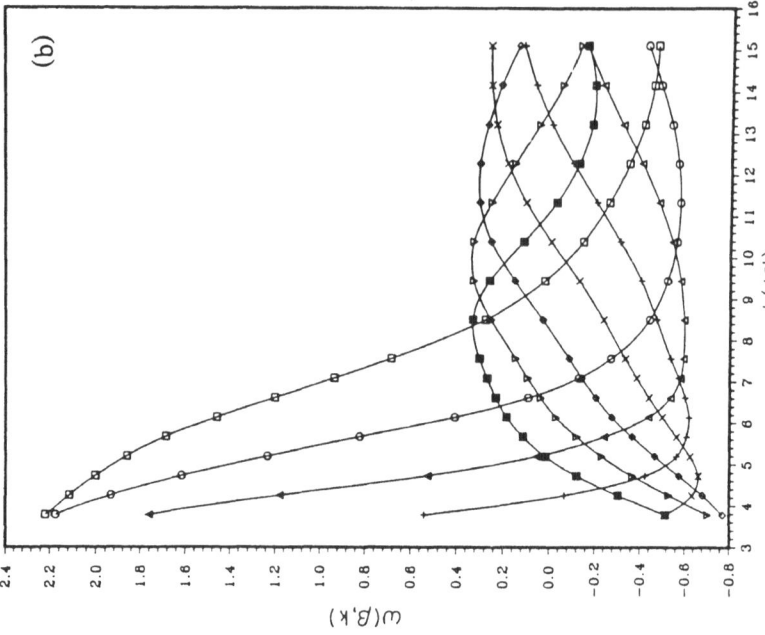

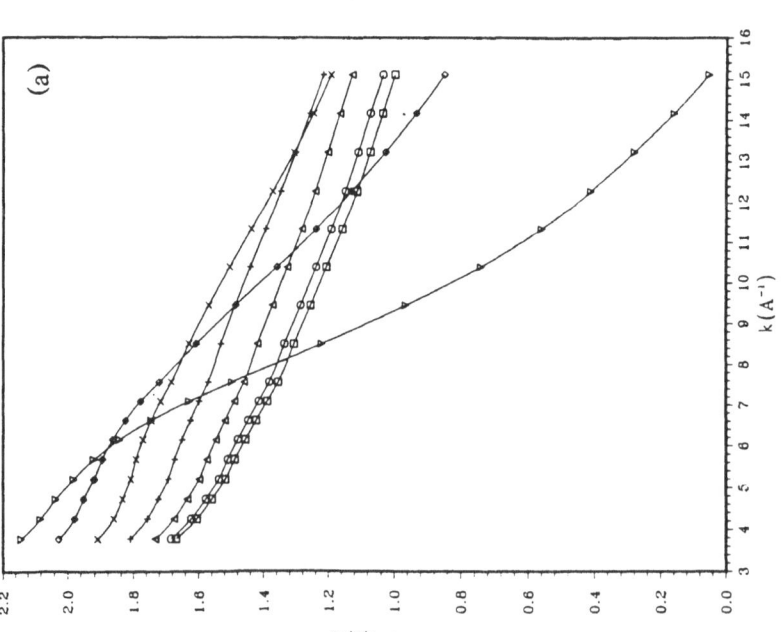

Fig. 8.10. Phase modification factor $\omega(\beta, k)$ in radian *vs* photoelectron wave vector k in Å^{-1} for oxygen with $r_{AB} = 1.95\text{Å}$, $r_{BC} = 1.28\text{Å}$ as a function of scattering angle β where $\beta = 0°$ (□), 5° (○), 10° (△), 15° (+), 20° (×), 25° (◇), 30° (▽), for (a) and $\beta = 35°$ (□), 40° (○), 45° (△), 50° (+), 55° (×), 60° (◇), 65° (▽), 70° (⊠) for (b). (Reproduced from Teo, 1981).

(amplitude enhancement) at low k values and ~ 0.4 (amplitude reduction) at high k values. For higher values of β, we find that $\Omega(\beta,k)$ exhibits characteristic features in k space for each β value, with its maxima and minima progressing smoothly as β changes. The systematic progression of these extrema as β changes stems from the composite effect of $F(\beta,k)$ and $k(\Delta r)$ as they vary systematically with β.

In contrast, the β dependence of the phase modification factor $\omega(\beta,k)$ is less dramatic. At $\beta \approx 30°$ where the strongest β dependence is observed, the maximum slope of ~ -0.20 rad./Å^{-1} will give rise to a distance correction of $\sim 0.10\text{Å}$. For $\beta \gtrsim 50°$, $\omega(\beta,k)$ is once again a weak function of both β and k. For example, at $\beta \approx 55°$, a slope of ~ 0.10 rad./Å^{-1} will cause a correction in distance of $\sim -0.05\text{Å}$. It should be noted that these phase shift corrections are in general smaller than that caused by the corresponding backscatter phase shift ($\beta = 180°$) and are much smaller than those caused by central atom phase shifts.

Examination of the three components of $\Omega(\beta,k)$ revealed (cf. Fig. 8.11) that at low scattering angles ($\beta \leq 30°$), the relative importance of the three scattering pathways follows the order: III $>>$ II $\geq$ I. That is, multiple scattering via the intermediary atom B on both forward and returning trips is the dominant scattering process such that the functional form (shape) of the amplitude modification function $\Omega(\beta,k)$ resembles that of $(\bar{r}F(\beta,k))^2$. This is the origin of the focusing effect. In fact, strictly speaking, at $\beta \equiv 0°$, only this term (pathway III) need to be considered. At higher β values, the relative magnitude of the three scattering pathways follows the order: I $\geq$ II $>>$ III. That is, at high scattering angles, ($\beta \geq 45°$), direct backscattering (pathway I) and the "round-the-triangle" multiple scattering (pathway II) are the most important scattering processes with the latter attenuating further as β increases. $\Omega(\beta,k)$ is now determined mainly by that of $2\,\bar{r}F(\beta,k)\cos\bar{\theta}$ since pathway I gives rise to only a constant 1 in Eq. 8.19. Even at scattering angles as high as $90-120°$, we can still find amplitude modification of $ca \pm 25\%$ due to multiple scattering pathway II at low k regions. Hence amplitude transferability must be treated with caution.

While focusing effect can, at low scattering angles (large bond angles), cause an order of magnitude enhancement in amplitude, its impact on the phase is, fortunately, less dramatic (cf. Fig. 8.12). In most cases, the corrections in

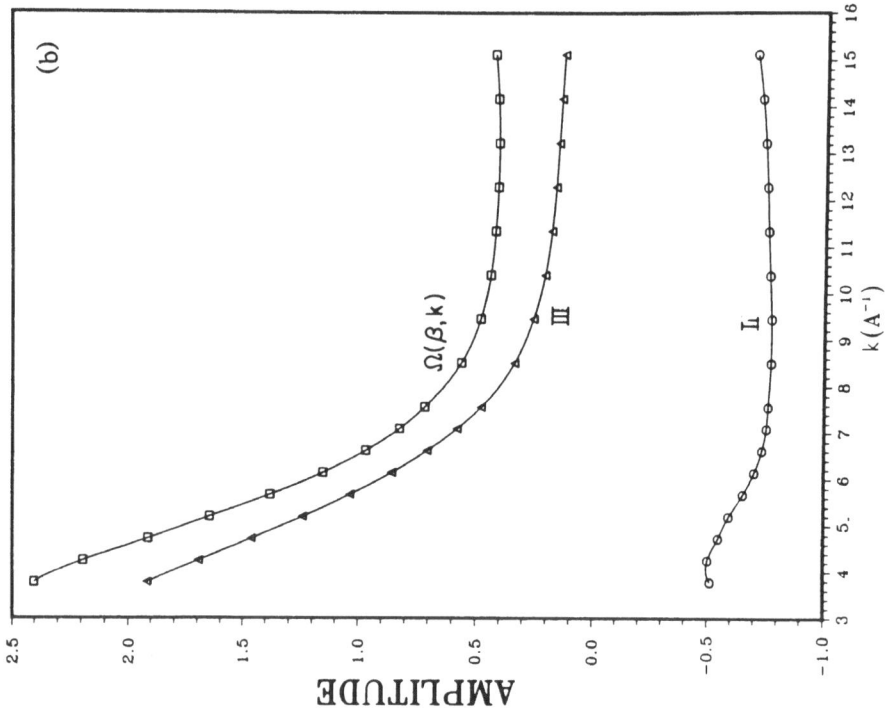

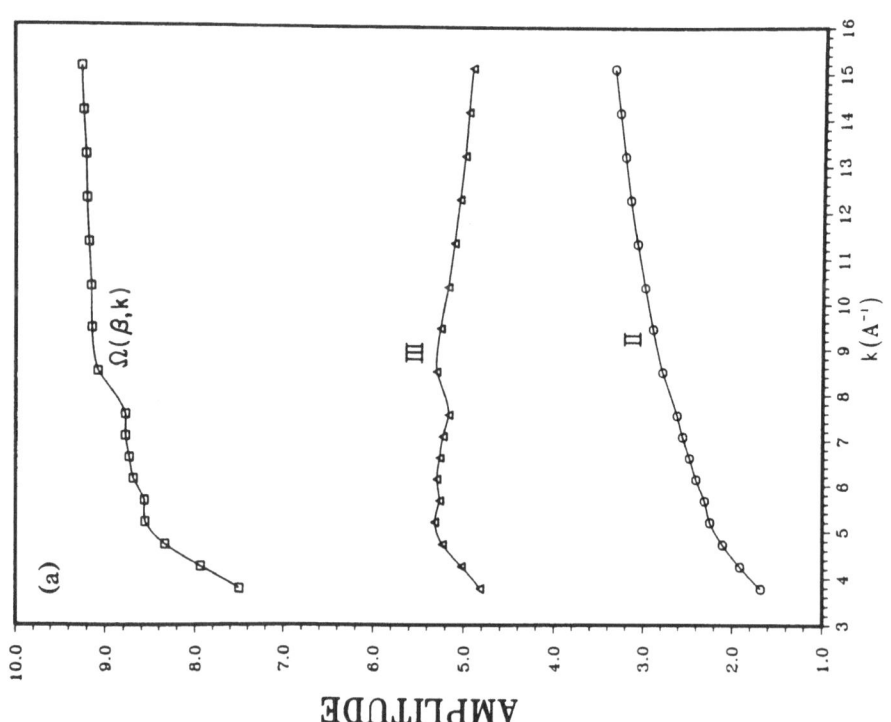

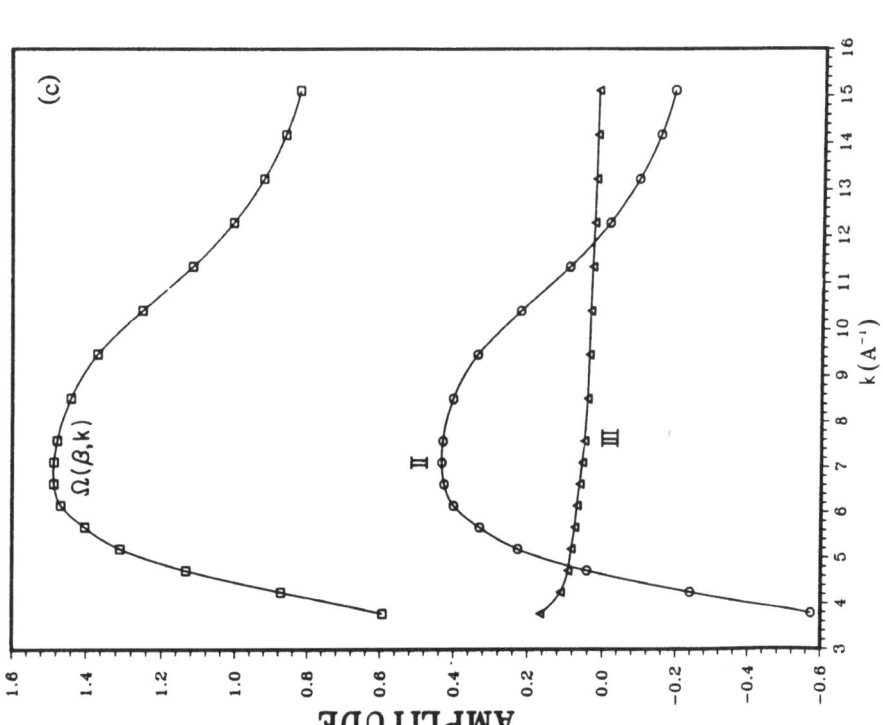

Fig. 8.11. Multiple scattering contributions to the amplitude modification factor $\Omega(\beta, k)$ ($\square$) *vs* k in Å^{-1} due to pathways II ($\triangle$) and III ($\bigcirc$) in Fig. 8.8 for oxygen at different scattering angles β where $\beta = 0°$ (a), 30° (b), and 60° (c). Curves II and III correspond to the functions $2\bar{r}F(\beta, k)$ $\text{Cos}\tilde{\theta}(\beta, k)$ and $[\bar{r}F(\beta, k)]^2$, respectively, defined in the text. (Reproduced from Teo, 1981).

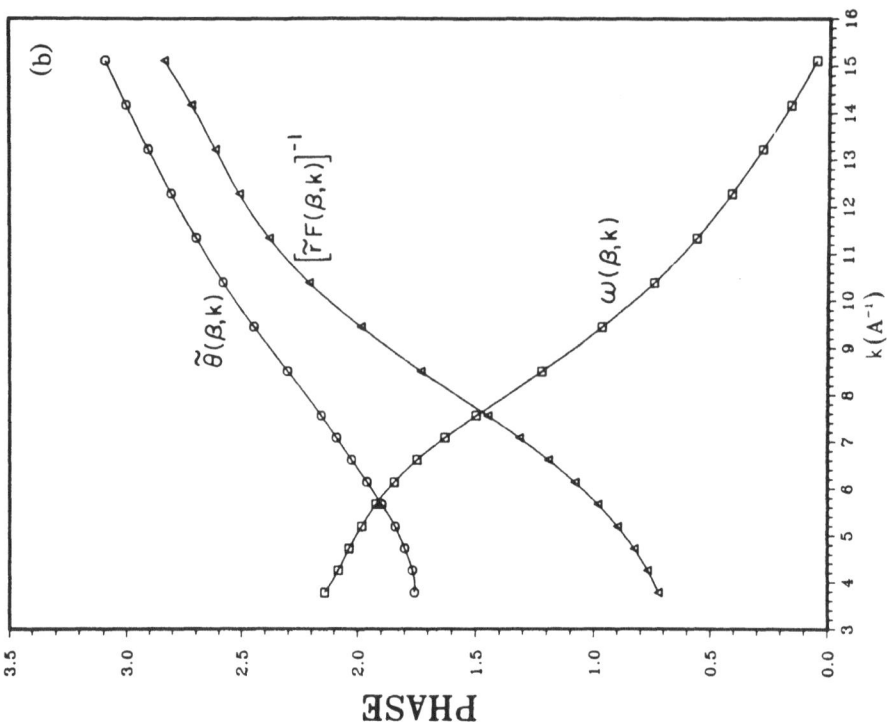

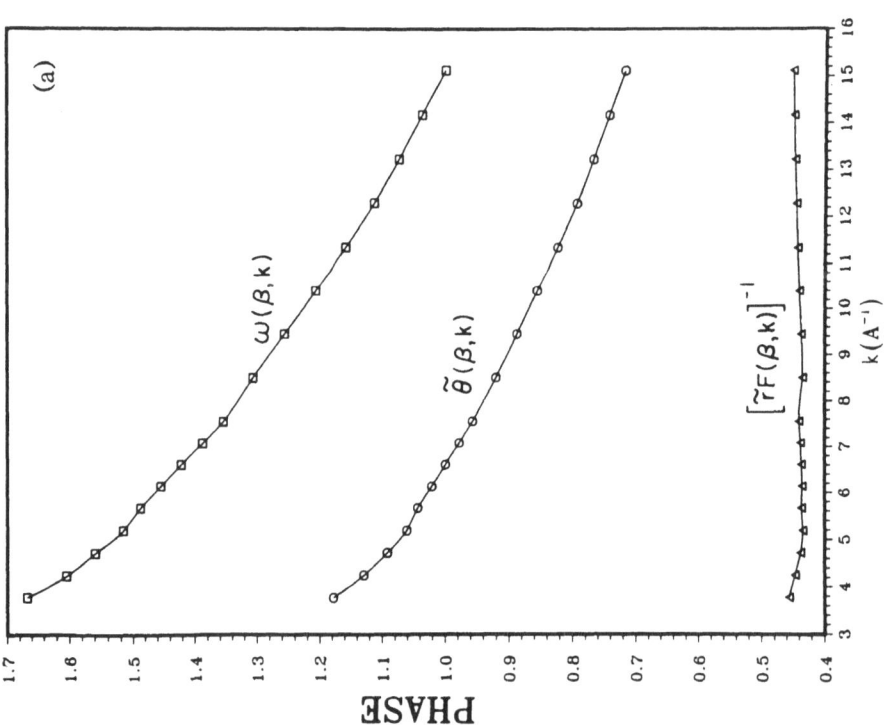

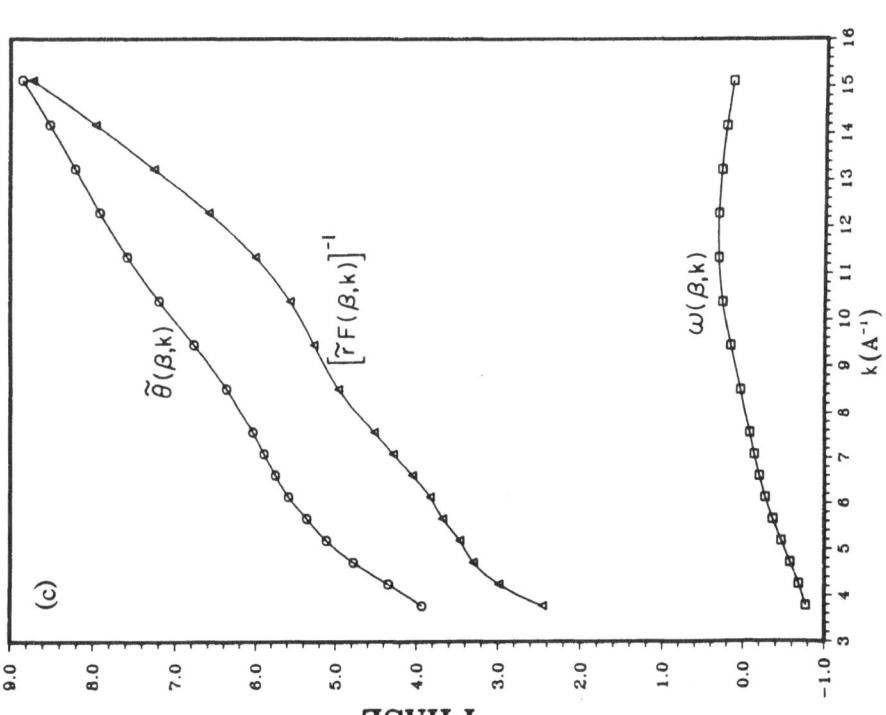

Fig. 8.12. Components of the multiple scattering phase modification factor $\omega(\beta, k)$ ($\square$) in radian vs k in Å^{-1} for oxygen at different scattering angles β where $\beta = 0°$ (a), $30°$ (b), and $60°$ (c). The curves are $\tilde{\theta}(\beta, k)$ (O) and $[\tilde{r}F(\beta, k)]^{-1}$ ($\triangle$) defined in the text. (Reproduced from Teo, 1981).

distance determination due to multiple scattering processes amount to less than ~0.10Å, especially when the intermediary atoms are low Z elements.

To summarize, the relative importance of the three scattering pathways follows the order III >> II $\geq$ I and I $\geq$ II >> III at low (high) and high (low) scattering (bond) angles, respectively. Fig. 8.13 shows schematically the relative importance of the three scattering components and their peak positions in distance space at two scattering angles(β). Here we ignore the phase shifts for the sake of clarity. The peaks are separated by $\frac{1}{2}(\Delta r) = \frac{1}{2}(r_{AB} + r_{BC} - r_{AC})$. At low β, $\Delta r \approx 0$, the three peaks merge at a distance corresponding to r_{AC} with intensities III >> II $\geq$ I. As β increases,

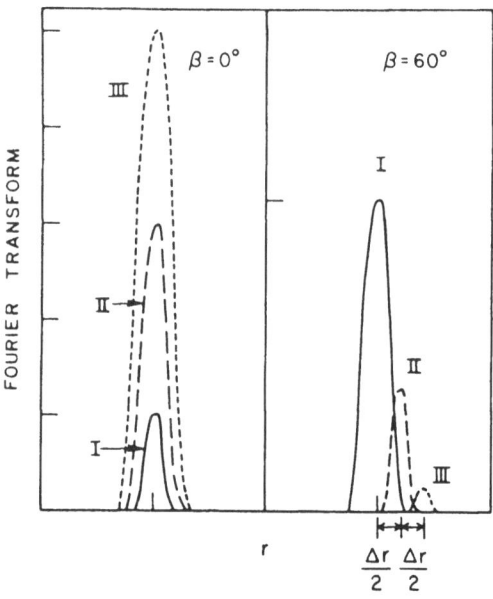

Fig. 8.13. Schematic illustrations of the relative importance of the three scattering components I, II, III (cf. Fig. 8.8) in distance r space in terms of their Fourier transform magnitudes. For the sake of clarity, the effects of phase shifts have been omitted. At $\beta = 0°$, the three peaks merge at a distance corresponding to r_{AC} with relative intensities III > II > I. As β increases, III and II move to distances corresponding to $r_{AB} + r_{BC}$ and $(r_{AB}+r_{BC}+r_{AC})/2$, respectively, with a concomitant drastic decrease in intensity, resulting in a relative ordering of I > II > III. The spacing between the peaks corresponds to $\Delta r/2 = (r_{AB} + r_{BC} - r_{AC})/2$. The vertical axis for the $\beta = 60°$ case is enlarged with respect to that for the $\beta = 0°$ case. (Reproduced from Teo, 1983).

$\Delta r > 0$, peaks III and II move to distances corresponding to $r_{AB} + r_{BC}$ and $\frac{1}{2}(r_{AB} + r_{BC} + r_{AC})$, respectively, with a concomitant dramatic decrease in intensities. The resulting peak positions therefore follow the order $I < II < III$ (separated by $\frac{1}{2}(\Delta r)$) in distance whereas the peak magnitudes follow the order $I \geq II >> III$. In this context, it should be cautioned that in Fourier filtering, one should not filter out the multiple scattering pathways II and III by choosing too narrow a filtering window for the A-C peak.

8.2.1.2 Anisotropic ABC Systems

It is also possible to derive the multiple scattering formalism of EXAFS using the full T operator in the Lippmann-Schwinger equation as demonstrated by Boland, Crane, and Baldeschwieler (1982). Details have already been given in Chapter 4, *Section 4.2.3*. (See also Rennert and Vasvári, 1983). Their equation takes into account polarization and geometrical factors (hence is applicable to anisotropic systems) but ignores the angular dependence of the scattering phase due to the neighboring atoms. (Note that ignoring the angular dependence of the phase function can cause errors of up to *ca* -0.20Å in distance determination). Taking into consideration polarization and geometrical factors and the angular dependence of the phase functions due to the neighboring atoms (but ignoring the Debye-Waller and electron mean free path terms), the EXAFS expression for the three pathway formalism of an *ABC* system can now be written as

$$\chi(k) = \frac{(-1)^l}{k} \left\{ 3(\hat{\epsilon}\cdot\hat{r}_{AC})^2 \frac{F_C(\pi,k)}{r_{AC}^2} \sin[2kr_{AC} + 2\delta_l'(k) + \theta_C(\pi,k)] \right.$$

$$+ \frac{6(\hat{\epsilon}\cdot\hat{r}_{AB})(\hat{\epsilon}\cdot\hat{r}_{AC})}{r_{AB}r_{BC}r_{AC}} F_B(\beta,k)F_C(\gamma,k)$$

$$\times \sin[k(r_{AB}+r_{BC}+r_{AC}) + 2\delta_l'(k) + \theta_B(\beta,k) + \theta_C(\gamma,k)]$$

$$+ \frac{3(\hat{\epsilon}\cdot\hat{r}_{AB})^2}{r_{AB}^2 r_{BC}^2} F_B(\beta,k)^2 F_C(\pi,k)$$

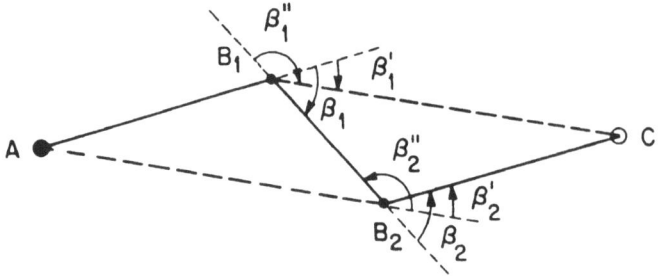

Fig. 8.14. A four-atom AB_1B_2C system where A is the absorber, C is the backscatterer, and B_1, B_2 are the intervening atoms. Also shown are the scattering angles β at the two intervening atoms. (Reproduced from Teo, 1983).

For unoriented samples, Eq. 8.21 must be spherically averaged over all polarization directions:

$$\int (\hat{\epsilon} \cdot \hat{r}_{AC})^2 \, \frac{d\Omega_e}{4\pi} = \frac{1}{3} \tag{8.22}$$

$$\int (\hat{\epsilon} \cdot \hat{r}_{AB})(\hat{\epsilon} \cdot \hat{r}_{AC}) \, \frac{d\Omega_e}{4\pi} = \frac{\hat{r}_{AB} \cdot \hat{r}_{AC}}{3}$$

$$= \frac{\cos(\gamma - \alpha)}{3} \tag{8.23}$$

where $(\gamma - \alpha)$ is the B-A-C angle. The net result is an equation similar to Eq. 8.8 (but without the exponential damping terms).

8.2.2 Multiple Scattering: AB_1B_2C Systems

Multiple scatterings involving more than three atoms are likely to be of less importance due to the large effective total path lengths involved; except, perhaps, in cases where all the bond angles are close to linearity. In this section, a general multiple-scattering theory will be formulated for a four-atom AB_1B_2C system and a systematic method for determining the bond angles developed.

Considering a four-atom framework AB_1B_2C shown in Fig. 8.14 and following the notations used in the previous section, it is obvious that there are two predominant groups of multiple scattering processes as shown in Fig. 8.15: those involving three atoms (a and b) and those involving four atoms (c, d, and e). Other more complex four-atom multiple scattering pathways necessarily involve larger effective total path length as well as larger scattering angles such that they are likely to be of less importance. In this context, it should be emphasized that the three-atom pathways (a, b) are in general more important

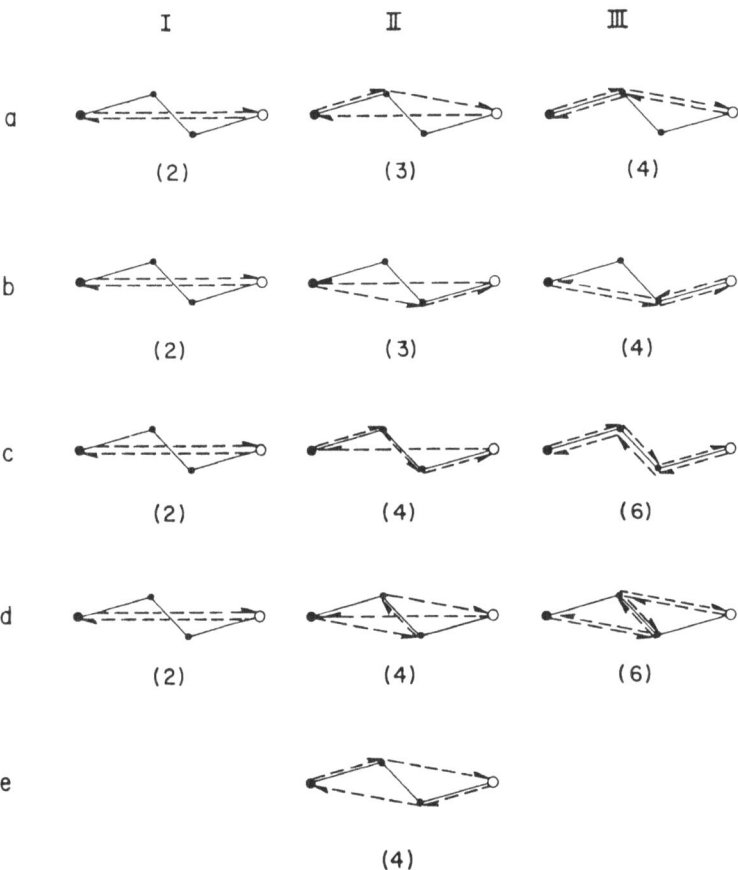

Fig. 8.15. Schematic representation of scattering pathways for the four-atom AB_1B_2C system shown in Fig. 8.14. The absorber A, the intervening atoms B, and the backscatterer C are represented by filled circles, dots, and open circles, respectively. Pathway I is the direct backscattering whereas pathways II and III are multiple scatterings involving three (a and b) or four (c, d, and e) atoms. The numbers in the parentheses indicate number of distances traveled. Pathway II should be counted twice. (Reproduced from Teo, 1983).

than the four-atom pathways (c,d,e) due to the less number of distances involved. Within each group, the relative importance of the multiple scattering pathways also decreases with increasing number of path lengths involved. Hence, for the three-atom pathways a and b, I (2 distances) $\geq$ II (3 distances) $>>$ III (4 distances) whereas for the four-atom pathways c, d, and e, I (2 distances) $\geq$ II (4 distances) $>>$ III (6 distances) (*cf.* Fig. 8.15). Obviously, the exact ordering also depends heavily on the scattering angles at each intervening atoms. In fact, when the atoms are arranged in a linear or nearly collinear fashion, the particular pathway(s) which have small scattering angle(s) will dominate. Under these conditions, the ordering may be reversed itself to give III $>>$ II $\geq$ I.

Generally speaking, for a given pathway involving a $AB_1...B_nC$ fragment which is detectable and separable via Fourier filtering, it can be analyzed with a single term model:

$$\chi(k) = g \frac{(-1)^l}{k} \times \frac{F_{B_1(\beta_1,k)}F_{B_2(\beta_2,k)}...F_{B_a(\beta_a,k)}F_{C(\gamma,k)}}{r_{AB_1}r_{B_1B_2}...r_{B_aC}r_{AC}} e^{-2r_{ef}/\lambda} e^{-2\sigma_{eff}^2k^2}$$

$$\times \sin[2kr_{eff}+2\delta_l'(k)+\theta_{B_1}(\beta_1,k)+\theta_{B_2}(\beta_2,k)+...+\theta_{B_a}(\beta_n,k)+\theta_C(\gamma,k)] \quad (8.24)$$

Here g is the geometrical factor and the photoelectron travels through distances $r_{AB_1}, r_{B_1B_2}, ...r_{B_aC}$, being scattered by atoms $B_1, B_2, ... B_n$, and C with scattering angles $\beta_1, \beta_2, ... \beta_n$, and γ, respectively. The effective pathlength, which corresponds to the distance observed in the Fourier transform after phase shift corrections, is $r_{eff} = (r_{AB_1}+r_{B_1B_2}+...+r_{B_aC}+r_{AC})/2$ and the effective Debye-Waller factor is $\sigma_{eff}^2 \approx (\sigma_{AB_1}^2+\sigma_{B_1B_2}^2+...+\sigma_{B_aC}^2+\sigma_{AC}^2)/2$.

8.2.3 Multiple Scattering: Linear or Nearly Linear $AB_1B_2...B_nC$ Systems

Owing to the short-range nature of EXAFS, higher shells can in general be analyzed using the multiple scattering formalism developed so far for three- and four-atom systems. In most cases, the signals from the third and fourth shells would also have reached the detection limit of EXAFS due to factors such as Debye-Waller factor, inelastic electron mean free path and $1/r^2$ damping. The obvious exception to these considerations is the case where the atoms are arranged in a linear or nearly linear fashion. In such a case, significant amplitude enhancement due to forward scattering at small scattering angle *via*

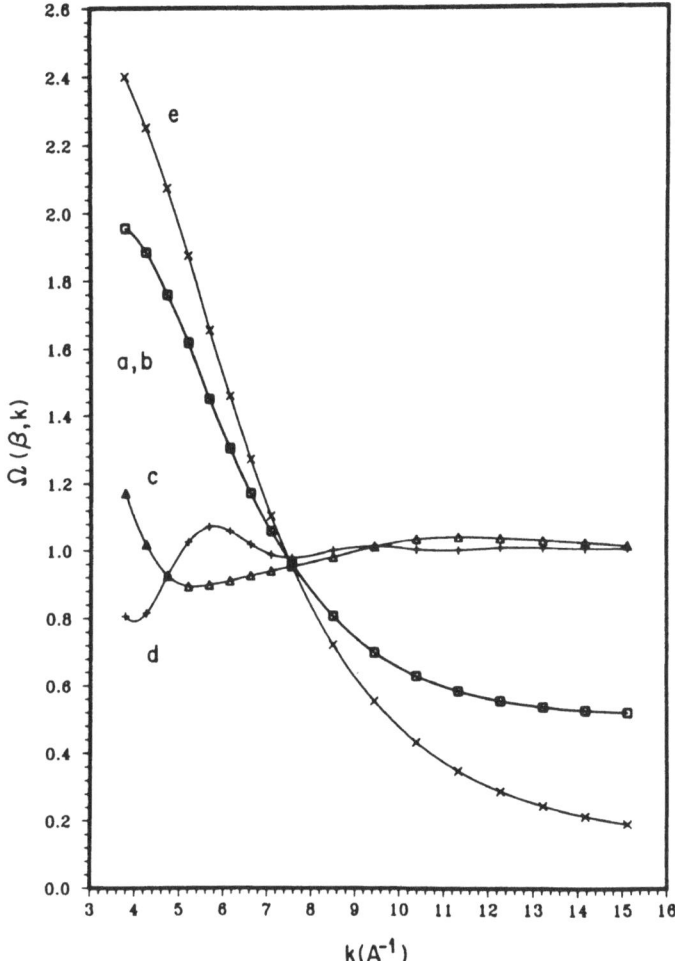

Fig. 8.16. The amplitude modification functions for various scattering pathways (a-e) for a zig-zag A-B_1-B_2-C system shown in Fig. 8.15. Here B_1, B_2 are oxygens and the structural parameters are given in the text. (Reproduced from Teo, 1983).

the intervening atoms can occur and EXAFS signals beyond the second or third shells can be observed. For these systems, the multiple scattering processes related to that involving the stepwise propagation of the photoelectron waves along the chemical bonds dominate and one can write (assuming $f_c(\gamma, k) \approx f_c(\pi, k)$):

$$\Omega e^{i\omega} = [1 + \tilde{r} f_{B_1}(\beta_1, k) \cdots f_{B_n}(\beta_n, k) e^{-\Delta r/\lambda} e^{ik(\Delta r)}]^2 \qquad (8.25)$$

where

$$\bar{r} = \frac{r_{AC}}{r_{AB_1} r_{B_1 B_2} \cdots r_{B_n C}} \tag{8.26}$$

$$\Delta r = r_{AB_1} + r_{B_1 B_2} + \cdots + r_{B_n C} - r_{AC} \tag{8.27}$$

If one defines

$$\bar{F} = F_{B_1}(\beta_1, k) \cdots F_{B_n}(\beta_n, k) \tag{8.28}$$

$$\tilde{\theta} = \theta_{B_1}(\beta_1, k) \cdots \theta_{B_n}(\beta_n, k) + k(\Delta r) \tag{8.29}$$

it is straightforward to show that

$$\Omega(\beta_1, \cdots \beta_n, k) = 1 + 2\bar{r}\bar{F}e^{-\Delta r/\Lambda}\cos\tilde{\theta} + [\bar{r}\bar{F}e^{-\Delta r/\Lambda}]^2 \tag{8.30}$$

$$\omega(\beta_1, \cdots \beta_n, k) = 2\tan^{-1}\frac{\bar{r}\bar{F}e^{-\Delta r/\Lambda}\sin\tilde{\theta}}{1+\bar{r}\bar{F}e^{-\Delta r/\Lambda}\cos\tilde{\theta}} \tag{8.31}$$

8.3 Comparison of Theory and Experiment

The theoretical amplitude and phase functions tabulated in Appendix VI and depicted in Appendix VII for various elements at differing scattering angles have been compared with experimental results and used in bond angle determinations (*vide infra*). We shall discuss a few examples here.

Biebesheimer, Marques, Sandstrom, Lytle, and Greegor (1984), using angle dependent theoretical phase and amplitude functions listed in Appendix VI and the multiple scattering theory discussed in the previous sections, calculated the L_{III} edge EXAFS of Pt metal (fcc structure). They found good agreement with experimental data for photon energy $E > 50$ eV above the absorption edge. This is not unexpected since the theory assumes plane wave approximation which is least accurate at low photoelectron energies (*cf.* Chapter 4). Moreover, the theory assumes that the photoelectron was at sufficiently high energy (approximately three times the plasma frequency or $E > 70$ eV in Pt) so that the attractive potential of the central atom nucleus became negligible. These authors also found that the forward scattering amplitude is not as strong as the calculated amplitude functions. That is, for $\beta = 0°$ or $\alpha = 180°$, the theoretical amplitude function needs to be multiplied by a factor of 0.6 to get reasonable agreement with experimental data. This is again expected since the theoretical amplitude functions are not as reliable in the forward scattering as in the back scattering (cf. Section 8.1). Finally, these authors concluded that, as in the case of Cu metal (Lee and Pendry, 1975; Ashley and Doniach, 1975), the effects of multiple scattering in Pt metal are only important in the fourth shell region. Multiple scattering contributions corresponding to second and third shell neighbor distances are very weak and can be neglected. At total scattering distances corresponding to the fourth shell, the major contribution to the EXAFS comes from the forward scattering by the second nearest neighbor and backscattering by the fourth shell.

As another example, Motta, DeCrescenzi, and Balzarotti (1983) measured the EXAFS of Fe (bcc) and TiFe (CsCl structure) and found that, for both materials the fifth and sixth coordination shells show clear evidence of multiple-scattering effects due to the intervening first- and second-shell atoms. Using Eq. 17-20 and the amplitude and phase functions tabulated by Fink and Ingram (1972), these authors found that such effects can be largely accounted for.

Finally, a recent report by Lengeler (1984) found evidence for multiple scattering effect of hydrogen as the intervening atom (which by itself shows no observable EXAFS signal) in nickel and chromium hydrides.

8.4 Angle Determination

We shall now describe a general method for interatomic angle determination by EXAFS on unoriented (spherically averaged) materials based on the multiple-scattering theory and calculations described in the previous sections.

8.4.1 ABC Systems

To determine the bond angle A-B-C ($\alpha = 180° - \beta$), the distance r_{AC}, and the Debye-Waller factor σ_C, one simply fits the experimental curve with Eq. 8.17 by least-squares refining β, r_{AC}, σ_C, ΔE_{0_C}, η_C, and S_C. The distance r_{AB} and the energy correction ΔE_{0_B} are taken from the AB (first shell) fit and held constant throughout the curve fitting process. The distance r_{BC} can be calculated from r_{AC}, r_{AB}, and β in each cycle of refinement. The chi-squares Σ^2, which is the sum of squares of the residuals, are then plotted as a function of β to determine the parameters for the best fit (minimum). The bond angle α is then $180° - \beta$.

The usefulness and the accuracy of this method have been demonstrated by Teo (1981) for three compounds: (1) the virtually linear metal carbonyl Fe-C-O in $Na_2Fe(CO)_4(1)$; (2) the bent Pt-O-C in $K_2Pt(C_2O_4)_2(2)$ and (3) the bent Pt-O-S in $K_2Pt_2(SO_4)_4(H_2O)_2(3)$ shown in Fig. 8.17. The known bond angles of each of these three A-B-C systems are 177° (Teller, *et al*, 1977), 113° (Mattes and Krogmann, 1964), and 124° (Muraveiskaya, *et al*, 1976).

Fig. 8.18 shows the chi-square (Σ^2) plots for the angle A-B-C. The minimum (best fit) occurs at $\beta = 0°$, 70°, 58°, or $\alpha = 180°$, 110°, 122° and for **1, 2, 3** respectively.

In Fig. 8.17, we compare the mean structural parameters derived from EXAFS spectroscopy with those from X-ray diffraction studies.

The accuracy for angle determination is better than *ca* 6% for low Z ($Z \leqslant 10$) and *ca* 3% for high Z scatterers, if η_C is allowed to vary (i.e., least-squares refined). In most cases, it amounts to an accuracy of *ca* 5°, which is comparable to the scattered range of crystallographically independent bond angles often observed in diffraction studies. Since the B-C distance is determined indirectly from the A-B, AC distances and the A-B-C angle, it is not surprising that it has the least accuracy.

It is interesting to consider why angle determination is feasible by EXAFS for unoriented samples (e.g., noncrystalline or polycrystalline materials). Even more puzzling is the question that at bond angles $\alpha \leq 120°$ where the focusing effect is virtually disappearing, such angular determination is still possible. If, as mentioned in previous sections, the phase correction factor ω due to multiple scatterings is normally quite small, it might be concluded that the well-defined Σ^2 minima observed for the angle must be related to the amplitude modification. This rationale is understandable for bond angle α approaching 180° ($\alpha \geq 150°$) where both the magnitude and the shape of the scattering amplitude change dramatically with the angle. For $90° \leqslant \alpha \leq 120°$, the scattering amplitude changes only slightly. Yet the amplitude modification factor Ω exhibits characteristic shape for each α. The reason is that at these angles, Δr becomes

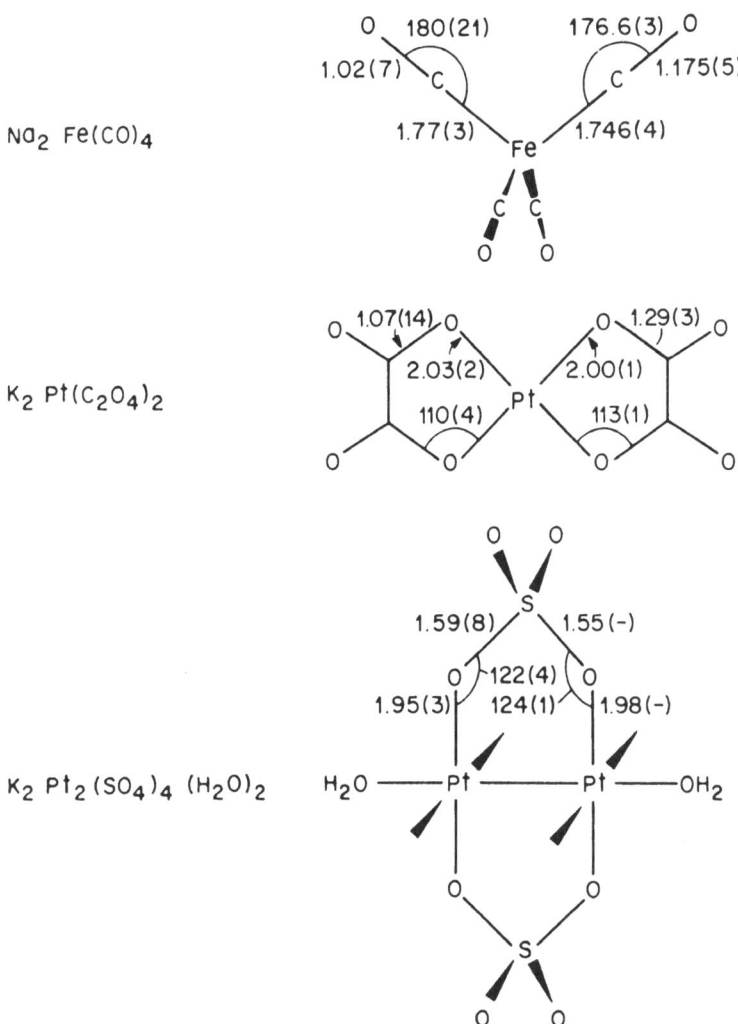

Fig. 8.17. The structures of $Na_2Fe(CO)_4$ (1), $K_2Pt(C_2O_4)_2$ (2), and $K_2Pt_2(SO_4)_4(H_2O)_2$ (3). For 3, two SO_4 bridges are omitted for clarity. The EXAFS results (left) are also compared with the crystallographical results (right).

large such that the second term $(k(\Delta r))$ in Eq. 8.14, which is now a strong function of k, becomes the dominant factor. It is therefore not surprising to find Ω (cf. Fig. 8.9), as well as ω (cf. Fig. 8.10), to be highly dependent upon (i.e. fast varying) the bond angle α. As α deviates far from the correct value, in

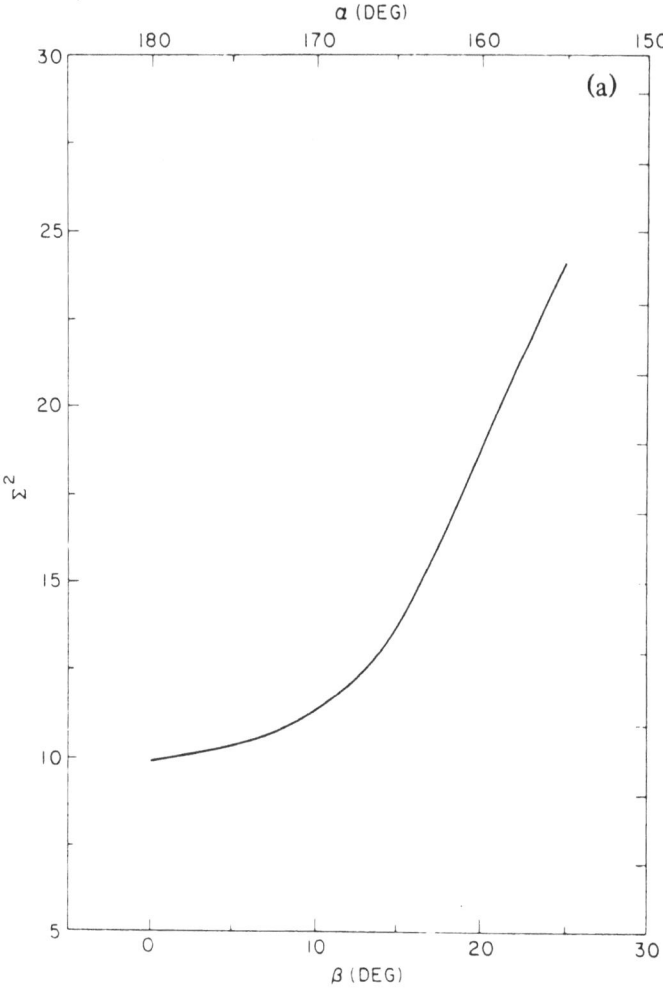

Fig. 8.18. The chi-squares Σ^2 (the sum of squares of the residuals) *vs* the A-B-C bond angle α in degree (or the scattering angle $\beta = 180° - \alpha$) for the ABC fits of (a) $Na_2Fe(CO)_4$; (b) $K_2Pt(C_2O_4)_2$; and (c) $K_2Pt_2(SO_4)_4(H_2O)_2$. (Adapted from Teo, 1981).

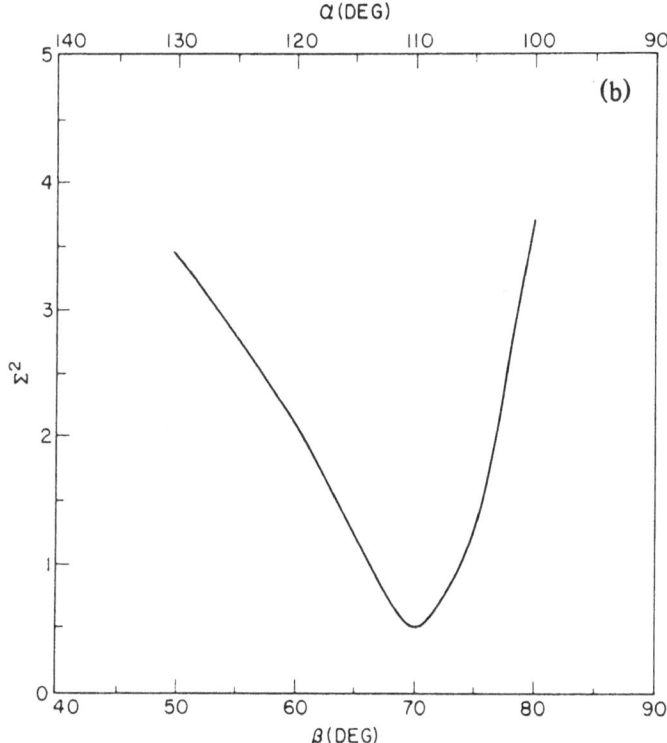

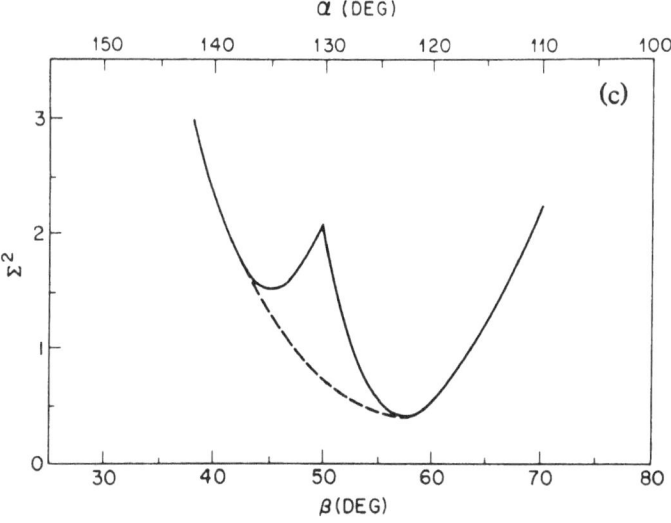

either direction (viz., beyond the correlation limits of other parameters such as ΔE_0), the wrong distances give rise to progressively worsening phase mismatch which causes the quality of the fit to deteriorate rapidly.

Furthermore, it is evident that multiple scattering processes involving three atoms (including the absorber and two neighboring atoms) are important in determining the EXAFS of the distant shells, especially at large bond angles α. Multiple scattering involving more than three atoms are likely to be of less importance due to the large effective total path length $2r_{eff}$ (*vide supra*); except, perhaps, in cases where all the bond angles are close to linearity. Yet the single-scattering theory has so far been successful in providing accurate distance, but not angle, information because of two facts. First, the effect in distance due to multiple scatterings is no more than 0.1Å; in fact, in most cases ($\alpha \geq 130°$), less than *ca* 0.03 Å. A large part of such phase effect can be compensated for by varying r and ΔE_o. Second, the effect in amplitude due to multiple scatterings is generally less than *ca* 70% for $\alpha \leq 150°$. The envelope of such amplitude effect (i.e. shape of Ω) is such that a large portion of which can be compensated for by varying σ and λ. For $\alpha \geq 150°$, the large amplitude enhancement Ω is accompanied by a relatively weak k dependence such that the effect can largely be compensated by the overall scale factor S. Therefore, in a sense, the single-scattering theory, which contains no angle information, corresponds to one section of the multiple-scattering formalism in which $\Omega(\beta, k)$ and $\omega(\beta, k)$ are rapidly varying (*cf.* Fig. 8.9 and 8.10). The latter characteristics of the multiple-scattering effects make possible the accurate bond angle determination by EXAFS.

It should also be cautioned that in Fourier filtering, one should not filter out the multiple scattering pathways II and III which correspond to effective distances of $(r_{AB} + r_{BC} + r_{AC})/2$ (II in Fig. 8.13) and $(r_{AB} + r_{BC})$ (III in Fig. 8.13), respectively, by choosing too narrow a window for the A-C peak. For systems with large $(r_{AB} + r_{BC} - r_{AC})/2$, each of these peaks may be analyzed separately using the corresponding term in Eq. 8.7. Fortunately, for most systems with large α, III $>>$ II $>$ I and $r_{AB} + r_{BC} \approx r_{AC}$ whereas for most systems with small α, I $>$ II $>>$ III and $(r_{AB} + r_{BC} + r_{AC})/2 \geq r_{AC}$. That is, for both limits of α, the dominant term(s) are more likely than not to be buried under (or lie close to) the A-C peak.

It should be emphasized that in order to determine the bond angle one must have a set of reliable amplitude and phase functions for each scattering angle for the intermediary atom and a consistent set of backscattering amplitude and phase functions for the terminal atom (backscatterer). One must also combine structural information obtained from AB and ABC fits in a consistent way to uniquely define the triangle. Finally, the accuracy of bond angle determinations is critically dependent upon the quality of the EXAFS data.

Using the theory described in Section 8.2.1.1, Alberding and Crozier (1983) investigated the temperature-dependent EXAFS of two μ-oxo-bridged iron compounds in order to determine the two different bridging geometries. Two extreme curve-fitting models assuming different correlations of Fe-Fe distance with bond angle were applied to the data: a *mean multiple-scattering* model in which the bond angle was rigid and an *integrated multiple-scattering* model in which the bond angle was permitted to flex. The latter model produced more accurate results for (Fe-HEDTA)$_2$O giving a bond angle of 171.0^{+9}_{-17} deg. and Fe-Fe separation of $3.579^{+0.02}_{-0.08}$Å relative to the crystallographically determined values of 163.8° and 3.554Å, respectively. The compound (Fe-dipic-OH)$_2$ has an equilibrium bond angle of 103.4°; therefore, collinear arrangements of atoms were less important and both models converged to about the same result, 105.4° ± 2.0°. These authors found that in at least these compounds, allowance must be made for large variations in multiple-scattering effects arising from dynamic fluctuations in bond angles due to thermal motion. These authors also pointed out that suggestions by some investigators to use empirical phase and amplitude functions derived from well-characterized model systems rather than taking theoretical functions in bond angle determination may cause problems in some cases. That is, where multiple scattering exists, the total amplitude factor and phase shifts depend on the relative positions of the atoms as well as their chemical nature. Therefore, "unless the geometry of the model system exactly matches that of the unknown system, a direct transfer of phase and amplitude functions is not correct. A possible solution to the problem would be to use the theoretical multiple-scattering functions of Teo (1981) to correct empirical functions for changes in geometry of the unknown system as well as differences in dynamic and static disorder."

8.4.1.1 Empirical Approach

Co, Hendrickson, Hodgson, and Doniach (1983) measured and analyzed a series of dihydroxo-, dialkoxo- or μ-oxo-bridged dimeric and trimeric iron complexes with Fe-O-Fe bridging angles ranging from 103° to 180°. Their results indicate that where the Fe-O-Fe bridging angle is greater than ca. 150°

the intervening oxygen atom in the absorber-to-scatterer path enhances the Fe-Fe EXAFS scattering amplitude and also causes additional phase shifts. Both the magnitude and the phase shifts of the Fe-Fe wave are correlated with the Fe-O-Fe bridging angle. The maximum effect was observed in the linear case (bond angle of $180°$) where an amplitude enhancement by a factor of 4 and an "extra" phase shift (due to the bridging oxygen) of $-0.21 Å$ were observed.

It is apparent that in cases where such a *working curve* can be obtained through measurements of model compounds with varying bond angles, it may be possible to determine bond angle in an unknown system involving the same molecular fragment $(A-B-C)$. For example, Hendrickson, *et al* (1982), have applied this technique to determine the bridging angle for a μ-oxo-bridged dimeric iron protein, hemerythrin, and calculate the bridging angle to be *ca.* $165°$. In general, however, it is not always possible to construct such a working curve since compounds containing the same molecular fragments often adopt the same or a few well-defined configurations (linear, trigonal or tetrahedral), rather than exhibit a wide range of bond angles. It is recommended that theoretical functions be used for bond angle determinations (*cf.*, previous section). Furthermore, curve fitting (in k space) is more accurate than Fourier analysis (in r space).

8.4.2 $AB_1...B_nC$ Systems

To obtain structural parameters for the nearly linear $AB_1...B_nC$ system, the procedure outlined for an ABC system can be repeated for each of the preceding (shorter) fragments $AB_1...B_jC$ where j runs from 1 to n, using Eq. 8.30 and 8.31. Note the similarity between Eq. 8.30 and 8.31 for the nearly linear $AB_1...B_nC$ system and Eq. 8.19 and 8.20 the triangular ABC system.

In each step, one distance $(r_{B_jB_{j+1}})$ and one angle (β_j) will be sought (*i.e.*, refined) while the previously obtained structural $(r_{B_1B_2},...r_{B_{j-1}B_j}, \beta_1,...\beta_{j-1})$ and energy $(\Delta E_{o_{B_1}},...\Delta E_{o_{B_{j-1}}})$ parameters will be fixed.

8.5 Conclusion

In this chapter, we have described an EXAFS formulation which takes into account the effect of multiple scatterings. Some of the results of theoretical calculations on the scattering amplitude and phase shift, both as functions of the scattering angle, are tabulated in Appendix VI. The relative importance of various multiple scattering processes is assessed as a function of the scattering angle which helps unravel the "focusing" effect. Based on these, a method for interatomic angle determinations by EXAFS is devised and its usefulness and accuracy illustrated with a few examples. The ultimate accuracy of such structural determination depends heavily on the quality of the EXAFS data, especially the amplitude.

Extension of the application of the angle determination technique to a wide variety of systems is obvious. Examples include: stereochemical information beyond the first coordination shell for inorganic and biological systems, especially when the ligand atoms are arranged in a nearly collinear fashion; solution structural determinations such as those pertaining to the ligand fluxionality and kinetics; ligation of simple well-defined molecules to biological molecules in solution; etc. It is clear that such structural information involving the distant, normally nonbonding, shells can greatly facilitate either interpretation of the mode of binding of known ligands or identification of the unknown ligands.

BIBLIOGRAPHY

1. Adachi, H.; Fujima, K.; Taniguchi, K.; Miyake, C.; Imoto, S. *Jpn. J. Appl. Phys.*, **1981**, *20*, L612.

2. Adhyapak, S. V.; Nigavekar, A. S. *J. Phys. Chem. Solids*, **1976**, *37*, 1037.

3. Agarwal, B. K. "*X-ray Spectroscopy*," Springer-Verlag, Heidelberg, **1979**, p. 215.

4. Agarwal, B. K.; Agarwal, B. R. K. *J. Phys. C*, **1978**, *11*, 4223.

5. Agarwal, B. K.; Balakrishnan, V. *J. Phys. F*, **1982**, *12*, 1519.

6. Agarwal, B. K.; Balakrishnan, V. *Phys. Rev. B*, **1983**, *28*, 2852.

7. Agarwal, B. K.; Bhargava, C. B.; Vishnoi, A. N.; Seth, V. P. *J. Phys. Chem. Solids*, **1976**, *37*, 725.

8. Agarwal, B. K.; Johri, R. K. *J. Phys. C*, **1977a**, *10*, 3213.

9. Agarwal, B. K.; Johri, R. K. *J. Phys. F*, **1977b**, *7*, 1607.

10. Agarwal, B. K.; Johri, R. K. *Phys. Status Solidi B*, **1978**, *88*, 309.

11. Ageev, A. L.; Babanov, Yu. A.; Vasin, V. V.; Ershov, N. V.; Serikov, A. V. *Phys. Status Solidi B*, **1983**, *117*, 345.

12. Ahlberg, J. H.; Nilson, E. N.; Walsh, J. L. "*The Theory of Splines and Their Applications*," Academic, New York, **1967**.

13. Akimov, V. N.; Vinogradov, A. S.; Zimkina, T. M. *Opt. Spektrosk.*, **1982a**, *53*, 109.

14. Akimov, V. N.; Vinogradov, A. S.; Zimkina, T. M. *Opt. Spektrosk.*, **1982b**, *53*, 476.

15. Akimov, V. N.; Vinogradov, A. S.; Zimkina, T. M. *Opt. Spektrosk.*, **1982c**, *53*, 918.

16. Alagna, L.; Hasnain, S. S.; Piggott, B.; Williams, D. J. *Biochem. J.*, **1984**, *220*, 591.

17. Alagna, L.; Prosperi, T.; Tomlinson, A. G. *Springer Ser. Chem. Phys.*, **1983**, *27*, 303.

18. Alagna, L.; Prosperi, T.; Ferragina, C.; La Ginestra, A.; Tomlinson, A. A. G. *J. Chem. Soc., Faraday Trans. 1*, **1983**, *79*, 1039.

19. Alagna, L.; Prosperi, T.; Tomlinson, A. A. G.; Vlaic, G. *J. Chem. Soc., Dalton Trans.*, **1983**, 645.

20. Alagna, L.; Tomlinson, A. A. G. *J. Chem. Soc., Faraday Trans.*, **1982**, *78*, 3009.

21. Alagna, L.; Tomlinson, A. A. G.; Ferragina, C.; La Ginestra, A. *Daresbury Lab. Rep. DL/SCI/R17*, **1981**, 98.

22. Alberding, N.; Crozier, E. D. *Phys. Rev. B*, **1983**, *27*, 3374.

23. Aliotta, F.; Galli, G.; Maisano, G.; Migliardo, P.; Vasi, C. *Springer Ser. Chem. Phys.*, **1983**, 27, 268.

24. Aliotta, F.; Galli, G.; Maisano, G.; Migliardo, P.; Vasi, C.; Wanderlingh, F. *Nuovo Cimento D.*, **1983**, *2*, 103.

25. Alema, S.; Bianconi, A.; Castellani, L.; Fasella, P. *Prog. Clin. Biol. Res.*, **1982**, *102*, 47.

26. Aliotta, F.; Galli, G.; Maisano, G.; Migliardo, P.; Vasi, C. *Springer Ser. Chem. Phys.*, **1983**, *27*, 268.

27. Aliotta, F.; Galli, G.; Maisano, G.; Migliardo, P.; Vasi, C.; Wanderlingh, F. *Nuovo Cimento D*, **1983**, *2*, 103.

28. Allred, A. L.; Rochow, D. E. G. *J. Inorg. Nucl. Chem.*, **1958**, *5*, 264.

29. Amus'ya, M. Ya.; Pavlychev, A. A.; Vinogradov, A. S.; Onopko, D. E.; Titov, S. A. *Opt. Spektrosk.*, **1982**, *53*, 157.

30. Anderson, S.; Hammarqvist, H.; Nyberg, C. *Rev. Sci. Instrum.*, **1974**, *45*, 877.

31. Antonangeli, F.; Apicella, M. L.; Balzarotti, A.; Incoccia, L.; Piacentini, M. *Physica B+C*, **1981**, *105*, 25.

32. Antonangeli, F.; Balzarotti, A.; Motta, N.; Piacentini, M.; Kisiel, A.; Zimnal-Starnawska, M.; Giriat, W. *Springer Ser. Chem. Phys.*, **1983**, *27*, 224.

33. Antonangeli, F.; Bellini, C.; De Crescenzi, M.; Rosei, R. *Springer Ser. Chem. Phys.*, **1983**, 27, 397.

34. Antonini, M.; Caprile, C.; Merlini, A.; Petiau, J.; Thornley, F. R. *Springer Ser. Chem. Phys.*, **1983**, *27*, 261.

35. Antonio, M. R.; Averill, B. A.; Moura, I.; Moura, J. J. G.; Orme-Johnson, W. H., Teo, B. K.; Xavier, A. V. *J. Biol. Chem.*, **1982**, *257*, 6646.

36. Antonio, M. R.; Teo, B. K.; Averill, B. A. *J. Am Chem. Soc.*, **1985**, *107*, 3583.

37. Antonio, M. R.; Teo, B. K.; Cleland, W. E.; Averill, B. A. *J. Am. Chem. Soc.*, **1983**, *105*, 3477.

38. Antonio, M. R.; Teo, B. K.; Orme-Johnson, W. H.; Nelson, M. J.; Groh, S. E.; Lindahl, P. A.; Kauzlarich, S. M.; Averill, B. A. *J. Am. Chem. Soc.*, **1982**, *104*, 4703.

39. Apai, G.; Hamilton, J. F.; Stohr, J.; Thompson, A. *Phys. Rev. Lett.*, **1979**, *43*, 165.

40. Apte, M. Y.; Mande, C. *J. Phys. Chem. Solids*, **1980**, *41*, 307.

41. Apte, M. Y.; Mande, C. *J. Phys. Chem. Solids*, **1981**, *42*, 605.

42. Apte, M. Y.; Mande, C. *J. Phys. C*, **1982**, *15*, 607.

43. Apte, M. Y.; Mande, C.; Suchet, J. P. *J. Chim. Phys. Phys.-Chim. Biol.*, **1982**, *79*, 325.

44. Ashley, C. A.; Doniach, S. *Phys. Rev. B*, **1975**, *11*, 1279.

45. Avery, J. *"The Quantum Theory of Atoms, Molecules and Photons,"* McGraw Hill, New York, **1972**.

46. Azaroff, L. V. *Rev. Mod. Phys.*, **1963**, *35*, 1012.

47. Azaroff, L. V., ed., *"X-ray Absorption Spectra,"* McGraw-Hill, New York, **1974**.

48. Azoulay, J.; Stern, E. A.; Shaltiel, D.; Grayevski, A. *Phys. Rev. B*, **1982**, *25*, 5627.

49. Babanov, Y. A.; Vasin, V. V.; Ageev, A. L.; Ershov, N. V. *Springer Ser. Chem. Phys.*, **1983**, *27*, 112.

50. Babanov, Y. A.; Vasin, V. V.; Ageev, A. L.; Ershov, N. V. *Phys. Status Solidi B*, **1981**, *105*, 747.

51. Bachelet, G. B.; Schluter, M. *Phys. Rev. B*, **1983**, *28*, 2302.

52. Bachrach, R. Z.; Hansson, G. V.; Bauer, R. S. *Surf. Sci.*, **1981**, *109*, L560.

53. Bachrach, R. Z.; Lindau, I. *Springer Ser. Chem. Phys.*, **1983**, *27*, 415.

54. Badiali, J. P.; Bosio, L.; Cortes, R.; Bondot, P; Loupias, G.; Petiau, J. *J. Phys. Colloq.*, **1980**, *C8*, 211.

55. Bair, R. A.; Goddard, W. A., III *Phys. Rev. B*, **1980**, *22*, 2767.

56. Ballhausen, C. J. *"Introduction to Ligand Field Theory,"* McGraw-Hill, New York, **1962**.

57. Balzarotti, A. *Springer Ser. Chem. Phys.*, **1983**, *27*, 135.

58. Balzarotti, A.; Comin, F.; Incoccia, L.; Piacentini, M.; Mobilio, S.; Savoia, A. *Solid State Commun.*, **1980**, *35*, 145.

59. Balzarotti, A.; Czyzyk, M.; Kisiel, A.; Motta, N.; Podgorny, M.; Zimnal-Starnawska, M. *Phys. Rev. B*, **1984**, *30*, 2295.

60. Balzarotti, A.; Menuschenkov, A. P.; Motta, N.; Purans, J. *Solid State Commun.*, **1984**, *49*, 887.

61. Bambynek, W.; Crasemann, B.; Fink, R. W.; Freund, H. U.; Mark, H.; Swift, C. D.; Price, R. E.; Rao, P. V. *Rev. Mod. Phys.*, **1972**, *44*, 716.

62. Barhate, A. V.; Pendharkar, A. V.; Sapre, V. B.; Mande, C. *Solid State Commun.*, **1980**, *36*, 473.

63. Barhate, A. V.; Sapre, V. B.; Mande, C. *Indian J. Phys.*, **1982**, *56A*, 81.

64. Bartlett, N.; McQuillan, B.; Robertson, A. S. *Mater. Res. Bull.*, **1978**, *13*, 1259.

65. Bassi, I. W.; Lytle, F. W.; Parravano, G. *J. of Catal.*, **1976**, *42*, 139.

66. Batmanian, S.; Garner, C. D.; Blackburn, N. J.; Bordas, J.; Diakun, G. P.; Hasnain, S. S.; Pantos, E.; Knowles, P. F. *Daresbury Lab. Rep. DL/SCI/R17*, **1981**, 91.

67. Batson, P. E. *"Proceedings of a Specialist Workshop on Analytical Electron Microscopy,"* Feyes, P. L., ed., Cornell University, Ithaca, N.Y., **1978a**, p. 31.

68. Batson, P. E. *Ultramicroscopy*, **1978b**, *3*, 367.

69. Batson, P. E.; Craven, A. J. *Phys. Rev. Lett.*, **1979**, *893*, 42.

70. Batterman, B. W. in *"EXAFS Spectroscopy: Techniques and Applications,"* Teo, B. K., Joy, D. C., ed., Plenum, New York, **1981**, p. 197.

71. Batterman, B. W.; Ashcroft, N. W. *Science*, **1979**, *206*, 157.

72. Baxter, D. V.; Williams, A.; Johnson, W. L. *Report DOE/ER/10870-142*, Order No. DE84002022; NTIS: *Energy Res. Abstr.*, **1984**, *9*, No. 769.

73. Baxter, D. V.; Williams, A.; Johnson, W. L. *J. Non-Cryst. Solids*, **1984**, 409.

74. Beagley, B.; Gahan, B.; Greaves, G. N.; McAuliffe, C. A. *J. Chem. Soc., Chem. Commun.*, **1983**, 1265.

75. Beaumont, J. H.; Hart, M. *J. Phys. E.*, **1974**, *7*, 823.

76. Beaurepaire, E.; Krill, G.; Kappler, J. P.; Roehler, J. *Solid State Commun.*, **1984**, *49*, 65.

77. Beinert, H.; Emptage, M. H.; Dreyer, J. L.; Scott, R. A.; Hahn, J. E.; Hodgson, K. O.; Thomson, A. J. *Proc. Natl. Acad. Sci. U. S. A.*, **1983**, *80*, 393.

78. Bellessa, J.; Gors, C.; Launois, P.; Quillec, M.; Launois, H. *Conf. Ser.- Inst. Phys.*, **1983**, *65*, 529.

79. Belli, M.; Bianconi, A.; Burattini, E.; Mobilio, S.; Natoli, C. R.; Palladino, L.; Reale, A.; Scafati, A. *Springer Ser. Chem. Phys.*, **1983**, *27*, 345.

80. Belli, M.; Scafati, A.; Bianconi, A.; Burattini, E.; Mobilio, S.; Natoli, C. R.; Palladino, L.; Reale, A. *Nuovo Cimento D*, **1983**, *2*, 1281.

81. Belli, M.; Bianconi, A.; Mobilio, S.; Palladino, L.; Reale, A.; Scafati, A. *Ital. J. Biochem.*, **1980**, *29*, 77.

82. Belli, M.; Scafati, A.; Bianconi, A.; Burattini, E.; Mobilio, S.; Natoli, C.; Palladino, L.; Reale, A. in *"Inn. -Shell X-Ray Phys. At. Solids,"* Fabian, D. J., Kleinpoppen, H., Watson, L. M., ed., Plenum, N.Y., **1981**, p. 731.

83. Belli, M.; Scafati, A.; Bianconi, A.; Mobilio, S.; Palladino, L.; Reale, A.; Burattini, E. *Solid State Commun.*, **1980**, *35*, 355.

84. Bellissent, R.; Chenevas-Paule, A.; Lagarde, P.; Bazin, D.; Raoux, D. *J. Non-Cryst. Solids*, **1983**, *59-60*, 237.

85. Beni, G.; Platzman, P. M. *Phys. Rev B*, **1976**, *14*, 1514.

86. Bernieri, E.; Burattini, E.; Cappuccio, G.; Dalba, G.; Fornasini, P. *Report ISS-FI-83/4*, Order No. N83-34820; NTIS: *Gov. Rep. Announce. Index (U.S.)*, **1984**, *84*, 125.

87. Bernieri, E.; Burattini, E.; Cappuccio, G.; Dalba, G.; Fornasini, P.; Grandolfo, M.; Vecchia, P.; Efendiev, S. M. *Ferroelectrics*, **1984**, *56*, 257.

88. Bernieri, E.; Burattini, E.; Dalba, G.; Fornasini, P.; Rocca, F. *Solid State Commun.*, **1983**, *48*, 421.

89. Bernieri, E.; Burattini, E.; Cavallo, N.; Masullo, M. R.; Morone, A.; Rinizivillo, R.; Natoli, C.; Palladino, L.; Reale, A. *Springer Ser. Chem. Phys.*, **1983**, *27*, 62.

90. Besson, B.; Moraweck, B.; Smith, A. K.; Basset, J. M.; Psaro, Rinaldo; Fusi, A.; Ugo, R. *J. Chem. Soc., Chem. Commun.*, **1980**, 569.

91. Bhat, N. V.; Salvi, S. V. *Spectrochim. Acta B*, **1983**, *38*, 481.

92. Bhattacharya, P.; Chetal, A. R. *Phys. Status Solidi B*, **1983**, *119*, 735.

93. Bianconi, A. *Surf. Sci.*, **1979**, *89*, 41.

94. Bianconi, A. *Appl. Surf. Sci.*, **1980**, *6*, 392.

95. Bianconi, A. *Daresbury Lab. Rep. DL/SCI/R17*, **1981**, 13.

96. Bianconi, A. *Phys. Rev. B*, **1982**, *26*, 2741.

97. Bianconi, A. *Springer Ser. Chem. Phys.*, **1983**, *27*, 118.

98. Bianconi, A.; Bachrach, R. Z. *Phys. Rev. Lett.*, **1979**, *42*, 104.

228

99. Bianconi, A.; Bachrach, R. Z.; Flodström, S. A. *Phys. Rev. B*, **1979**, *19*, 3879.

100. Bianconi, A.; Bachrach, R. Z.; Hagstrom, S. B. M.; Flodström, S. A. *Phys. Rev. B*, **1979**, *19*, 2837.

101. Bianconi, A.; Bauer, R. S. *Surf. Sci.*, **1980**, *99*, 76.

102. Bianconi, A.; Campagna, M.; Stizza, S. *Phys. Rev. B*, **1982**, *25*, 2477.

103. Bianconi, A.; Campagna, M.; Stizza, S.; Davoli, I. *Phys. Rev. B*, **1981**, *24*, 6139.

104. Bianconi, A.; Dell'Ariccia, M.; Giovannelli, A.; Burattini, E.; Cavallo, N.; Patteri, P.; Pancini, E.; Carlini, C.; Ciardelli, F.; et al. *Chem. Phys. Lett.*, **1982**, *90*, 257.

105. Bianconi, A.; Dell'Ariccia, M.; Durham, P. J.; Pendry, J. B. *Phys. Rev. B*, **1982**, *26*, 6502.

106. Bianconi, A.; Dell'Ariccia, M.; Gargano, A.; Natoli, C. R. *Springer Ser. Chem. Phys.*, **1983**, *27*, 57.

107. Bianconi, A.; Dell'Ariccia, M.; Gargano, A.; Natoli, C. R. *Springer Ser. Chem. Phys.*, **1983**, *27*, 57.

108. Bianconi, A.; Giovannelli, A.; Ascone, I.; Alema, S.; Durham, P. J.; Fasella, P. *Springer Ser. Chem. Phys.*, **1983**, *27*, 355.

109. Bianconi, A.; Giovannelli, A.; Davoli, I.; Stizza, S.; Palladino, L.; Gzowski, O.; Murawski, L. *Solid State Commun.*, **1982**, *42*, 547.

110. Bianconi, A.; Incoccia, L.; Stipcich, S., ed., *"EXAFS and Near Edge Structure,"* Springer-Verlag, Berlin, Heidelberg, *Springer Ser. Chem. Phys.*, **1983**.

111. Bianconi, A.; Jackson, D.; Monahan, K. *Phys. Rev. B*, **1978**, *17*, 2021.

112. Bianconi, A.; Modesti, S.; Campagna, M.; Fischer, K.; Stizza, S. *J. Phys. C*, **1981**, *14*, 4737.

113. Bianconi, A.; Natoli, C. R. *Solid State Commun.*, **1978**, *27*, 1177.

114. Biebesheimer, V. A.; Marques, E. C.; Sandstrom, D. R.; Lytle, F. W.; Greegor, R. B. *J. Chem. Phys.*, **1984**, *81*, 2599.

115. Bienenstock, A. in *"EXAFS Spectroscopy: Techniques and Applications,"* Teo, B. K., Joy, D. C., ed., Plenum, New York, **1981**, p. 185.

116. Bienenstock, A. in *"Structure of Non-Crystalline Materials,"* Gaskell, P. H.; Davis, E. A., ed., Taylor and Francis, London, **1977**, p. 5.

117. Bigler, E.; Polack, F.; Lowenthal, S. *Nucl. Instrum. Methods Phys. Res.*, **1983**, *208*, 387.

118. Binsted, N.; Hasnain, S. S.; Hukins, D. W. L. *Biochem. Biophys. Res. Commun.*, **1982**, *107*, 89.

119. Blackburn, N. J.; Hasnain, S. S.; Binsted, N.; Diakun, G. P.; Garner, C. D.; Knowles, P. F. *Biochem. J.*, **1984**, *219*, 985.

120. Blackburn, N. J.; Hasnain, S. S.; Diakun, G. P.; Knowles, P. F.; Binsted, N.; Garner, C. D. *Biochem. J.*, **1983**, *213*, 765.

121. Blumberg, W. E.; Eisenberger, P.; Peisach, J.; Shulman, R. G. *Adv. Exp. Med. Biol.*, **1976**, *74*, 389.

122. Blumberg, W. E.; Peisach, J.; Eisenberger, P.; Fee, J. A. *Biochemistry*, **1978**, *17*, 1842.

123. Böhmer, W.; Rabe, P. *J. Phys. C*, **1979**, *12*, 2465.

124. Boland, J. J.; Baldeschwieler, J. D. *J. Chem. Phys.*, **1984a**, *80*, 3005.

125. Boland, J. J.; Baldeschwieler, J. D. *J. Chem. Phys.*, **1984b**, *81*, 1145.

126. Boland, J. J.; Crane, S. E.; Baldeschwieler, J. D. *J. Chem. Phys.*, **1982**, *77*, 142.

127. Boland, J. J.; Halaka, F. G.; Baldeschwieler, J. D. *Phys. Rev. B*, **1983**, *28*, 2921.

128. Bomben, K. D.; Bahl, M. K.; Gimzewski, J. K.; Chambers, S. A.; Thomas, T. D. *Phys. Rev. A*, **1979**, *20*, 2405.

129. Bonnelle, C.; Karnatak, R. C.; Spector, N. *J. Phys. B*, **1977**, *10*, 795.

130. Bonse, U.; Hartmann-Lotsch, I.; Lotsch, H. *Springer Ser. Chem. Phys.*, **1983**, *27*, 376.

131. Bonse, U.; Hartmann-Lotsch, I.; Lotsch, H.; Olthoff-Muenter, K. *Z. Phys. B*, **1982**, *47*, 297.

132. Bonse, U.; Materlik, G.; Schröder, W. *J. Appl. Crystallogr.*, **1976**, *9*, 223.

133. Bordas, J.; Bray, R. C.; Garner, C. D.; Gutteridge, S.; Hasnain, S. S. *J. Inorg. Biochem.*, **1979**, *11*, 181.

134. Bordas, J.; Bray, R. C.; Garner, C. D.; Gutteridge, S.; Hasnain, S. S.; Pettifer, R. *Daresbury Lab. Rep. DL/SCI/R13*, **1979**, 116.

135. Bordas, J.; Bray, R. C.; Garner, C. D.; Gutteridge, S.; Hasnain, S. S. *Biochem. J.*, **1980**, *191*, 499.

136. Bordas, J.; Dodson, G. G.; Grewe, H.; Koch, M. H. J.; Krebs, B.; Randall, J. *Proc. R. Soc. London, B*, **1983**, *219*, 21.

137. Bordas, J.; Koch, M. H. J.; Hartmann, H. J.; Weser, U. *FEBS Lett.*, **1982**, *140*, 19.

138. Bosio, L.; Cortes, R.; Defrain, A.; Froment, M.; Lebrun, A. M. in *"Passivity Met. Semicond.,"* Froment, M., ed., Elsevier, Amsterdam, **1983**, p. 131.

139. Boudart, M.; Sanchez Arrieta, J.; Dalla Betta, R. *J. Am. Chem. Soc.,* **1983**, *105*, 6501.

140. Bouldin, C.; Stern, E. A. *Phys. Rev. B,* **1982**, *25*, 3462.

141. Bouldin, C. E.; Stern, E. A.; Von Roedern, B.; Azoulay, J. *J. Non-Cryst. Solids,* **1984**, *66*, 105.

142. Bouldin, C. E.; Stern, E. A.; Von Roedern, B.; Azoulay, J. *Phys. Rev. B,* **1984**, *30*, 4462.

143. Bourdillon, A. J.; Pettifer, R. F.; Marseglia, E. A. *J. Phys. C,* **1979**, *12*, 3889.

144. Bourdillon, A. J.; Pettifer, R. F.; Marseglia, E. A. *Physica B+C,* **1980**, *99*, 64.

145. Boudreaux, D. S.; Reidinger, F. in *"Amorphous Mater.: Model. Struct. Prop.,"* Vitek, V., ed., Metall. Soc. AIME, Warrendale, Pa., 1983, p. 65.

146. Bowen, D. K.; Stock, S. R.; Davies, S. T.; Pantos, E.; Birnbaum, H. K.; Chen, H. *Nature,* **1984**, *309*, 336.

147. Bowen, D. K.; Stock, S. R.; Davies, S. T.; Pantos, E.; Birnbaum, H. K. *Daresbury Lab. Rep. DL/SCI/P399E,* Order No. AD-A137 11610; NTIS: *Gov. Rep. Announce. Index (U.S.),* **1984**, *84*, 303.

148. Boyce, J. B.; Baberschke, K. *Solid State Commun.,* **1981**, *39*, 781.

149. Boyce, J. B.; Carter, W. L.; Geballe, T. H.; Claeson, T. *Phys. Scr.,* **1982**, *25*, 749.

150. Boyce, J. B.; Hayes, T. M. in *"Fast Ion Transp. Solids: Electrodes Electrolytes,"* Proc. Int. Conf., Vashishta, P., Mundy, J. N., Shenoy, G. K., ed., Elsevier, N. Holland, N.Y., **1979**, p. 535.

151. Boyce, J. B.; Hayes, T. M. *Top. Curr. Phys.,* **1979**, *15*, 5.

152. Boyce, J. B.; Hayes, T. M. in *"EXAFS Spectroscopy: Techniques and Applications,"* Teo, B. K., Joy, D. C., ed., Plenum, New York, **1981**, p. 103.

153. Boyce, J. B.; Hayes, T. M.; Mikkelsen, J. C., Jr. *Phys. Rev. B,* **1981**, *23*, 2876.

154. Boyce, J. B.; Hayes, T. M.; Mikkelsen, J. C., Jr. *Solid State Ionics,* **1981**, *5*, 497.

155. Boyce, J. B.; Hayes, T. M.; Mikkelsen, J. C., Jr.; Stutius, W. *Solid State Commun.,* **1980**, *33*, 183.

156. Boyce, J. B.; Hayes, T. M.; Stutius, W.; Mikkelsen, J. C., Jr. *Phys. Rev. Lett.,* **1977,** *38,* 1362.

157. Boyce, J. B.; Martin, R. M.; Allen, J. W. *Springer Ser. Chem. Phys.,* **1983,** *27,* 187.

158. Boyce, J. B.; Martin, R. M.; Allen, J. W.; Holtzberg, F. in *"Valence Fluctuations Solids,"* St. Barbara Inst. Theor. Phys. Conf., Falicov, L. M., Hanke, W., Maple, M. B., ed., North-Holland, Amsterdam, Netherlands, **1981,** p. 427.

159. Boyd, R. J. *J. Phys. B,* **1976,** *9,* L69.

160. Braun, W.; Bradshaw, A. M. *Europhys. News,* **1984,** *15,* 11.

161. Braun, W.; Petersen, H.; Feldhaus, J.; Bradshaw, A. M.; Dietz, E.; Haase, J.; McGovern, I. T.; Puschmann, A.; Reimer, A.; et al., *Proc. SPIE-Int. Soc. Opt. Eng.,* **1983,** *447,* 117.

162. Breinig, M.; Chen, M. H.; Ice, G. E.; Parente, F.; Crasemann, B.; Brown, G. S. *Phys. Rev. A,* **1980,** *22,* 520.

163. Bremner, I.; Hasnain, S. S.; Garner, C. D.; Bordas, J. *Daresbury Lab. Rep. DL/SCI/R17,* **1981,** 111.

164. Brennan, S.; Stöhr, J.; Jaeger, R. *Phys. Rev. B,* **1981,** *24,* 4871.

165. Briand, J. P.; Touati, A.; Frilley, M.; Chevallier, P.; Johnson, A.; Rozet, J. P.; Tavernier, M.; Shafroth, S.; Krause, M. O. *J. Phys. B,* **1976,** *9,* 1055.

166. Brillouin, L. *"Science and Information Theory,"* 2nd Ed., Academic, **1962.**

167. Brown, F. C.; Bachrach, R. Z.; Bianconi, A. *Chem. Phys. Lett.,* **1978,** *54,* 425.

168. Brown, G. S. in *"Synchrotron Radiation Research,"* Winick, H., Doniach, S., ed., Plenum, New York, **1980,** p. 387.

169. Brown, G. S.; Doniach, S. in *"Synchrotron Radiation Research,"* Winick, H., Doniach, S., ed., Plenum, New York, **1980,** p. 353.

170. Brown, G. S.; Eisenberger, P.; Schmidt, P. *Solid State Commun.,* **1977,** *24,* 201.

171. Brown, G. S.; Navon, G.; Shulman, R. G. *Proc. Natl. Acad. Sci. U.S.A.,* **1977,** *74,* 1794.

172. Brown, G. S.; Testardi, L. R.; Wernick, J. H.; Hallak, A. B.; Geballe, T. H. *Solid State Commun.,* **1977,** *23,* 875.

173. Brown, J. M.; Powers, L.; Kincaid, B.; Larrabee, J. A.; Spiro, T. G. *J. Am. Chem. Soc.,* **1980,** *102,* 4210.

174. Brown, N. M. D.; McMonagle, J. B.; Greaves, G. N. *J. Chem. Soc., Faraday Trans. 1*, **1984**, *80*, 589.

175. Bruck, M. A.; Korte, H.-J.; Bau, R.; Hadjiliadis, N.; Teo, B. K. *ACS Symp. Series*, **1983**, *209*, 245.

176. Brunschwig, B. S.; Creutz, C.; Macartney, D. H.; Sham, T. K.; Sutin, N. *Faraday Discuss. Chem. Soc.*, **1982**, *74*, 113.

177. Bunker, B. A.; Stern, E. A. *Phys. Rev. B*, **1983**, *27*, 1017.

178. Bunker, G. *Nucl. Instrum. Methods Phys. Res.*, **1983**, *207*, 437.

179. Bunker, G.; Stern, E. A.; Blankenship, R. E.; Parson, W. W. *Biophys. J.*, **1982**, *37*, 539.

180. Burattini, E.; Cappuccio, G.; Dalba, G.; Fornasini, P.; Grandolfo, M.; Vecchia, P.; Efendiev, Sh. M. *Ferroelectrics*, **1984**, *55*, 675.

181. Burattini, E.; Cavallo, N.; Cappuccio, G.; Dalba, G.; Fornasini, P.; Grandolfo, M.; Vecchia, P.; Efendiev, Sh. M. *Springer Ser. Chem. Phys.*, **1983**, *27*, 216.

182. Burattini, E.; Cavallo, N.; Cappuccio, G.; Efendiev, Sh. M.; Grandolfo, M.; Vecchia, P. *Ferroelectrics*, **1982**, *43*, 211.

183. Bylander, D. M.; Kleinman, L. *Phys. Rev. B*, **1984**, *39*, 2997.

184. Calas, G.; Levitz, P.; Petiau, J.; Bondot, P.; Loupias, G. *Rev. Phys. Appl.*, **1980**, *15*, 1161.

185. Calas, G.; Petiau, J. *Struct. Non-Cryst. Mater., Proc. 2nd Int. Conf.*, Gaskell, P. H.; Parker, J. M.; Davis, E. A., ed., Taylor & Francis: London, **1983**, p. 18.

186. Camilloni, R.; Fainelli, E.; Petrocelli, G.; Stefani, G. *Springer Ser. Chem. Phys.*, **1983**, *27*, 174.

187. Cargill, G. S., III *Proc. 4th Int. Conf. Rapidly Quenched Met., Volume 1*, Masumoto, T.; Suzuki, K., ed., *Jpn. Inst. Met*: Sendai, Japan. **1982**, p. 389.

188. Cargill, G. S., III *J. Non-Cryst. Solids*, **1984**, *61-62*, 261.

189. Cargill, G. S., III; Boehme, R. F.; Weber, W. *Phys. Rev. Lett.*, **1983**, *50*, 1391.

190. Cargill, G. S., III; Weber, W.; Boehme, R. F. *Springer Ser. Chem. Phys.*, **1983**, *27*, 277.

191. Carlson, T. A. *Phys. Rev.*, **1963**, *131*, 676.

192. Carlson, T. A.; Nestor, C. W., Jr.; Tucker, T. C.; Malik, T. B. *Phys. Rev.*, **1968**, *169*, 27.

193. Carlson, T. A. *Physics Today,* **1972,** *25,* 30.

194. Carlson, T. A. *"Photoelectron and Auger Spectroscopy,"* Plenum, New York, **1975,** p. 83.

195. Caswell, N.; Solin, S. A.; Hayes, T. M.; Hunter, S. J. *Physica B+C,* **1980,** *99,* 463.

196. Catlow, C. R. A.; Chadwick, A. V.; Greaves, G. N.; Moroney, L. M.; Worboys, M. R. *Solid State Ionics,* **1983,** *9-10,* 1107.

197. Catlow, C. R. A.; Chadwick, A. V.; Greaves, G. N.; Moroney, L. M. *Springer Ser. Chem. Phys.,* **1983,** *27,* 200.

198. Catlow, C. R. A.; Moroney, L. M.; Chadwick, A. V.; Greaves, G. N. *Radiat. Eff.,* **1983,** *75,* 159.

199. Chafekar, V. D.; Joshi, Pankaj; Sen, S. C. *Proc. Nucl. Phys. Solid State Phys. Symp.,* **1980,** *23,* 898.

200. Chan, S. I.; Gamble, R. C. *Methods Enzymol.,* **1978,** *54,* 323.

201. Chan, S. I.; Hu, V. W.; Gamble, R. C. *J. Mol. Struct.,* **1978,** *45,* 239.

202. Chance, B.; Kumar, C.; Powers, L.; Ching, Y. C. *Biophys. J.,* **1983,** *44,* 353.

203. Chance, B.; Fischetti, R.; Powers, L. *Biochemistry,* **1983,** *22,* 3820.

204. Chance, B.; Kumar, C.; Korszun, Z. R.; Legallis, V.; Pennie. W.; Sorge, J.; Khalid, S. *Nucl. Instrum. Methodsds Phys. Res. A,* **1984,** *222,* 180.

205. Chance, B.; Pennie, W.; Carman, M.; Legallais, V.; Powers, L. *Anal. Biochem.,* **1982,** *124,* 248.

206. Chance, B.; Powers, L.; Ching, Y. in *"Mitochondria Microsomes,"* Lee, C. P., Schatz, Gottfried, Dallner, Gustav, ed., **1981,** p. 271.

207. Chandra, S.; Sharma, Y.; Sharma, B. K.; Garg, K. B. *Proc. Nucl. Phys. Solid State Phys. Symp.,* **1983,** *23,* 73.

208. Chappert, J. *J. Phys., Colloq.,* **1980,** *C1,* 9.

209. Chattopadhyay, S.; Sapre, V. B.; Mande, C. *X-Ray Spectrom.,* **1984,** *13,* 153.

210. Chen, F. Y.; Huang, H. W. *Chin. J. Phys. (Taipei),* **1983,** *21,* 44.

211. Chen, H. *J. Phys. Chem. Solids,* **1980,** *41,* 641.

212. Chen, H. S.; Teo, B. K.; Wang, R. *J. Phys., Colloq.,* **1980,** *C8,* 254.

213. Chermashentsev, V. M.; Mazalov, L. N.; Gel'mukhanov, F. Kh. *Zh. Strukt. Khim.,* **1979,** *20,* 209.

214. CHESS Users Manual for EXAFS Measurements.

234

215. Chiarello, G.; Colavita, E.; De Crescenzi, M.; Nannarone, S. *Phys. Rev. B*, **1984**, *29*, 4878.

216. Chiu, N. S.; Bauer, S. H.; Johnson, M. F. L. *J. Catal.*, **1984**, *89*, 226.

217. Cho, Z. H.; Tsai, C. M.; Eriksson, L. A. *IEEE Trans. Nucl. Sci.*, **1975**, *22*, 72.

218. Chougule, B. K.; Patil, R. N. *Indian J. Pure Appl. Phys.*, **1979**, *17*, 38.

219. Citrin, P. H.; Eisenberger, P.; Hewitt, R. C. *J. Vac. Sci. Technol.*, **1977**, *15*, 449.

220. Citrin, P. H.; Eisenberger, P.; Hewitt, R. C. *Phys. Rev. Lett.*, **1978**, *44*, 309.

221. Citrin, P. H.; Eisenberger, P.; Hewitt, R. C. *Phys. Rev. Lett.*, **1980**, *45*, 1948.

222. Citrin, P. H.; Eisenberger, P.; Hewitt, R. C. *Nucl. Instrum. Methods*, **1978**, *152*, 330.

223. Citrin, P. H.; Eisenberger, P.; Hewitt, R. C. *Surf. Sci.*, **1979a**, *89*, 28.

224. Citrin, P. H.; Eisenberger, P.; Kincaid, B. M. *Phys. Rev. Lett.*, **1976**, *36*, 1346.

225. Citrin, P. H.; Hamman, D. R.; Mattheiss, L. F.; Rowe, J. E. *Phys. Rev. Lett.*, **1983**, *50*, 1824.

226. Citrin, P. H.; Rowe, J. E. *Surf. Sci.*, **1983**, *132*, 205.

227. Citrin, P. H.; Rowe, J. E.; Eisenberger, P. *Phys. Rev. B*, **1983**, *28*, 2299.

228. Citrin, P. H.; Rowe, J. E.; Eisenberger, P.; Comin, F. *Physica B+C*, **1983**, *117-118*, 786.

229. Claeson, T.; Boyce, J. B. *Phys. Rev. B*, **1984**, *29*, 1551.

230. Claeson, T.; Boyce, J. B.; Lowe, W. P.; Geballe, T. H. *Phys. Rev. B*, **1984**, *29*, 4969.

231. Claeson, T.; Boyce, J. B.; Geballe, T. H. *Phys. Rev. B*, **1982**, *25*, 6666.

232. Clausen, B. S.; Lengeler, B.; Candia, R.; Als-Nielsen, J.; Topsoee, H. *Bull. Soc. Chim. Belg.*, **1981**, *90*, 1249.

233. Clausen, B. S.; Topsoee, H.; Candia, R.; Lengeler, B. *ACS Symp. Ser.*, **1984**, *248*, 71.

234. Clausen, B. S.; Topsoe, H.; Candia, R.; Villadsen, J.; Lengeler, B.; Als-Nielsen, J.; Christensen, F. *J. Phys. Chem.*, **1981**, *85*, 3868.

235. Clausen, B. S.; Topsoe, H.; Candia, R.; Villadsen, J.; Lengeler, B.; Als-Nielsen, J.; Christensen, F. *Report DESY-SR-81/02, Atomindex*, **1982**, 13, No. 658543.

236. Clementi, E.; Roetti, C. *At. Data Nucl. Data Tables*, **1974**, *14*, 177.

237. Clout, P. N.; Ridley, P. A. *Nucl. Instrum. Methods*, **1978**, *152*, 145.

238. Co, M. S. *Report SSRL-84/02*, Order No. DE84010930; NTIS: *Energy Res. Abstr.*, **1983**, *9*, No. 27102.

239. Co, M. S.; Hendrickson, W. A.; Hodgson, K. O.; Doniach, S. *J. Am. Chem. Soc.*, **1983**, *105*, 1144.

240. Co, M. S.; Hendrickson, W. A.; Hodgson, K. O.; Doniach, S. *J. Am. Chem. Soc.*, **1983**, *105*, 1144.

241. Co, M. S.; Hodgson, K. O. *J. Am. Chem. Soc.*, **1981**, *103*, 3200.

242. Co, M. S.; Hodgson, K. O. in *"Front. Biochem. Stud. Proteins Membr.,"* *Proc. Int. Conf.*, Liu, T.-Y., ed., Elsevier, New York, p. 351.

243. Co, M. S.; Hodgson, K. O. in *"Copper Proteins Copper Enzymes,"* Vol. *1*, Lontie, R., ed., CRC, Boca Raton, Fla., **1984**, *1*, p. 93.

244. Co, M. S.; Hodgson, K. O.; Eccles, T. K.; Lontie, R. *J. Am. Chem. Soc.*, **1981**, *103*, 984.

245. Co, M. S.; Scott, R. A.; Hodgson, K. O. *J. Am. Chem. Soc.*, **1981**, *103*, 986.

246. Cocco, G.; Enzo, S.; Fagherazzi, G.; Schiffini, L.; Bassi, I. W.; Vlaic, G.; Galvagno, S.; Parravano, G. *J. Phys. Chem.*, **1979**, *83*, 2527.

247. Cocco, G.; Enzo, S.; Incoccia, L.; Mobilio, S. *Z. Naturforsch., A: Physics. Chem., Kosmophys.*, **1983**, *38*, 1391.

248. Codling, K.; Houlgate, R. G.; West, J. B.; Woodruff, P. R. *J. Phys. B*, **1976**, *9*, L83.

249. Cohen, G. G.; Deslattes, R. D. *Nucl. Instrum. Methods Phys. Res.*, **1982**, *193*, 33.

250. Cohen, G. G.; Fischer, D. A.; Colbert, J.; Shevchik, N. J. *Rev. Sci. Instrum.*, **1980**, *51*, 273.

251. Cohen, P. I.; Einstein, T. L.; Elam, W. T.; Fukuda, Y.; Park, R. L. *Appl. Surf. Sci.*, **1978**, *1*, 538.

252. Cohen, R. L.; Feldman, L. C.; West, K. W.; Kincaid, B. M. *Phys. Rev. Lett.*, **1982**, *49*, 1416.

253. Colavita, E.; De Crescenzi, M.; Papagno, L.; Caputi, L. S.; Chiarello, G.; Scarmozzino, R.; Rosei, R. *Solid State Commun.*, **1982**, *41*, 545.

254. Colliex, C.; Jouffrey, B. *Philos. Mag.*, **1972**, *25*, 191.

255. Colosimo, A.; Brunori, M.; Andreasi, F.; Mobilio, S. *J. Inorg. Biochem.*, **1981**, *15*, 179.

256. Comin, F.; Incoccia, L.; Mobilio, S. *J. Phys. E*, **1983**, *16*, 83.

257. Comin, F.; Incoccia, L.; Mobilio, S.; Motta, N. *Springer Ser. Chem. Phys.*, **1983**, *27*, 292.

258. Comin, F.; Rowe, J. E.; Citrin, P. H. *Proc. SPIE-Int. Soc. Opt. Eng.*, **1983**, *447*, 107.

259. Comin, F.; Rowe, J. E.; Citrin, P. H. *Phys. Rev. Lett.*, **1983**, *51*, 2402.

260. Conradson, S. D. *Report SSRL-83/04*, Order No. DE84012741; NTIS: *Energy Res. Abstr.*, **1984**, *9, No. 29346.*

261. Cook, J. W., Jr.; Sayers, D. E. *J. Appl. Phys.*, **1981**, *52*, 5024.

262. Cook, S. L.; Evans, J.; Greaves, G. N. *J. Chem. Soc., Chem. Commun.*, **1983**, 1287.

263. Cook, S. L.; Evans, J.; Greaves, G. N.; Johnson, B. F. G.; Lewis, J.; Raithby, P. R.; Wells, P. B.; Worthington, P. *J. Chem. Soc., Chem. Commun.*, **1983**, 777.

264. Cooper, M. J.; Sakata, M. *Acta Crystallogr. A*, **1979**, *35*, 989.

265. Cordts, B.; Pease, D.; Azaroff, L. V. *Phys. Rev. B*, **1981**, *24*, 538.

266. Cotton, F. A.; Meyers, M. D. *J. Am. Chem. Soc.*, **1960**, *82*, 5023.

267. Cox, A. D. in *"Charact. Catal."*, Thomas, J. M., Lambert, R. M., ed., Wiley, Chichester, United Kingdom, **1980**, p. 254.

268. Cox, A. D. *Daresbury Lab Rep. DL/SCI/R17*, **1981**, 51.

269. Cox, A. D.; Beaumont, J. H. *Philos. Mag. B*, **1980**, *42*, 115.

270. Cox, A. D.; McMillan, P. W. *J. Non-Cryst. Solids*, **1981**, *44*, 257.

271. Craievich, A.; Dartige, E.; Fontaine, A.; Raoux, D. *Springer Ser. Chem. Phys.*, **1983**, *27*, 274.

272. Cramer, S. P. *Daresbury Lab Rep. DL/SCI/R17*, **1981**, 47.

273. Cramer, S. P. *NATO Adv. Study Inst. Ser. A*, **1979**, *25*, 291.

274. Cramer, S. P. *Adv. Inorg. Bioinorg. Mech.*, **1983**, *2*, 259.

275. Cramer, S. P.; Dawson, J. H.; Hodgson, K. O.; Hager, L. P. *J. Am. Chem. Soc.*, **1978**, *100*, 7282.

276. Cramer, S. P.; Eccles, T. K.; Kutzler, F.; Hodgson, K. O.; Doniach, S. *J. Am. Chem. Soc.*, **1976**, *98*, 8059.

277. Cramer, S. P.; Eccles, T. K.; Kutzler, F. W.; Hodgson, K. O.; Mortenson, L. E. *J. Am. Chem. Soc.*, **1976**, *98*, 1287.

278. Cramer, S. P.; Eidem, P. K.; Paffett, M. T.; Winkler, Z. D.; Dori, Z.; Gray, H. B. *J. Am. Chem. Soc.*, **1983**, *105*, 799.

279. Cramer, S. P.; Eidem, P. K.; Paffett, M. T.; Winkler, J. R.; Dori, Z.; Gray, H. B. *J. Am. Chem. Soc.,* **1983**, *105,* 799.

280. Cramer, S. P.; Gillum, W. O.; Hodgson, K. O.; Mortenson, L. E.; Stiefel, E. I.; Chisnell, J. R.; Brill, W. J.; Shah, V. K. *J. Am. Chem. Soc.,* **1978**, *100,* 3814.

281. Cramer, S. P.; Gray, H. B.; Dori, Z.; Bino, A. *J. Am. Chem. Soc.,* **1979**, *101,* 2770.

282. Cramer, S. P.; Gray, H. B.; Rajagopalan, K. V. *J. Am. Chem. Soc.,* **1979**, *101,* 2772.

283. Cramer, S. P.; Gray, H. B.; Scott, N. S.; Barber, M.; Rajagopalan, K. V. in *"Molybdenum Chem. Biol. Significance,"* Newton, W. E., Otsuka, S., ed., Plenum, New York, **1980**, p. 157.

284. Cramer, S. P.; Hodgson, K. O. *Prog. Inorg. Chem.,* **1979**, *25,* 1.

285. Cramer, S. P.; Hodgson, K. O.; Gillum, W. O.; Mortenson, L. E. *J. Am. Chem. Soc.,* **1978**, *100,* 3398.

286. Cramer, S. P.; Hodgson, K. O.; Stiefel, E. I.; Newton, W. E. *J. Am. Chem. Soc.,* **1978**, *100,* 2748.

287. Cramer, S. P.; Liang, K. S.; Jacobson, A. J.; Chang, C. H.; Chianelli, R. R. *Inorg. Chem.,* **1984**, *23,* 1215.

288. Cramer, S. P.; Moura, J. J. G.; Xavier, A. V.; LeGall, J. *J. Inorg. Biochem.,* **1984**, *20,* 275.

289. Cramer, S. P.; Scott, R. A. *Rev. Sci. Instrum.,* **1981**, *52,* 395.

290. Cramer, S. P.; Solomonson, L. P.; Adams, M. W. W.; Mortenson, L. E. *J. Am. Chem. Soc.,* **1984**, *106,* 1467.

291. Cramer, S. P.; Wahl, R.; Rajagopalan, K. V. *J. Am. Chem. Soc.,* **1981**, 103, 7721.

292. Crespin, M.; Levitz, P.; Gatineau, L. *Springer Ser. Chem. Phys.,* **1983**, *27,* 228.

293. Crozier, E. D. in *"EXAFS Spectroscopy: Techniques and Applications,"* Teo, B. K., Joy, D. C., ed., Plenum, New York, **1981**, p. 89.

294. Crozier, E. D.; Alberding, N.; Sundheim, B. R. *J. Chem. Phys.,* **1983**, *79,* 939.

295. Crozier, E. D.; Lytle, F. W.; Sayers, D. E.; Stern, E. A. *Can. J. Chem.,* **1977**, *55,* 1968.

296. Crozier, E. D.; Seary, A. J. *Can. J. Phys.,* **1980**, *58,* 1388.

297. Crozier, E. D.; Seary, A. J. *Can. J. Phys.,* **1981**, *59,* 876.

238

298. Csillag, S.; Johnson, D. E.; Stern, E. A. in *"EXAFS Spectroscopy: Techniques and Applications,"* Teo, B. K., Joy, D. C., ed., Plenum, New York, **1981**, p. 241.

299. Cyrin, S. J. *"Molecular Vibrations and Mean Square Amplitudes,"* Elsevier, Amsterdam, **1968**, p. 77.

300. Dalba, G.; Ferrari, F.; Fornasini, P.; Burattini, E.; Cavallo, N.; Foresti, M.; Mencuccini, C.; Pancini, E.; Patteri, P.; Rinzivillo, R. *Daresbury Lab. Rep. DL/SCI/R17,* **1981**, 104.

301. Dalba, G.; Fontana, A.; Fornasini, O.; Mariotto, G.; Rocca, F.; Bernieri, E.; Masullo, M. R.; Morone, A. *Springer Ser. Chem. Phys.,* **1983**, *27*, 290.

302. Dalba, G.; Fornasini, P.; Burattini, E. *J. Phys. C,* **1983**, *16*, L165.

303. Darshan, B.; Padalia, B. D.; Nagrajan, R.; Sampathkumaran, E. V.; Gupta, L. C.; Vijayaraghvan, R. *Proc. Nucl. Phys. Solid State Phys. Symp.,* **1983**, *23*, 562.

304. Dartyge, E.; Flank, A. M.; Fontaine, A.; Jucha, A. *J. Phys. Colloq.,* **1984**, *C2*, 275.

305. Dartyge, E.; Fontaine, A. *J. Phys. F,* **1984**, *14*, 721.

306. Dartyge, E.; Fontaine, A.; Mimault, J. *Springer Ser. Chem. Phys.,* **1983**, *27*, 80.

307. Davies, B. M.; Brown, F. C. *Phys. Rev. B,* **1982**, *25*, 2997.

308. Davis, L. A.; Mac Donald, N. C.; Palmberg, P. W.; Riach, G. E.; Weber, R. E., ed., *"Handbook for Auger Electron Spectroscopy,"* 2nd Ed., Phys. Elect. Ind., Eden Prairie, Minn., **1976**.

309. Davoli, I.; Palladino, L.; Stizza, S.; Bianconi, A. *Solid State Commun.,* **1982**, *44*, 1585.

310. Davoli, I.; Stizza, S.; Benfatto, M.; Gzowski, O.; Murawski, L.; Bianconi, A. *Springer Ser. Chem. Phys.,* **1983**, *27*, 161.

311. Dawson, J. H.; Andersson, L. A.; Davis, I. M.; Hahn, J. E. *Dev. Biochem.,* **1980**, *13*, 565.

312. Dawson, J. H.; Andersson, L. A.; Hodgson, K. O.; Hahn, J. E. *Dev. Biochem.,* **1982**, *23*, 589.

313. deBoor, C. *J. Approx. Th.,* **1968**, *1*, 219.

314. deBoor, C. *J. Approx. Th.,* **1972**, *6*, 50.

315. deBoor, C. *"A Practical Guide to Splines,"* Springer, New York, **1978**.

316. De Grescenzi, M. *Springer Ser. Chem. Phys.,* **1983**, *27*, 382.

317. De Crescenzi, M.; Balzarotti, A.; Comin, F.; Incoccia, L.; Mobilio, S.; Bacci, D. *J. Phys., Colloq.*, **1980**, *C8*, 238.

318. De Crescenzi, M.; Balzarotti, A.; Comin, F.; Incoccia, L.; Mobilio, S.; Motta, N. *Daresbury Lab. Rep. DL/SCI/R17*, **1981a**, 119.

319. De Crescenzi, M.; Balzarotti, A.; Comin, F.; Incoccia, L.; Mobilio, S.; Motta, N. *Solid State Commun.*, **1981b**, *37*, 921.

320. De Crescenzi, M.; Chiarello, G.; Colavita, E.; Memeo, R. *Phys. Rev. B*, **1984**, *29*, 3730.

321. Defrain, A.; Bosio, L.; Cortes, R.; Gomes da Costa, P. *J. Non-Cryst. Solids*, **1984**, *61-62*, 439.

322. Dehmer, J. L.; Dill, D. in *Proc. 2nd Int. Conf. Inn. Shell Ioniz. Phenom.*, Mehlhorn, W., Brenn, R., ed., **1976**, p. 221.

323. Del Cueto, J. A.; Shevchik, N. J. *J. Phys. F*, **1977**, *7*, L215.

324. Del Cueto, J. A.; Shevchik, N. J. *J. Phys. C*, **1978a**, *11*, L829.

325. Del Cueto, J. A.; Shevchik, N. J. *J. Phys. C*, **1978b**, *11*, L833.

326. Del Cueto, J. A.; Shevchik, N. J. *J. Phys. E*, **1978c**, *11*, 616.

327. den Boer, M. L.; Cohen, P. I.; Park, R. L. *J. Vac. Sci. Tech.*, **1978**, *15*, 502.

328. Einstein, T. L.; Mehl, M. J.; Morar, J. F.; Park, R. L.; Laramore, G. E. *Springer Ser. Chem. Phys.*, **1983**, *27*, 391.

329. den Boer, M. L.; Einstein, T. L.; Elam, W. T.; Park, R. L.; Roelofs, L. D.; Laramore, G. E. *Phys. Rev. Lett.*, **1980**, *44*, 496.

330. Denley, D.; Perfetti, P.; Williams, R. S.; Shirley, D. A.; Stöhr, J. *Phys. Rev. B*, **1980**, *21*, 2267.

331. Denley, D. R.; Raymond, R. H.; Tang, S. C. *Springer Ser. Chem. Phys.*, **1983**, *27*, 325.

332. Denley, D. R.; Raymond, R. H.; Tang, S. C. *J. Catal.*, **1984**, *87*, 414.

333. Denley, D.; Williams, R. S.; Perfetti, P.; Shirley, D. A.; Stöhr, J. *Phys. Rev. B*, **1979**, *19*, 1762.

334. Deshmukh, P.; Deshmukh, P.; Mande, C. *J. Phys. C*, **1981**, *14*, 531.

335. Desideri, A.; Comin, F.; Morpurgo, L.; Cocco, D.; Calabrese, L.; Mondovi, B.; Maret, W.; Rotilio, G. *Biochim. Biophys. Acta*, **1981**, *670*, 312.

336. Desideri, A.; Rotilio, G. *Springer Ser. Chem. Phys.*, **1983**, *27*, 342.

337. Dev., B. N.; Mishra, K. C.; Gibson, W. M.; Das, T. P. *Phys. Rev. B*, **1984**, *29*, 1101.

338. Dexpert, H.; Lagarde, P.; Bournonville, J. P. *J. Mol. Catal.*, **1984**, *25*, 347.

339. Diamond, H.; Pan, H. K.; Knapp, G. S.; Horwitz, E. P. *Solvent Extr. Ion Exch.*, **1983**, *1*, 515.

340. Dill, D., Dehmer, J. L. *J. Chem. Phys.*, **1974**, *61*, 692.

341. Ding, Y. S.; Yarusso, D. J.; Pan, H. K. D.; Cooper, S. L. *J. Appl. Phys.*, **1984**, *56*, 2396.

342. Doebler, U.; Baberschke, K.; Haase, J.; Puschmann, A. *Phys. Rev. Lett.*, **1984**, *52*, 1437.

343. Donato, E.; Giuliano, E. S.; Ruggeri, R.; Ginatempo, B.; Stancanelli, A. *Nuovo Cimento D*, **1982**, *1*, 351.

344. Doniach, S.; Eisenberger, P.; Hodgson, K. O. in "Synchrotron Radiation Research," Winick, H., Doniach, S., ed., Plenum, New York, **1980**, p. 425.

345. Dreier, P.; Rabe, P.; Malzfeldt, W.; Nieman, W. *Springer Ser. Chem. Phys.*, **1983**, *27*, 378.

346. Drezdzom, M. A.; Tessier-Youngs, C.; Woodcock, C.; Blonsky, P. M.; Leal, O.; Teo, B. K.; Burwell, R. L., Jr.; Shriver, D. F. *Inorg. Chem.*, **1985**, *24*, 2349.

347. Dubois, J. M.; Goulon, J.; Le Caer, G.; Lagarde, P. *Springer Ser. Chem. Phys.*, **1983**, *27*, 284.

348. Dubois, J. M.; Le Caer, G.; Chieux, P.; Goulon, J. *Nucl. Instrum. Methods Phys. Res.*, **1982**, *199*, 315.

349. Dukhnyakov, A. Yu.; Vinogradov, A. S. *Opt. Spektrosk.*, **1982**, *53*, 841.

350. Durham, P. J. *Springer Ser. Chem. Phys.*, **1983**, *27*, 37.

351. Durham, P. J.; Pendry, J. B.; Hodges, C. H. *Solid State Commun.*, **1981**, *38*, 159.

352. Durham, P. J.; Pendry, J. B.; Hodges, C. H. *Comput. Phys. Commun.*, **1982**, *25*, 193.

353. Durham, P. J.; Pendry, J. B.; Norman, D. *Daresbury Lab. Rep. DL/SCI/R17*, **1981**, 108.

354. Durham, P. J.; Pendry, J. B.; Norman, D. *J. Vac. Sci. Technol.*, **1982**, *20*, 665.

355. Dutta, C. M.; Huang, H. W. *Phys. Rev. Lett.*, **1980**, *44*, 643.

356. Dyson, N. A. "*X-rays in Atomic and Nuclear Physics*," Longman Group, London, **1973**, p. 69.

357. Eanes, E. D.; Costa, J. L.; MacKenzie, A.; Warburton, W. K. *Rev. Sci. Instrum.*, **1980**, *51*, 1579.

358. Eanes, E. D.; Powers, L.; Costa, J. L. *Cell Calcium*, **1981**, *2*, 251.

359. Eason, R. W.; Bradley, D. K.; Kilkenny, J. D.; Greaves, G. N. *J. Phys. C*, **1984**, *17*, 5067.

360. Eastman, D. E.; Freeouf, J. L. *Phys. Rev. Lett.*, **1974**, *33*, 1601.

361. Eberle, H. G.; Schnuerer, E. *Wiss. Ber. - Akad. Wiss. D. D. R., Zentralinst. Festkoerperphys. Werkstofforsch*, **1980**, *20*, 42.

362. Einstein, T. L. *Appl. Surf. Sci.*, **1982**, *11-12*, 42.

363. Einstein, T. L.; denBoer, M. L.; Morar, J. F.; Park, R. L. *J. Vac. Sci. Technol.*, **1981**, *18*, 490.

364. Einstein, T. L.; Mehl, M. J.; Morar, J. F.; Park, R. L.; Laramore, G. E. *Springer Ser. Chem. Phys.*, **1983**, *27*, 391.

365. Eisenberger, P. *Daresbury Lab. Rep. DL/SCI/R17*, **1981a**, 1.

366. Eisenberger, P. *Hyperfine Interact.*, **1981b**, *10*, 915.

367. Eisenberger, P.; Brown, G. S. *Solid State Commun.*, **1979**, *29*, 481.

368. Eisenberger, P.; Citrin, P.; Hewitt, R.; Kincaid, B. *CRC Crit. Rev. Solid State Mater. Sci.*, **1981**, *10*, 191.

369. Eisenberger, P. M.; Kincaid, B. M. *Chem. Phys. Lett.*, **1975**, *36*, 134.

370. Eisenberger, P.; Kincaid, B. M. *Science*, **1978**, *200*, 1441.

371. Eisenberger, P.; Kincaid, B. M.; Shulman, R. G. in "*Front. Biol. Energ.,*" *Vol. 1*, Dutton, P. L., Leigh, J. S., Scarpa, A., ed., Academic, New York, **1978**, p. 652.

372. Eisenberger, P.; Lengeler, B. *Phys. Rev. B*, **1980**, *22*, 3551.

373. Eisenberger, P.; Okamura, M. Y.; Feher, G. *Biophys. J.*, **1982**, *37*, 523.

374. Eisenberger, P.; Shulman, R. G.; Brown, G. S.; Ogawa, S. *Proc. Natl. Acad. Sci., U.S.A.*, **1976**, *73*, 491.

375. Eisenberger, P.; Shulman, R. G.; Kincaid, B. M.; Brown, G. S.; Ogawa, S. *Nature* (London), **1978**, *274*, 30.

376. Ekardt, W.; Thoai, D. B. *Solid State Commun.*, **1981**, *40*, 939.

377. Ekardt, W.; Thoai, D. B. *Solid State Commun.*, **1983**, *45*, 1083.

378. Elam, W. T.; Cohen, P. I.; Roelofs, L.; Park, R. L. *Appl. Surf. Sci.*, **1979**, *2*, 636.

379. Elam, W. T.; Stern, E. A.; McCallum, J. D.; Sanders-Loehr, J. *J. Am. Chem. Soc.*, **1982**, *104*, 6369.

380. Elam, W. T.; Stern, E. A.; McCallum, J. D.; Sanders-Loehr, J. *J. Am. Chem. Soc.*, **1983**, *105*, 1919.

381. Elder, R. C.; Eidsness, M. K.; Heeg, M. J.; Tepperman, K. G.; Shaw, C. F., III; Schaeffer, N. *ACS Symp. Ser.*, **1983**, *209*, 385.

382. El-Mashri, S. M.; Forty, A. J.; Jones, R. G., *Scan. Elect. Microsc.*, **1983**, *2*, 569.

383. El-Mashri, S. M.; Jones, R. G.; Forty, A. J. *Philos. Mag. A.*, **1983**, *48*, 665.

384. Emrich, R. J.; Katzer, J. R. in *"Laboratory EXAFS Facilities-1980,"* AIP Conf. Proc., **1980**, *64*, 131.

385. Enderby, J. E. *Can. J. Chem.*, **1977**, *55*, 1961.

386. Enderby, J. E.; Biggin, S. *Adv. Molten Salt Chem.*, **1983**, *5*, 1.

387. Engel, T.; Rieder, K. H. *Springer Tracts in Mod. Phys.*, **1982**, *91*, 55.

388. Epstein, H. M.; Schwerzel, R. E.; Mallozzi, P. J.; Campbell, B. E. *J. Am. Chem. Soc.*, **1983**, *105*, 1466.

389. Eriksson, L. A.; Tsai, C. M.; Cho, A. H.; Hurlbut, C. R. *Nucl. Instrum. Methodsds*, **1974**, *122*, 373.

390. Ershov, N. V.; Ageev, A. L.; Vasin, V. V.; Babanov, Yu. A. *Phys. Status Solidi B*, **1981**, *108*, 103.

391. Ershov, N. V.; Babanov, Y. A.; GValakhov, V. R. *Phys. Status Solidi B*, **1983**, *117*, 749.

392. Esteva, J. M.; Karnatak, R. C. *J. Phys. Colloq.*, **1984**, C2, 279.

393. Evangelisti, F.; Proietti, M. G.; Balzarotti, A.; Comin, F.; Incoccia, L.; Mobilio, S. *Solid State Commun.*, **1981**, *37*, 413.

394. Fang, S. S.; Chen, H. *J. Phys. Chem. Solids*, **1983**, *44*, 521.

395. Fang, S. S.; Chen, H. *J. Phys. Chem. Solid.*, **1983**, *44*, 521.

396. Fano, U.; Cooper J. W. *Rev. Mod. Phys.*, **1968**, *40*, 441.

397. Feibelman, P. J.; Knotek, M. L. *Phys. Rev. B*, **1978**, *18*, 6531.

398. Feldman, J. L.; Skelton, E. F.; Ehrlich, A. C.; Dominguez, D. D.; Elam, W. T.; Qudri, S. B.; Lytle, F. W. *Ext. Abstr. Program — Bienn. Conf. Carbon, 16th*, **1983**, 209.

399. Felton, R. H.; Barrow, W. L.; May, S. W.; Sowell, A. L.; Goel, S.; Bunker, G.; Stern, E. A. *J. Am. Chem. Soc.*, **1982**, *104*, 6132.

400. Figgis, B. N. *"Introduction to Ligand Fields,"* John Wiley & Sons, London, **1966**, p. 203.

401. Fink, J.; Mueller-Heinzerling, T.; Pflueger, J.; Bubenzer, A.; Koidl, P.; Crecelius, G. *Solid State Commun.*, **1983**, *47*, 687.

402. Fischer, D. A.; Cohen G. G.; Shevchik, N. J. *J. Phys. F*, **1980**, *10*, L139.

403. Fichtner-Schmittler, H. *Cryst. Res. Technol.*, **1984**, *19*, 1225.

404. Flank, A. M.; Fontaine, A.; Jucha, A.; Lemonnier, M.; Raoux, D.; Williams, C. *Nucl. Instrum. Methods Phys. Res.*, **1983**, *208*, 651.

405. Flank, A. M.; Fontaine, A.; Jucha, A.; Lemonnier, M.; Williams, C. *J. Phys. Lett.*, **1982**, *43*, 315.

406. Flank, A. M.; Fontaine, A.; Jucha, A.; Lemonnier, M.; Williams, C. *Springer Ser. Chem. Phys.*, **1983**, *27*, 405.

407. Flank, A. M.; Fontaine, A.; Lagarde, P.; Lemonnier, M.; Mimault, J.; Raoux, D.; Sadoc, A. *Daresbury Lab. Rep. DL/SCI/R17*, **1981**, 70.

408. Flank, A. M.; Lagarde, P.; Raoux, D.; Rivory, J.; Sadoc, A. *Proc. 4th Int. Conf. Rapidly Quenched Met.*, **1981**, *1*, 393.

409. Flank, A. M.; Raoux, D.; Naudon, A.; Sadoc, S. F. *J. Non-Cryst. Solids*, **1984**, *61-62*, 445.

410. Fleet, M. E.; Herzberg, C. T.; Henderson, G. S.; Crozier, E. D.; Osborne, M. D.; Scarfe, C. M. *Geochim. Cosmochim. Acta*, **1984**, *48*, 1455.

411. Fontaine, A.; Lagarde, P.; Naudon, A.; Raoux, D.; Spnjaard, D. *Philos. Mag. B*, **1979**, *40*, 17.

412. Fontaine, A.; Lagarde, P.; Raoux, D.; Esteva, J. M. *J. Phys. F*, **1979**, *9*, 2143.

413. Fontaine, A.; Lagarde, P.; Raoux, D.; Fontana, M. P.; Maisano, G.; Migliardo, P.; Wanderlingh, F. *Phys. Rev. Lett.*, **1978**, *41*, 504.

414. Fontana, M. P.; Lottici, P. P.; Razzetti, C.; Bianchi, D.; Antonioli, G.; Emiliani, U. *Solid State Commun.*, **1982**, *43*, 561.

415. Fontana, M. P.; Maisano, G.; Migliardo, P.; Wanderlingh, F. *J. Chem. Phys.*, **1978**, *69*, 676.

416. Fox, R.; Gurman, S. J. *J. Phys. C*, **1980**, *13*, L249.

417. Fox, R.; Gurman, S. J. *Phys. Chem. Glasses*, **1981** 22, 32.

418. Fox, P. A.; Hall, A. D.; Schryer, N. L. *"The PORT Library Mathematical Subroutine Library," Bell Laboratories Computing Science Technical Report*, **1976**, No. 47.

419. Frahm, R.; Haensel, R.; Rabe, P. *Springer Ser. Chem. Phys.*, **1983**, *27*, 107.

420. Frahm, R.; Haensel, R.; Rabe, P. *J. Phys. F*, **1984**, *14*, 1029.

244

421. Frahm, R.; Haensel, R.; Rabe, P. *J. Phys. F*, **1984**, *14*, 1333.

422. Franchy, R.; Menzel, D. *Phys. Rev. Lett.*, **1979**, *43*, 865.

423. Frank, K. H.; Kaindl, G.; Feldhaus, J.; Wortmann, G.; Krone, W.; Materlik, G.; Bach, H. in *"Valence Instab.,"* Wachter, P.; Boppart, H., ed., North-Holland: Amsterdam, **1982**, p. 189.

424. Fujikawa, T. *J. Electron Spectrosc. Relat. Phenom.*, **1981**, *22*, 353.

425. Fujikawa, T. *J. Electron Spectrosc. Relat. Phenom.*, **1982a**, *26*, 79.

426. Fujikawa, T. *J. Phys. Soc. Jpn.*, **1982b**, *51*, 2619.

427. Fukamachi, T.; Hosoya, S. *Acta Cryst. Allogr. A*, **1975**, *31*, 215.

428. Fukamachi, T.; Kawamura, T. *Bunseki*, **1981**, *4*, 221.

429. Fukushima, T.; Katzer, J. R.; Sayers, D. E.; Cook, J. *Stud. Surf. Sci. Catal.*, **1981**, *7*, 79.

430. Fussa, O.; Kauzlarich, S.; Dye, J. L.; Teo, B. K. *J. Am Chem Soc.*, **1985**, *107*, 3727.

431. Gallezot, P.; Weber, R.; Dalla Betta, R. A.; Boudart, M. *Z. Naturforsch A*, **1979**, *34*, 40.

432. Galli, G.; Maisano, G.; Migliardo, P.; Vasi, C.; Wanderlingh, F.; Fontana, M. P. *Solid State Commun.*, **1982**, *42*, 213.

433. Gamble, R. C. in *"Laboratory EXAFS Facilities-1980,"* AIP Conf. Proc., **1980**, *64*, 123.

434. Garcia-Iniguez, L.; Powers, L; Chance, B,; Sellin, S.; Mannervik, B.; Mildvan, A. S. *Biochemistry*, **1984**, *23*, 685.

435. Garner, C. D.; Hasnain, S. S.; Ed. *"EXAFS (Extended X-Ray Absorption Fine Structure) for Inorganic Systems,"* Daresbury Lab. Rep. *DL/SCI/R17*, **1981b**.

436. Garner, C. D.; Hasain, S. S.; Bremner, I.; Bordas, J. *J. Inorg. Biochem.*, **1982**, *16*, 253.

437. Gaskell, P. H. *J. Phys. C*, **1979**, *12*, 4337.

438. Gaskell, P. H. *Daresbury Lab. Rep. DL/SCI/R19*, **1983**, 28.

439. Gaskell, P. H.; Glover, D. M.; Livesey, A. K.; Durham, P. J.; Greaves, G. N. *J. Phys. C*, **1982**, *15*, L597.

440. Gaskell, P. H.; Glover, D. M.; Livesey, A. K.; Durham, P. J.; Greaves, G. N. *Springer Ser. Chem. Phys.*, **1983**, *27*, 157.

441. Gaskell, P. H.; Glover, D. M.; Livesey, A. K.; Durham, P. J.; Greaves, G. N. in *"Struct. Non-Cryst. Mater. 1982,"* Gaskell, P. H.; Parker, J. M.; Davis, E. A., ed., Taylor & Francis, London, **1983**, p. 29.

442. Gaspar, R. *Acta Phys. Hung.,* **1954**, *3*, 263.

443. Gavrila, M.; Hansen, J. E. *J. Phys. B,* **1978**, *11*, 1353.

444. Geere, R. G.; Gaskell, P. H.; Greaves, G. N.; Greengrass, J.; Binstead, N. *Springer Ser. Chem. Phys.,* **1983**, *27*, 256.

445. Georgopoulos, P.; Knapp, G. S. *J. Appl. Crystallogr.,* **1981**, *14*, 3.

446. Ghatikar, M. N. *Phys. Status Solidi B,* **1983**, *120*, 445.

447. Ghatikar, M. N.; Hatwar, T. K.; Padalia, B. D.; Sampathkumaran, E. V.; Gupta, L. C.; Vijayaraghavan, R. *Phys. Status Solidi B,* **1981**, *106*, K89.

448. Ghatikar, M. N.; Padalia, B. D. *J. Phys. C,* **1978**, *11*, 1941.

449. Ghosh, D.; Furey, W.; O'Donnell, S.; Stout, D. *J. of Biol. Chem.,* **1981**, *256*, 4185.

450. Godart, C.; Achard, J. C.; Krill, G.; Ravet-Krill, M. F. *J. Less-Common Met.,* **1983**, *94*, 177.

451. Godart, C.; Gupta, L. C.; Ravet-Krill, M. F. *J. Less-Common Met.,* **1983**, *94*, 187.

452. Gohshi, Y.; Fukushima, T. *Shokubai,* **1980**, *22*, 396.

453. Goldman, A. I.; Canova, E.; Kao, Y. H.; Fitzpatrick, B. J.; Bhargava, R. N.; Phillips, J. C. *Appl. Phys. Lett.,* **1983**, *43*, 836.

454. Goulding, F. S.; Jaklevic, J. M.; Thompson, A. C. *Lawrence Berkeley Lab. Report, LBL-7542, CONF-780497-2, Energy Res. Abstr.,* **1979**, *4*, No. 27635.

455. Goulon, J.; Friant, P.; Goulon-Ginet, C.; Coutsolelos, A.; Fuilard, R. *Chem. Phys.,* **1984**, *83*, 367.

456. Goulon, J.; Friant, P.; Poncet, J. L.; Guilard, R.; Fischer, J.; Ricard, L. *Springer Ser. Chem. Phys.,* **1983**, *27*, 100.

457. Goulon, J.; Goulon-Ginet, C. *Pure Appl. Chem.,* **1982**, *54*, 2307.

458. Goulon, J.; Goulon-Ginet, C.; Chabanel, M. *J. Solution Chem.,* **1981**, *10*, 649.

459. Goulon, J.; Georges, E.; Goulon-Ginet, C.; Chauvin, Y.; Commereuc, D.; Dexpert, H.; Freund, E. *Chem. Phys.,* **1984**, *83*, 357.

460. Goulon, J.; Goulon-Ginet, C.; Cortes, R.; Dubois, J. M. *J. Phys.,* **1982**, *43*, 539.

461. Goulon, J.; Goulon-Ginet, C.; Cortes, R.; Dubois, J. M. *Springer Ser. Chem. Phys.,* **1983**, *27*, 96.

462. Goulon, J.; Goulonk, C.; Niedercorn, F.; Selve, C.; Castro, B. *Tetrahedron,* **1981**, *37*, 3707.

463. Goulon, J.; Lemonnier, M.; Cortes, R.; Retournard, A.; Raoux, D. *Nucl. Instrum. Methods Phys. Res.*, **1983**, *208*, 625.

464. Goulon, J.; Retournard, A.; Friant, P.; Goulon-Ginet, C.; Berthe, C.; Muller, J. F.; Poncet, J. L.; Guilard, R.; Escalier, J. C.; Neff, B. *J. Chem. Soc., Dalton Trans.*, **1984**, 1095.

465. Goulon, J.; Tola, P.; Lemonnier, M.; Dexpert-Ghys, J. *Chem. Phys.*, **1983**, *78*, 347.

466. Goulon-Ginet, C.; Goulon, J.; Battioni, J. P.; Mansuy, D.; Chottard, J. C. *Springer Ser. Chem. Phys.*, **1983**, *27*, 349.

467. Goulon-Ginet, C.; Goulon, J. *Stud. Phys. Theor. Chem.*, **1983**, *24*, 169.

468. Greaves, G. N. *J. Phys. Colloq.*, **1981**, *C4*, 225.

469. Greaves, G. N. *Springer Ser. Chem. Phys.*, **1983**, *27*, 248.

470. Greaves, G. N.; Diakun, G. P.; Quinn, P. D.; Hart, M.; Siddons, D. P. *Nucl. Instrum. Methods Phys. Res.*, **1983**, *208*, 335.

471. Greaves, G. N.; Durham, P. J.; Diakun, G.; Quinn, P. *Nature (London)*, **1981**, *294*, 139.

472. Greaves, G. N.; Fontaine, A.; Lagarde, P.; Raoux, D. *Daresbury Lab. Rep. DL/SCI/R17*, **1981**, 115.

473. Greaves, G. N.; Fontaine, A.; Lagarde, P.; Raoux, D.; Gurman, S. J. *Nature (London)*, **1981**, *293*, 611.

474. Greaves, G. N.; Fontaine, A.; Lagarde, P.; Raoux, D.; Gurman, S. J.; Parke, S. in *"Recent Dev. Condens. Matter Phys.,"* Vol. 2, Devreese, J. T., ed., **1981**, p. 225.

475. Greaves, G. N.; Raoux, D. in *"Struct. Non-Cryst. Mater. 1982,"* Gaskell, P. H.; Parker, J. M.; Davis, E. A., ed., Taylor & Francis: London, **1983**, p. 55.

476. Greaves, G. N.; Simkiss, K.; Taylor, M.; Binsted, N. *Biochem. J.*, **1984**, *221*, 855.

477. Greaves, N. *Recherche*, **1982**, *13*, 1184.

478. Greegor, R. B.; Lytle, F. W. *Phys. Rev. B*, **1979**, *20*, 4902.

479. Greegor, R. B.; Lytle, F. W. *J. Catal.*, **1980**, *63*, 476.

480. Greegor, R. B.; Lytle, F. W.; Chin, R. L.; Hercules, D. M. *J. Phys. Chem.*, **1981**, *85*, 1232.

481. Greegor, R. B.; Lytle, F. W.; Ewing, R. C.; Haaker, R. F. *Mater. Res. Soc. Symp. Proc.*, **1982**, *11*, 409.

482. Greegor, R. B.; Lytle, F. W.; Ewing, R. C.; Haaker, R. F. *Nucl. Instrum. Methods Phys. Res. B*, **1984**, *229*, 587.

483. Greegor, R. B.; Lytle, F. W.; Sandstrom, D. R.; Wong, J.; Schultz, P. *J. Non-Cryst. Solids.*, **1983**, *55*, 27.

484. Green, G. K. *BNL Report 50522*, **1977**, 90; *BNL Report 50595*, **1977**, **Vol. II.**

485. Grosso, G.; Pastori, P. G. *J. Phys. C*, **1980**, *13*, L919.

486. Grunes, L. A. *Phys. Rev. B*, **1983**, *27*, 2111.

487. Gudat, W.; Kunz, C. *Phys. Rev. Lett.*, **1972**, *29*, 169.

488. Gupta, S. N.; Vijayavargiya, V. P.; Padalia, B. D.; Tripathi, B. C.; Ghatikar, M. N. *Phys. Status Solidi B*, **1977**, *82*, 603.

489. Gurman, S. J. *J. Mater. Sci.*, **1982**, *17*, 1541.

490. Gurman, S. J.; Binsted, N.; Ross, I. *J. Phys. C.*, **1984**, *17*, 143.

491. Gurman, S. J.; Pettifer, R. F. *Phil. Mag. B*, **1979**, *40*, 345.

492. Gusatinskii, A. N.; Bunin, M. A.; Blokhin, M. A.; Minin, V. I.; Prochukhan, V. D.; Averkieva, G. K. *Phys. Status Solidi B*, **1980**, *100*, 739.

493. Haensel, R. in *"Laboratory EXAFS Facilities-1980,"* AIP Conf. Proc., **1980a**, *64*, p. 73.

494. Haensel, R. in *"High Pressure Sci. Technol.,"* Vol. *1*, Vodar, B., Marteau, P., ed., **1980b**, p. 54.

495. Haensel, R.; Rabe, P.; Tolkiehn, G.; Werner, A. *Report DESY-SR-80/06*, **1980**, 20; NTIS: *Energy Res. Abstr.*, **1982**, *7*, No. 22459.

496. Hahn, J. E.; Hodgson, K. O. *ACS Symp. Ser.*, **1983**, *211*, 431.

497. Hahn, J. E.; Hodgson, K. O.; Andersson, L. A.; Dawson, J. H. *J. Biol. Chem.*, **1982**, *257*, 10934.

498. Hahn, J. E.; Scott, R. A.; Hodgson, K. O.; Doniach, S.; Desjardins, S. R.; Solomon, E. I. *Chem. Phys. Lett.*, **1982**, *88*, 595.

499. Halaka, F. G.; Boland, J. F.; Baldeschwieler, J. D. *J. Am. Chem. Soc.*, **1984**, *106*, 5408.

500. *"Handbook for Auger Electron Spectroscopy,"* see Davis, L. A., *et al.*, **1976**.

501. Hanus, M. J.; Gilberg, E. *J. Phys. B*, **1976**, *9*, 137.

502. Hartree, D. R.; deL. Kronig, R.; Petersen, H. *Physia*, **1934**, *1*, 895.

503. Hasnain, S. S. *Daresbury Lab. Rep. DL/SCI/R17*, **1981**, 23.

504. Hasnain, S. S. *Springer Ser. Chem. Phys.*, **1983**, *27*, 330.

505. Hasnain, S. S.; Piggott, B. *Biochem. Biophys. Res. Commun.*, **1983**, *112*, 279.

506. Hasnain, S. S.; Piggott, B. *Springer Ser. Chem. Phys.*, **1983**, *27*, 358.

507. Hasnain, S. S.; Diakun, G. P.; Knowles, P. F.; Binsted, N.; Garner, C. D.; Blackburn, N. J. *Biochem. J.*, **1984**, *221*, 545.

508. Hasnain, S. S.; Quinn, P. D.; Diakun, G. P.; Wardell, E. M.; Garner, C. D. *J. Phys. E*, **1984**, *17*, 40.

509. Hastings, J. B. in *"EXAFS Spectroscopy: Techniques and Applications,"* Teo, B. K., Joy, D. C., ed., Plenum, New York, **1981**, p. 171 and p. 205.

510. Hastings, J. B.; Eisenberger, P.; Lengeler, B.; Perlman, M. L. *Phys. Rev. Lett.*, **1979**, *43*, 1807.

511. Hatwar, T. K.; Ghatikar, M. N.; Padalia, B. D.; Malik, S. K.; Vijayaraghavan, R. *Phys. Status Solidi B*, **1981**, *103*, 159.

512. Hatwar, T. K.; Malik, S. K.; Ghatikar, M. N.; Padalia, B. D. *Phys. Status Solidi B*, **1979**, *95*, 621.

513. Hatwar, T. K.; Nayak, R. M.; Padalia, B. D.; Ghatikar, M. N.; Sampathkumaran, E. V.; Gupta, L. C.; Vijayaraghavan, R. *Solid State Commun.*, **1980**, *34*, 617.

514. Hayes, T. M. in *"Phys. Non-Cryst. Solids,"* Frischat, G. H., ed., Trans. Tech. Publ., Aedermannsdorf, Switzerland, **1977**, p. 108.

515. Hayes, T. M. *J. Non-Cryst. Solids*, **1978**, *31*, 57.

516. Hayes, T. M. *Nuovo Cimento D*, **1984a**, *3*, 816.

517. Hayes, T. M. *Nuovo Cimento D*, **1984b**, *3*, 803.

518. Hayes, T. M.; Allen, J. W.; Tauc, J.; Giessen, B. C.; Hauser, J. J. *Phys. Rev. Lett.*, **1978**, *40*, 1282.

519. Hayes, T. M.; Boyce, J. B. *J. Phys. C*, **1980**, *13*, L731.

520. Hayes, T. M.; Boyce, J. B. in *"EXAFS Spectroscopy: Techniques and Applications,"* Teo, B. K., Joy, D. C., ed., Plenum, New York, **1981**, p. 81.

521. Hayes, T. M.; Boyce, J. B. *Solid State Phys.*, **1982**, *37*, 173.

522. Hayes, T. M.; Boyce, J. B. *Springer Ser. Chem. Phys.*, **1983**, *27*, 182.

523. Hayes, T. M.; Boyce, J. B.; Beeby, J. L. *J. Phys. C*, **1978**, *11*, 2931.

524. Hayes, T. M.; Hunter, S. H. in *"Struct. Non-Cryst. Mater.,"* Gaskell, P. H.; Davis, E. A., ed., Taylor and Francis, London, **1977**, p. 69.

525. Hayes, T. M.; Knights, J. C.; Mikkelsen, J. C., Jr. in *"Amorphous Liq. Semicond.,"* Spear, W. E., ed., Univ. Edinburgh, Edinburgh, Scotland, **1977**, p. 73.

526. Hayes, T. M.; Sen, P. N.; Hunter, S. H. *J Phys. C.*, **1976**, *9*, 4357.

527. Hayes, T. M.; Sen, P. N. *Phys. Rev. Lett.*, **1975**, *34*, 956.

528. Hayes, T. M.; Wright, A. C. in "*Struct. Non-Cryst. Mater.*, *1982*" Gaskell, P. H.; Parker, J. M.; Davis, E. A., ed., Taylor & Francis, London, **1983**, p. 108.

529. Heald, S. M. in "*Laboratory EXAFS Facilities-1980*," AIP Conf. Proc., **1980**, *64*, p. 31.

530. Heald, S. M. *Springer Ser. Chem. Phys.*, **1983**, *27*, 98.

531. Heald, S. M. *Nucl. Instrum. Methods Phys. Res. A*, **1984**, *222*, 160.

532. Heald, S. M.; Keller, E.; Stern, E. A. *Phys. Lett. A*, **1984**, *103*, 155.

533. Heald, S. M.; Stern, E. A. *Phys. Rev. B*, **1977**, *16*, 5549.

534. Heald, S. M.; Stern, E. A. *Phys. Rev. B*, **1978**, *17*, 4069.

535. Heald, S. M.; Stern, E. A. *Synth. Met.*, **1980**, *1*, 249.

536. Heald, S. M.; Stern, E. A.; Bunker, B.; Holt, E. M.; Holt, S. L. *J. Am. Chem. Soc.*, **1979**, *101*, 67.

537. Heinz, K., Müller, K. *Springer Tracts in Mod. Phys.*, **1982**, *91*, 1.

538. Hermes, C.; Gilberg, E.; Koch, M. H. J. *Nucl. Instrum. Methods Phys. Res., A*, **1984**, *222*, 207.

539. Hendrickson, W. A.; Co, M. S.; Smith, J. L.; Hodgson, K. O.; Klippenstein, G. L. *Proc. Natl. Acad. Sci. U.S.A.*, **1982**, *79*, 6255.

540. Henke, B. L.; Lee, P.; Tanaka, T. J.; Shimabukuro, R. L.; Fujikawa, B. K. *At. Data Nucl. Data Tables*, **1982**, *27*, 1.

541. Herman, F.; Skillman, S. "*Atomic Structure Calaculations*," Prentice-Hall, Englewood Cliffs, N.J., **1963**.

542. Hershfield, S. P.; Einstein, T. L. *Phys. Rev. B*, **1984**, *29*, 1048.

543. Hida, M.; Maeda, H.; Kamijo, N.; Terauchi, H. *Phys. Status Solidi A*, **1982**, *69*, 297.

544. Hida, M.; Maeda, H.; Kamijo, N.; Tanabe, K.; Terauchi, H.; Tsu, Y.; Watanabe, S. *J. Non-Cryst. Solids*, **1984**, *61-62*, 415.

545. Hirota, S.; Fujikawa, T. *J. Electron Spectrosc. Relat. Phenom.*, **1982**, *28*, 95.

546. Hitchcock, A. P.; Lock, C. J. L.; Pratt, W. M. C. *Inorg. Chim. Acta*, **1982**, *66*, L45.

547. Hodgson, K. O.; Hedman, B.; Penner-Hahn, J. E., ed., "*EXAFS and Near Edge Structure III*," Springer-Verlag, Berlin, **1985**.

548. Hoffman, R. W. in *"Passivity Met. Semicond.,"* Froment, M., ed., Elsevier, Amsterdam, **1983**, p. 147.

549. Holland, B. W.; Pendry, J. B.; Pettifer, R. F.; Bordas, J. *J. Phys. C,* **1978**, *11*, 633.

550. Holt, C.; Hasnain, S. S.; Hukins, D. W. L. *Biochim. Biophys. Acta,* **1982**, *719*, 299.

551. Horowitz, P.; Howell, J. A. *Science,* **1976**, *191*, 1172.

552. Horsley, J. A. *J. Chem. Phys.,* **1982**, *76*, 1451.

553. Hosoya, S.; Kawamura, T.; Fukamachi, T. *Oyo Butsuri,* **1978**, *47*, 708.

554. Hu, V. W.; Chan, S. I.; Brown, G. S. *Proc. Natl. Acad. Sci.,* U.S.A., **1977**, *74*, 3821.

555. Huang, H. W. *Gov. Rep. Announce. Index (U.S.),* **1979**, *79*, 93.

556. Huang, H. W.; Hunter, S. H.; Warburton, W. K.; Moss, S. C. *Science,* **1979**, *204*, 191.

557. Huang, H. W.; Liu, W. H.; Buchanan, J. A. *Nucl. Instrum. Methods Phys. Res.,* **1983**, *205*, 375.

558. Huang, H. W.; Liu, W. H.; Teng, T. Y.; Wang, X. F. *Rev. Sci. Instrum.,* **1983**, *54*, 1488.

559. Huang, H. W.; Williams, C. R. *Biophys. J.,* **1981**, *33*, 269.

560. Huffman, G. P.; Huggins, F. E.; Cuddy, L. J.; Lytle, F. W.; Greegor, R. B. *Scr. Metall.,* **1984**, *18*, 719.

561. Hunter, S. H. in *"EXAFS Spectroscopy: Techniques and Applications,"* Teo, B. K., Joy, D. C., ed., Plenum, New York, **1981**, p. 163.

562. Hunter, S.; Bienenstock, A. in *"Strukt. Svoistva Nekristallicheskikh Poluprovodn.,"* Kolomiets, B. T., ed., Nauka, Leningr. Otd., Leningrad, USSR, **1976**, p. 151.

563. Hunter, S. H.; Bienenstock, A.; Hayes, T. M. in *"Amorphous Liq. Semicond.,"* Spear, W. E., ed., Univ. Edinburgh, Edinburgh, Scotland, **1977a**, p. 78.

564. Hunter, S. H.; Bienenstock, A.; Hayes, T. M. in *"Struct. Non-Cryst. Mater.,"* Gaskell, P. H.; Davis, E. A., ed., Taylor and Francis, London, **1977b**, p. 73.

565. Huntley, D. R.; Parham, T. G.; Merrill, R. P.; Sienko, M. J. *Inorg. Chem.,* **1983**, *22*, 4114.

566. Hussain, Z.; Umbach, E.; Shirley, D. A.; Stöhr, J.; Feldhaus, J. *Nucl. Instrum. Methods Phys. Res.,* **1982**, *195*, 115.

567. Iizuka, T.; Uchida, K.; Ishimura, Y.; Ohyanago, H.; Hosoya, S. *Seibutsu Butsuri,* **1979**, *19*, 223.

568. Ikeda, S. in *"Recent Adv. Anal. Spectrosc.,"* Fuwa, K, ed., Pergamon, Oxford, U.K., **1982**, p. 201.

569. Ikeda, S. *Seisan to Gijutsu,* **1982**, *34*, 17.

570. Il'in, V. E.; Chermashentsev, V. M. *Zh. Strukt, Khim.,* **1982**, *23*, 148.

571. Il'in, V. E.; Chermashentsev, V. M.; Mazalov, L. N. *Zh. Strukt. Khim.,* **1982**, *23*, 161.

572. Incoccia, L.; Mobilio, S. *Nuovo Cimento D,* **1984**, *3*, 867.

573. Incoccia, L.; Mobilio, S. *Springer Ser. Chem. Phys.,* **1983**, *27*, 91.

574. Incoccia, L.; Mobilio, S.; Benfatto, M.; Davoli, I.; Stizza, S.; Bianconi, A. *Springer Ser. Chem. Phys.,* **1983**, *27*, 177.

575. Indrea, E.; Aldea, N. *Comput. Phys. Commun.,* **1980**, *21*, 91.

576. Ingalls, R.; Crozier, E. D.; Whitmore, J. E.; Seary, A. J.; Tranquada, J. M. *J. Appl. Phys.,* **1980**, *51*, 3158.

577. Ingalls, R.; Garcia, G. A.; Stern, E. A. *Phys. Rev. Lett.,* **1978**, *40*, 334.

578. Ingalls, R.; Tranquada, J. M.; Whitmore, J. E.; Crozier, E. D.; Seary, A. J. in *"EXAFS Spectroscopy: Techniques and Applications,"* Teo, B. K., Joy, D. C., ed., **1981a**, p. 127.

579. Ingalls, R.; Tranquada, J. M.; Whitmore, J. E.; Crozier, E. D. *Springer Ser. Chem. Phys.,* **1983**, *27*, 153.

580. Ingalls, R.; Tranquada, J. M.; Whitmore, J. E.; Crozier, E. D. in *"Phys. Solids High Pressure,"* Schilling, J. S., Shelton, R. N., ed., North-Holland, Amsterdam, **1981b**, p. 67.

581. Ingalls, R.; Whitmore, J. E.; Tranquada, J. M.; Crozier, E. D. in *"High Pressure Sci. Technol.,"* Vol. 1, Vodar, B., Marteau, P., ed., Pergamon, Oxford, England, **1980**, p. 528.

582. *"International Tables for X-Ray Crystallography,"* Vol. *III*, Kynoch Press, Birmingham, England, 1962:

(a) Rieck, G. D., "Tables Relating to the Production, Wavelengths and Intensities of X-rays," Sec. 2.2, p. 59.

(b) Koch, B., MacGillavry, C. H., "X-ray Absorption," Sec. 3.2.1, p. 157.

(c) Table 3.2.2C, p. 171.

(d) Table 3.2.2D, p. 174.

583. Isaacson, M. *J. Chem. Phys.,* **1972**, *56*, 1818.

252

584. Islam, M. S.; Mande, C. *Phys. Status Solidi A*, **1984**, *81*, 197.

585. Ismail, I. M.; Mazid, M. A.; Sadler, P. J.; Greaves, G. N. *Daresbury Lab. Rep. DL/SCI/R17*, **1981**, *95*.

586. Ito, M.; Iwasaki, H. *Jpn. J. Appl. Phys., Part 1*, **1983**, *22*, 357.

587. Ito, M.; Iwasaki, H.; Shiotani, N.; Narumi, H.; Mizoguchi, T.; Kawamura, T. *J. Non-Cryst. Solids*, **1984**, *61-62*, 303.

588. Ito, M.; Kawamura, T. *Philos. Mag. A*, **1984**, *49*, L9.

589. Iwata, M.; Oyanagi, H. *Kagaku to Kogyo*, **1979**, *32*, 298.

590. Jaeger, R.; Feldhaus, J.; Haase, J.; Stöhr, J.; Hussain, Z.; Menzel, D.; Norman, D. *Phys. Rev. Lett.*, **1980**, *45*, 1870.

591. Jain, D. C.; Chandra, U.; Garg, K. B.; Sharma, B. K. *J. Phys. D*, **1980**, *13*, 1113.

592. Jain, D.; Garg, K. B.; Sharma, B. K. *Proc. Nucl. Phys. Solid State Phys. Symp. C*, **1978**, *21*, 42.

593. Jain, D. C.; Sharma, B. K.; Garg, K. B. *J. Phys. D*, **1981**, *14*, L5.

594. Jaklevic, J.; Kirby, J. A.; Klein, M. P.; Robertson, A. S.; Brown, G. S.; Eisenberger, P. *Solid State Commun.*, **1977**, *23*, 679.

595. Jaklevic, J. M.; Kirby, J. A.; Ramponi, A. J.; Thompson, A. C. *Environ. Sci. Technol.*, **1980**, *14*, 437.

596. James, R. W. *"The Optical Principles of the Diffraction of X-Rays,"* Cornell University Press, Ithaca, N.Y., **1965**, p. 65.

597. Jenkins, R. *"An Introduction to X-Ray Spectrometry,"* Heyden, New York, **1974**. (a) p. 73. (b) p. 72. (c) p. 95. (d) p. 88.

598. Jerome, R.; Vlaic, G.; Williams, C. E. *J. Phys., Lett.*, **1983**, *44*, 717.

599. Johansson, L. I.; Stöhr, J. *Phys. Rev. Lett.*, **1979**, *43*, 1882.

600. Johansson, L. I.; Stöhr, J.; Brennan, S. *Appl. Surf. Sci.*, **1980**, *6*, 419.

601. Johansson, L. I.; Stöhr, J.; Brennan, S.; Hecht, M. H.; Miller, J. N. *Ned. Tijdschr. Vacuumtech.*, **1980**, *18*, 84.

602. Johnson, A. L.; Muetterties, E. L. *J. Am. Chem. Soc.*, **1983**, *105*, 7183.

603. Johri, R. K.; Agarwal, B. K. *J. Phys. F.*, **1978**, *8*, 555.

604. Jona, F.; Marcus, P. M. *J. Phys. C*, **1980**, *13*, L447.

605. Jona, F.; Marcus, P. M. *Phys. Rev. Lett.*, **1983**, *50*, 1823.

606. Joy, D. C.; Maher, D. M. *Science*, **1979**, *206*, 162.

607. Joyner, R. W. *Daresbury Lab Rep. DL/SCI/R13*, **1979**, 114.

608. Joyner, R. W. in *"Charact. Catal.,"* Thomas, J. M., Lambert, R. M., ed., Wiley, Chichester, U.K., **1980a**, p. 237.

609. Joyner, R. W. *Chem. Phys. Lett.,* **1980b**, *72*, 162.

610. Joyner, R. W. *J. Chem. Soc., Faraday Trans.,* **1980c**, *76*, 357.

611. Joyner, R. W. *Daresbury Lab. Rep. DL/SCI/R17,* **1981**, 65.

612. Joyner, R. W.; Meehan, P. *Vacuum,* **1983**, *33*, 691.

613. Joyner, R. W.; Van Veen, J. A. R.; Sachtler, W. M. H. *J. Chem. Soc., Faraday Trans.,* **1982**, *78*, 1021.

614. Jucha, A.; Bonin, D.; Dartyge, E.; Flank, A. M.; Fontaine, A.; Raoux, D. *Nucl. Instrum. Methods Phys. Res. A,* **1984**, *226*, 40.

615. Kaindl, G.; Brewer, W. D.; Kalkowski, G.; Holtzberg, F. *Phys. Rev. Lett.,* **1983**, *51*, 2056.

616. Kaindl, G.; Kalkowski, G.; Brewer, W. D.; Perscheid, B.; Holtzberg, F. *J. Appl. Phys.,* **1984**, *55*, 1910.

617. Jungfleisch, M. L.; Islam, M. S.; Sapre, V. B.; Mande, C. *Indian J. Phys., A,* **1983**, *57*, 250.

618. Kalb, A. J.; Stern, E. A.; Heald, S. M. *J. Mol. Biol.,* **1979**, *135*, 501.

619. Kambe, K.; Scheffler, M. *Surf. Sci.,* **1979**, *89*, 262.

620. Kaminaga, U.; Matsushita, T.; Kohra, K. *Jpn. J. Appl. Phys.,* **1981**, *20*, L355.

621. Karim, D. P.; Georgopoulos, P.; Knapp, G. S. *Nucl. Technol.,* **1980**, *51*, 162.

622. Kawamura, T. *Natl. Lab. High Energy Phys.,* KEK (Jpn.), **1980**, 156.

623. Kawamura, T.; Hosoya, S. *Seibutsu Butsuri,* **1979**, *19*, 213.

624. Kawamura, T.; Shimomura, O.; Fukamachi, T.; Fuoss, P. H. *Acta Crystallogr. A,* **1981**, *37*, 653.

625. Kennedy, O. J., Manson, S. T. *Phys. Rev A,* **1972**, *5*, 227.

626. Kevan, S. D.; Tobin, J. G.; Rosenblatt, D. H.; Davis, R. F.; Shirley, D. A. *Phys. Rev. B,* **1981**, *23*, 493.

627. Khalid, S.; Emrich, R.; Dujari, R.; Shultz, J.; Katzer, J. R. *Rev. Sci. Instrum.,* **1982**, *53*, 22.

628. Khasbardar, B. V.; Vaingankar, A. S.; Patil, R. N. *Indian J. Pure Appl. Phys.,* **1981**, *19*, 612.

629. Khristenko, S. V. *Phys. Lett. A,* **1976**, *59*, 202.

630. Kincaid, B. M., Ph.D. Thesis, Stanford University, **1975**.

631. Kincaid, B. M.; Eisenberger, P. *Phys. Rev. Lett.*, **1975**, *34*, 1361.

632. Kincaid, B. M.; Eisenberger, P.; Hodgson, K. O.; Doniach, S. *Proc. Natl. Acad. Sci. U.S.A.*, **1975**, *72*, 2340.

633. Kincaid, B. M.; Meixner, A. E.; Platzman, P. M. *Phys. Rev. Lett.*, **1978**, *40*, 1296.

634. Kincaid, B. M.; Shulman, R. G. *Adv. Inorg. Biochem.*, **1980**, *2*, 303.

635. Kirby, J. A.; Goodin, D. B.; Wydrzynski, T.; Robertson, A. S.; Klein, M. P. *J. Am. Chem. Soc.*, **1981**, *103*, 5537.

636. Kirby, J. A.; Jaklevic, J. M.; Kafka, N.; Klein, M. P.; Robertson, A. S.; Smith, J. P.; Thompson, A. C.; Walker, T. P. *Nucl. Instrum. Methods*, **1978**, *152*, 330.

637. Kirby, J. A.; Robertson, A. S.; Smith, J. P.; Thompson, A. C.; Cooper, S. R.; Klein, M. P. *J. Am. Chem. Soc.*, **1981**, *103*, 5529.

638. Kiyono, S.; Muranaka, T.; Watanabe, T. *Jpn. J. Appl. Phys.*, **1979**, *18*, 1865.

639. Klein, O.; Nishina, Y. *Z. Phys.*, **1929**, *52*, 853.

640. Knapp, G. S.; Chen, H.; Klippert, T. E. *Rev. Sci. Instrum.*, **1978**, *49*, 1658.

641. Knapp, G. S.; Fradin, F. Y. in *"Electron Positron Spectrosc. Mater. Sci. Eng.,"* Buck, O., Tien, J. K., Marcus, H. L., ed., Academic, New York, **1979**, p. 243.

642. Knapp, G. S.; Georgopoulos, P. in *"Laboratory EXAFS Facilities-1980,"* AIP Conf. Proc., **1980**, *64*, p. 2.

643. Knapp, G. S.; Georgopoulos, P. *Cryst.: Growth, Prop., Appl.*, **1982**, *7*, 75.

644. Knapp, G. S.; Georgopoulos, P. *Cryst.: Growth, Prop., Appl.*, **1982**, *7*, 75.

645. Knapp, G. S.; Kampwirth, R. T.; Georgopoulos, P.; Brown, B. S. in *"Supercond. d- f-Band Met.,"* Suhl, H., Maple, M. B., ed., Academic, New York, **1980**, p. 363.

646. Knapp, G. S.; Pan, H. K.; Georgopoulos, P.; Klippert, T. E. *Springer Ser. Chem. Phys.*, **1983**, *27*, 402.

647. Knapp, G. S.; Veal, B. W.; Lam, D. J.; Paulikas, A. P.; Pan, H. K. *Mater. Lett.*, **1984**, *2*, 253.

648. Knotek, M. L.; Feibelman, P. J. *Phys. Rev. Lett.*, **1978**, *40*, 964.

649. Knotek, M. L.; Jones, V. O.; Rehn, V. *Phys. Rev. Lett.*, **1979**, *43*, 300.

650. Kobayashi, S.; Takeuchi, S. *J. Phys. F*, **1982**, *12*, 1273.

651. Kordesch, M. E.; Hoffman, R. W. *Phys. Rev. B*, **1984**, *29*, 491.

652. Kordesch, M. E.; Hoffman, R. W. *Nucl. Instrum. Methods Phys. Res. A* **1984**, *222*, 347.

653. Koestner, R. J.; Stöhr, J.; Gland, J. L.; Horsley, J. A. *Chem. Phys. Lett.*, **1984**, *105*, 332.

654. Kohn, W.; Lee, T. K.; Lin-Liu, Y. R. *Phys. Rev. B*, **1982**, *25*, 3557.

655. Koningsberger, D. C.; Cook, J. W., Jr. *Springer Ser. Chem. Phys.*, **1983**, *27*, 412.

656. Koningsberger, D. C.; Huizinga, T.; van't Blik, H. F. J.; van Zon, J. B. A. D.; Prins, R.; Sayers, D. E. *Springer Ser. Chem. Phys.*, **1983**, *27*, 310.

657. Koningsberger, D. C.; Prins, R. *Trends Anal. Chem.*, **1981**, *1*, 16.

658. Koningsberger, D. C.; Prins, R. *Chem. Mag.*, **1982**, 33.

659. Korszun, Z. R.; Moffat, K.; Frank, K.; Cusanovich, M. A. *Biochemistry*, **1982**, *21*, 2253.

660. Kortright, J.; Warburton, W.; Bienenstock, A. *Springer Ser. Chem. Phys.*, **1983**, *27*, 362.

661. Kostarev, A. I. *Zh. Eksp. Theo. Fiz.*, **1941**, *11*, 60.

662. Kozlenkov, A. T. *Bull. Acad. Sci., USSR, Ser. Phys.*, **1961**, *25*, 968.

663. Kozlowski, R.; Pettifer, R. F.; Thomas, J. M. *J. Phys. Chem.*, **1983a**, *87*, 5172.

664. Kozlowski, R.; Pettifer, R. F.; Thomas, J. M. *J. Phys. Chem.*, **1983b** *87*, 5176.

665. Kozlowski, R.; Pettifer, R. F.; Thomas, J. M. *Springer Ser. Chem. Phys.*, **1983b**, *27*, 313.

666. Kozlowski, R.; Pettifer, R. F.; Thomas, J. M. *J. Chem. Soc., Chem. Commun.*, **1983**, 438.

667. Kohn, W.; Sham, L. J. *Phys. Rev. A*, **1965**, *140*, 1133.

668. Krasnoperova, A. A.; Gluskin, E. S.; Mazalov, L. N.; Kochubei, V. A. *Zh. Strukt. Khim.*, **1976**, *17*, 1113.

669. Krill, G.; Kappler, J. P.; Röhler, J. *Springer Ser. Chem. Phys.*, **1983**, *27*, 190.

670. Krill, G.; Kappler, J. P.; Röhler, J.; Ravet, M. F.; Leger, J. M.; Gautier, F. in *"Valence Instab.,"* Wachter, P., Boppart, H., ed., North-Holland: Amsterdam, **1982**, p. 155.

671. Krishna, V.; Prasad, J.; Nigam, H. L. *Indian J. Pure Appl. Phys.*, **1979**, *17*, 95.

672. Kronig, R. deL. *Z. Phys.*, **1931**, *70*, 317.

673. Kronig, R. deL. *Z. Phys.*, **1932**, *75*, 468; *ibid.*, *76*, 468.

674. Kruger, J.; Long, G. G.; Kuriyama, M.; Goldman, A. I. in *"Passivity Met Semicond.,"* Froment, Michel., ed., Elsevier, Amsterdam, **1983**, p. 163.

675. Kulipanov, G. N.; Skrinsky, A. N. *Usp. Fiz. Nauk*, **1977a**, *122*, 369; Eng. Trans., *Sov. Phys. Usp.*, **1977b**, *20*, 559.

676. Kumar, V.; Chetal, A. R.; Srivastava, K. S. *Phys. Status Solidi A*, **1982**, *70*, K107.

677. Kumar, A.; Nigam, A. N. *X-Ray Spectrom.*, **1979**, *8*, 135.

678. Kumar, A.; Nigam, A. N.; Srivastava, B. D. *J. Phys. C*, **1980**, *13*, 3523.

679. Kuroda, H.; Fujikawa, T. *Kagaku No Ryoiki*, **1979**, *33*, 276.

680. Kuroda, H.; Ikemoto, I.; Asakura, K.; Ishii, H.; Shirakawa, H.; Kobayashi, T.; Oyanagi, H.; Matsushita, T. *Solid State Commun.*, **1983**, *46*, 235.

681. Kutzler, F. W.; Hodgson, K. O.; Doniach, S. *Phys. Rev. A*, **1982**, *26*, 3020.

682. Kutzler, F. W.; Hodgson, K. O.; Misemer, D. K.; Doniach, S. *Chem. Phys. Lett.*, **1982**, *92*, 626.

683. Kutzler, F. W.; Natoli, C. R.; Misemer, D. K.; Doniach, S.; Hodgson, K. O. *J. Chem. Phys.*, **1980**, *73*, 3274.

684. Kutzler, F. W.; Scott, R. A.; Berg, J. M.; Hodgson, K. O.; Doniach, S.; Cramer, S. P.; Chang, C. H. *J. Am. Chem. Soc.*, **1981**, *103*, 6083.

685. Labhardt, A.; Yuen, C. *Nature*, **1979**, *277*, 150.

686. Laderman, S.; Bienenstock, A.; Liang, K. S. *Sol. Energy Mater.*, **1982**, *8*, 15.

687. Lagarde, P. *Phys Rev. B*, **1976**, *13*, 741.

688. Lagarde, P. *Nuovo Cimento D*, **1984**, *3*, 885.

689. Lagarde, P. *Springer Ser. Chem. Phys.*, **1983**, *27*, 296.

690. Lagarde, P. *Nucl. Instrum. Methods Phys. Res.*, **1983**, *208*, 621.

691. Lagarde, P.; Fontaine, A.; Raoux, D.; Sadoc, A.; Migliardo, P. *J. Chem. Phys.*, **1980**, *72*, 3061.

692. Lagarde, P.; Murata, T.; Vlaic, G.; Dexpert, H.; Freund, E.; Bournonville, J. P. *Springer Ser. Chem. Phys.*, **1983**, *27*, 319.

693. Lagarde, P.; Murata, T.; Vlaic, G.; Freund, E.; Dexpert, H.; Bournonville, J. P. *J. Catal.*, **1983**, *84*, 333.

694. Lagarde, P.; Raoux, D.; Fontaine, A. *Daresbury Lab. Rep. DL/SCI/R17*, **1981**, 122.

695. Lagarde, P.; Rivory, J.; Vlaic, G. *J. Non-Cryst. Solids*, **1983**, *57*, 275.

696. Landman, U.; Adams, D. L. *Proc. Nat. Acad. Sci. U.S.A.*, **1976**, *73*, 2550.

697. Lane, R. W.; Ibers, J. A.; Frankel, R. B.; Papaefthymiou, G. C.; Holm, R. H. *J. Am. Chem. Soc.*, **1977**, *99*, 84.

698. Lapeyre, C.; Petiau, J.; Calas, G. *Springer Ser. Chem. Phys.*, **1983a**, *27*, 265.

699. Lapeyre, C.; Petiau, J.; Calas, G. in *"Struct. Non-Cryst. Mater.,"* Proc 2nd Int. Conf., Gaskell, P. H.; Parker, J. M.; Davis, E. A., ed., Taylor & Francis: London, UK, **1983b**, p. 42.

700. Lapeyre, G. J.; Baer, A. D.; Anderson, J.; Hermanson, J. C.; Knapp, J. A.; Gobby, P. L. *Solid State Commun.*, **1974**, *15*, 1601.

701. Laramore, G. E. *Phys. Rev. B*, **1978**, *18*, 5254.

702. Laramore, G. E. *Surf. Sci.*, **1979**, *81*, 43.

703. Laramore, G. E. *Phys. Rev. A*, **1981**, *24*, 1904.

704. Laramore, G. E. *Phys. Rev. B*, **1983**, *28*, 4778.

705. Laramore, G. E.; Einstein, T. L.; Roelofs, L. D.; Park, R. L. *Phys. Rev. B*, **1980**, *21*, 2108.

706. Larue, J. F.; Moraweck, B.; Renouprez, A.; Lagarde, P. *Springer Ser. Chem. Phys.*, **1983**, *27*, 321.

707. Launois, H. *Recherche*, **1978**, *9*, 786.

708. Launois, H.; Rawiso, M.; Holland-Moritz, E.; Pott, R.; Wohlleben, D. *Phys. Rev. Lett.*, **1980**, *44*, 1271.

709. Le Calonnec, D.; Dürr, J.; Hannoyer, B.; Lenglet, M.; Calas, G.; Petiau, J. *Springer Ser. Chem. Phys.*, **1983**, *27*, 210.

710. Lea, K. *Phys. Rep.*, **1978**, *43*, 337.

711. Leapman, R. D.; Cosslet, V. E. *J. Phys. D*, **1976**, *9*, 25.

712. Leapnan, R. D.; Grumes, L. A.; Fejes, P. L.; Silcox, J. in *"EXAFS Spectroscopy: Techniques and Applications,"* Teo, B. K., Joy, D. C., ed., Plenum, New York, **1981**, p. 217.

713. Lee, P. A. *Phys. Rev. B*, **1976**, *13*, 5261.

714. Lee, P. A. *J. Phys.*, Colloq., **1978**, *C4*, 120.

715. Lee, P. A. in *"EXAFS Spectroscopy: Techniques and Applications,"* Teo, B. K., Joy, D. C., ed., Plenum, New York, **1981**, p. 5.

258

716. Lee, P. A.; Beni, G. *Phys. Rev. B*, **1977**, *15*, 2862.

717. Lee, P. A.; Citrin, P. H.; Eisenberger, P.; Kincaid, B. M. *Rev. Mod. Phys.*, **1981**, *53*, 769.

718. Lee, P. A.; Pendry, J. B. *Phys. Rev B*, **1975**, *11*, 2795.

719. Lee, P. A.; Teo, B. K.; Simons, A. L. *J. Am. Chem. Soc.*, **1977**, *99*, 3856.

720. Lee, P. L.; Boehm, F.; Vogel, P. *Phys. Lett. A*, **1977**, *63*, 251.

721. Lelieur, J. P.; Goulon, J.; Cortes, R.; Friant, P. *J. Phys. Chem.*, **1984**, *88*, 3730.

722. Lengeler, B.; Eisenberger, P. *Phys. Rev. B*, **1980**, *21*, 4507.

723. Lengeler, B.; Materlik, G.; Mueller, J. E. *Phys. Rev. B*, **1983a**, *28*, 2276.

724. Lengeler, B.; Materlik, G.; Mueller, J. E. *Springer Ser. Chem. Phys.*, **1983b**, *27*, 150.

725. Levitz, P.; Crespin, M.; Gatineau, L. *Springer Ser. Chem. Phys.*, **1983**, *27*, 219.

726. Levitz, P.; Crespin, M.; Gatineau, L. *J. Chem. Soc., Faraday Trans. 2.*, **1983**, *79*, 1195.

727. Licheri, G.; Paschina, G.; Piccaluga, G.; Pinna, G.; Vlaic, G. *Chem. Phys. Lett.*, **1981**, *83*, 384.

728. Licheri, G.; Paschina, G.; Piccaluga, G.; Pinna, G. *J. Chem. Phys.*, **1983**, *79*, 2168.

729. Licheri, G.; Pinna, G. *Springer Ser. Chem. Phys.*, **1983**, *27*, 240.

730. Licheri, G.; Pinna, G.; Navarra, G. *Z. Naturforsch. A*, **1983**, *38*, 559.

731. Lindahl, P. A.; Kojima, N.; Hausinger, R. P.; Fox, J. A.; Teo, B. K.; Walsh, C. T.; Orme-Johnson, W. H. *J. Am. Chem. Soc.*, **1984**, *106*, 3062.

732. Lindau, I.; Winick, H. *J. Vac. Sci. Technol.*, **1978**, *15*, 977.

733. Littke, W. *Nachr. Chem., Tech. Lab.*, **1979**, *27*, 318,320.

734. Liu, W. H.; Wang, X. F.; Teng, T. Y.; Huang, H. W., *Rev. Sci. Instrum.*, **1983**, *54*, 1653.

735. Lokhande, N. R. *Phys. Status Solidi B*, **1982**, *114*, K35.

736. Lokhande, N. R.; Mande, C. *Phys. Status Solidi B*, **1980**, *102*, K11.

737. Lokhande, N. R.; Mande, C. *J. Phys. Chem. Solids*, **1982**, *43*, 731.

738. Long, G. G.; Kruger, J.; Black, D. R.; Kuriyama, M. *J. Electrochem. Soc.*, **1983**, *130*, 240.

739. Long, G. G.; Kruger, J.; Black, D. R.; Kuriyama, M. *J. Electroanal. Chem. Interfacial Electrochem.*, **1983**, *150*, 603.

740. Long, G. G.; Kruger, J.; Kuriyama, M. in *"Passivity Met. Semicond.,"* Froment, M., ed., Elsevier, Amsterdam, **1983**, p. 139.

741. Lottici, P. P.; Rehr, J. J. *Solid State Commun.*, **1980**, *35*, 565.

742. Lu, K. Q.; Stern, E. A. *Nucl. Instrum. Methods Phys. Res.*, **1983**, *212*, 475.

743. Lukirskii, A. P.; Brytov, I. A. *Sov. Phys. Solid State*, **1964**, *6*, 33.

744. Lukirskii, A. P.; Ershov, O. A.; Zimkina, T. M.; Savinov, E. P. *Sov. Phys. Solid State*, **1966**, *8*, 1422.

745. Lukirskii, A. P.; Savinov, E. P.; Brytov, I. A.; Shepelev, Yu. F. *Bull. Acad. Sci. USSR Phys. Ser.*, **1964**, *28*, 774.

746. Lukirskii, A. P.; Zimkina, T. M. *Izv. Akad. Nauk. USSR, Ser. Fiz.*, **1964**, *28*, 765.

747. Lye, R. C.; Phillips, J. C.; Kaplan, D.; Doniach, S.; Hodgson, K. O. *Proc. Natl. Acad. Sci. U.S.A.*, **1980**, *77*, 5884.

748. Lynch, D. W. *J. Electroanal. Chem. Interfacial Electrochem.*, **1983**, *150*, 229.

749. Lytle, F. W. *J. Catal.*, **1976**, *43*, 376.

750. Lytle, F. W. *NBS Spec. Publ. (U.S.)*, **1977**, *475*, 34.

751. Lytle, F. W. *Prepr. — Am. Chem. Soc., Div. Pet. Chem.*, **1984**, *29*, 785.

752. Lytle, F. W.; Greegor, R. B.; Via, G. H.; Sinfelt, J. H. *Prepr. - Am. Chem. Soc., Div. Pet. Chem.*, **1981**, *26*, 400.

753. Lytle, F. W.; Sayers, D. E.; Moore, E. B. *Appl. Phys. Lett.*, **1974**, *24*, 45.

754. Lytle, F. W.; Sayers, D. E.; Stern, E. A. *Phys. Rev. B*, **1975**, *11*, 4825; **1975**, *11*, 4836.

755. Lytle, F. W.; Sayers, D. E.; Stern, E. A. *Phys. Rev. B*, **1977**, *15*, 2426.

756. Lytle, F. W.; Sayers, D. E.; Stern, E. A. in *"Adv. X-Ray Spectrosc.,"* Bonnelle, C.; Mande, C., ed., Pergamon, Oxford, **1982**, p. 267.

757. Lytle, F. W.; Via, G. H.; Sinfelt, J. H. *Prepr., Div. Pet. Chem., Am. Chem. Soc.*, **1976**, *21*, 366.

758. Lytle, F. W.; Via, G. H.; Sinfelt, J. H. *J. Chem. Phys.*, **1977**, *67*, 3831.

759. Lytle, F. W.; Via, G. H.; Sinfelt, J. H. in *"Synchrotron Radiation Research,"* Winick, H., Doniach, S., ed., Plenum, New York, **1980**, p. 401.

260

760. Lytle, F. W.; Wei, P. S. P.; Greegor, R. B.; Via, G. H.; Sinfelt, G. H. *J. Chem. Phys.*, **1979**, *70*, 4849.

761. Ma, L; Xue, Y. *Huaxue Tongbao*, **1982**, *7*, 394.

762. Macovei, D.; Pausescu, P.; Grigorovici, R.; Manaila, R. *J. Appl. Crystallogr.*, **1982**, *15*, 39.

763. Madey, T. E. *Surf. Sci.*, **1980**, *94*, 483.

764. Madey, T. E.; Stockbauer, R.; Van der Veen, J. F.; Eastman, D. E. *Phys. Rev. Lett.*, **1980**, *45*, 187.

765. Madey, T. E.; Yates, J. T., Jr. *Surf. Sci.*, **1977**, *63*, 203.

766. Maeda, H.; Tanimoto, T.; Terauchi, H.; Hida, M. *Phys. Status Solidi A.*, **1980**, *58*, 629.

767. Maeda, H.; Terauchi, H.; Kamijo, N.; Hida, M.; Osamura, K. in "*Proc. Int. Conf. Rapidly Quenched Met. 4th*," Vol. *1*, Masumoto, T.; Suzuki, K., ed., Jpn. Inst. Met., Sendai, Japan, **1982** p. 397.

768. Maeda, H.; Terauchi, H.; Tanabe, K.; Kamijo, N.; Hida, Moritaka; Kawamura, H. *Jpn. J. Appl. Phys. Part1*, **1982**, *21*, 1342.

769. Maire, G.; Hilaire, L.; Zahraa, O.; Ravet, M. F. *Springer Ser. Chem. Phys.*, **1983**, *27*, 316.

770. Mallozzi, P. J.; Schwerzel, R. E.; Epstein, H. M.; Campbell, B. E. *Science*, **1979**, *206*, 353.

771. Mallozzi, P. J.; Schwerzel, R. E.; Epstein, H. M.; Campbell, B. E. *Phys. Rev. A*, **1981**, *23*, 824.

772. Malzfeldt, W.; Niemann, W.; Rabe, P.; Schwentner, N. *Springer Ser. Chem. Phys.*, **1983**, *27*, 203.

773. Manaila, R.; Macovei, D. in "*Amorphous Semicond.*"--'82, Grigorovici, R.; Ciurea, M., ed., Cent. Inst. Phys., Bucharest, Rom., **1982**, p. 46.

774. Mande, C.; Apte, M. Y. *Bull. Mater. Sci.*, **1981a**, *3*, 193.

775. Mande, C.; Apte, M. Y. in "*Inn.-Shell X-Ray Phys. At. Solids*," (Proc. Int. Conf. X-Ray Processes Inn.-Shell Ioniz., Fabian, D. J., Kleinpoppen, H., Watson, L. M., ed., **1981b**, p. 691.

776. Mande, C.; Apte, M. Y.; Kondawar, V. K. *Trans. Indian Inst. Met.*, **1981**, *34*, 376.

777. Mansour, A. N.; Cook, J. W., Jr.; Sayers, D. E.; Emrich, R. J.; Katzer, J. R. *J. Catal.*, **1984**, *89*, 462.

778. Mansour, A. N.; Sayers, D. E.; Cook, J. W., Jr.; Short, D. R.; Shannon, R. D.; Katzer, J. R. *J. Phys. Chem.*, **1984**, *88*, 1778.

779. March, N. H. *J. Phys. B*, **1976**, *9*, L73.

780. Marcus, M. in *"EXAFS Spectroscopy: Techniques and Applications,"* Teo, B. K., Joy, D. C., ed., Plenum, New York, **1981a**, p. 181.

781. Marcus, M. *Solid State Commun.,* **1981b**, *38*, 251.

782. Marcus, M.; Powers, L. S.; Storm, A. R.; Kincaid, B. M.; Chance, B. *Rev. Sci. Instrum.,* **1980**, *51*, 1023.

783. Marcus, M.; Tsai, C. L., *Solid State Commun.,* **1984**, *52*, 511.

784. Margaritondo, G.; Stoffel, N. G. *Phys. Rev. Lett.,* **1979**, *42*, 1567.

785. Marquardt, D. W. *J. Soc. Ind. Appl. Math.,* **1963**, *11*, 443.

786. Marques, E. C.; Sandstrom, D. R.; Lytle, F. W.; Greegor, R. B. *J. Chem. Phys.,* **1982**, *77*, 1027.

787. Martens, G.; Rabe, P. in *J. Phys. C,* **1980a**, *13*, L913.

788. Martens, G.; Rabe, P. *Phys. Status Solidi A,* **1980b**, *57*, K31.

789. Martens, G.; Rabe, P. *Phys. Status Solidi A,* **1980c**, *58*, 415.

790. Martens, G.; Rabe, P. *J. Phys. C,* **1981a**, *14*, 1523.

791. Martens, G.; Rabe, P. in *"Inn.-Shell X-Ray Phys. At. Solids,"* Fabian, D. J., Kleinpoppen, H., Watson, L. M., ed., Plenum, New York, **1981**, p. 683.

792. Martens, G.; Rabe, P.; Schwentner, N.; Werner, A. *Phys. Rev. Lett.,* **1977**, *39*, 1411.

793. Martens, G.; Rabe, P.; Schwentner, N.; Werner, A. *J. Phys. C,* **1978a**, *11*, 3125.

794. Martens, G.; Rabe, P.; Schwentner, N.; Werner, A. *Phys. Rev. B,* **1978b**, *17*, 1481.

795. Martens, G.; Rabe, P.; Tolkiehn, G.; Werner, A. *Phys. Status Solidi A,* **1979**, *55*, 105.

796. Martin, R. L.; Davidson, E. R. *Phys. Rev. A,* **1977**, *16*, 1341.

797. Martin, R. M.; Boyce, J. B.; Allen, J. W.; Holtzberg, F. *Phys. Rev. Lett.,* **1980**, *44*, 1275.

798. Massey, H. S. W.; Burhop, E. H. S. *"Electronic and Ionic Impact Phenomena, Vol. 1: Collision of Electrons with Atoms,"* 2nd Ed., Oxford Univ. Press, New York, **1969**.

799. Mathey, Y.; Michalowicz, A.; Toffoli, P.; Vlaic, G. *Inorg. Chem.,* **1984**, *23*, 897.

800. Matsushita, T.; Phizackerley, R. P. *Jpn. J. Appl. Phys.,* **1981**, *20*, 2223.

801. Mauer, M.; Friedt, J. M.; Krill, G. *J. Phys. F,* **1983**, *13*, 2389.

802. Maylotte, D. H.; Wong, J.; Peters, R. L. S.; Lytle, F. W.; Greegor, R. B. *Science*, **1981**, *214*, 554.

803. Maylotte, D. H.; Wong, J.; St. Peters, R. L.; Lytle, F. W.; Greegor, R. B. in *"Proc. - Int. Kohlenwiss. Tag.,"* Springer Verlag, Heidelberg, **1981**, p. 756.

804. Mazid, M. A.; Razi, M. T.; Sadler, P. J.; Greaves, G. N.; Gurman, S. J.; Koch, M. H. J.; Phillips, J. C. *J. Chem. Soc., Chem. Commun.,* **1980**, *24*, 1261.

805. McMaster, W. H.; Kerr Del Grande, N.; Mallet, J. H.; Hubell, J. H. *"Compilation of X-Ray Cross Sections,"* Lawrence Radiation Laboratory, **1969**.

806. Meitzner, G.; Via, G. H.; Lytle, F. W.; Sinfelt, J. H. *J. Chem. Phys.,* **1983a**, *78*, 882.

807. Meitzner, G.; Via, G. H.; Lytle, F. W.; Sinfelt, J. H. *J. Chem. Phys.,* **1983b**, *78*, 2533.

808. Mehta, M.; Fadley, C. S.; Bagus, P. S. *Chem. Phys. Lett.,* **1976**, *37*, 454.

809. Menzel, D. *Top. Appl. Phys.,* **1975**, *4*, 101.

810. Messiah, A. *"Quantum Mechanics,"* Vol. I, John Wiley, N.Y., **1970**, p. 386.

811. Michalowicz, A.; Clement, R. *Inorg. Chem.,* **1982**, *21*, 3872.

812. Michalowicz, A.; Girerd, J. J.; Goulon, J. *Inorg. Chem.,* **1979**, *18*, 3004.

813. Michalowicz, A.; Huet, J.; Gaudemer, A. *Nouv. J. Chim.,* **1982**, *6*, 79.

814. Michalowicz, A.; Vlaic, G.; Clement, R.; Mathey, Y. *Springer Ser. Chem. Phys.,* **1983**, *27*, 222.

815. Mikkelsen, J. C., Jr.; Boyce, J. B. *Phys. Rev. B,* **1981**, *24*, 5999.

816. Mikkelsen, J. C., Jr.; Boyce, J. B. *Phys. Rev. B,* **1983**, *28*, 7130.

817. Mikkelsen, J. C., Jr.; Boyce, J. B. *Phys. Rev. Lett.,* **1982**, *49*, 1412.

818. Mikkelsen, J. C., Jr.; Boyce, J. B.; Allen, R. *Rev. Sci. Instrum,* **1980**, *51*, 388.

819. Miller, R. M.; Hukins, D. W. L.; Hasnain, S. S.; Lagarde, P. *Biochem. Biophys. Res. Commun.,* **1981**, *99*, 102.

820. Mills, D. M.; Lewis, A.; Harootunian, A.; Huang, J.; Smith, B. *Science,* **1984**, *223*, 811.

821. Mills, D.; Pollock, V. *Rev. Sci. Instrum.,* **1980**, *51*, 1664.

822. Minault, J.; Fontaine, A.; Lagarde, P.; Raoux, D.; Sadoc, A.; Spanjaard, D. *J. Phys. F,* **1981**, *11*, 1311.

823. Minomura, S.; Tsuji, K.; Wakagi, M.; Ishidate, T.; Inoue, K.; Shibuya, M. *J. Non-Cryst. Solids*, **1983**, *59-60*, 541.

824. Mobilio, S.; Incoccia, L. *Nuovo Cimento D*, **1984**, *3*, 846.

825. Mobilio, S.; Incoccia, L. *Springer Ser. Chem. Phys.*, **1983**, *27*, 87.

826. Modesti, S.; De Crescenzi, M.; Perfetti, P.; Quaresima, C.; Rosei, R.; Savoia, A.; Sette, F. *Springer Ser. Chem. Phys.*, **1983**, *27*, 394.

827. Moisy-Maurice, V. *Report CEA-N-2171, Atomindex*, **1981**, *12*, No. 581039.

828. Montano, P. A.; Shenoy, G. K. *Solid State Commun.*, **1980**, *35*, 53.

829. Montano, P. A.; Schulze, W.; Tesche, B.; Schenoy, G. K.; Morrison, T. I. *Phys. Rev. B*, **1984**, *30*, 672.

830. Moore, C. E. *"Atomic Energy Levels,"* Nat. Bur Stand. (U.S.), Cir. 467, **1952**, *Vol. II.*

831. Morante, S.; Cerdonio, M.; Vitale, S.; Congiu-Castellano, A.; Vaciago, A.; Giacometti, G. M.; Incoccia, L. *Springer Ser. Chem. Phys.*, **1983**, *27*, 352.

832. Moraweck, B.; Renouprez, A. J. *Surf. Sci.*, **1981**, *106*, 35.

833. Morawitz, H.; Bagus, P.; Clarke, T.; Gill, W.; Grant, P.; Street, G. B.; Sayers, D. *Synth. Met.*, **1980**, *1*, 267.

834. Morrison, T. I.; Reis, A. H., Jr.; Knapp, G. S.; Fradin, F. Y.; Chen, H.; Klippert, T. E. *J. Am. Chem. Soc.*, **1978**, *100*, 3262.

835. Morrison, T. I.; Shenoy, G. K.; Iton, L. E.; Stucky, G. D.; Suib, S. L. *J. Chem. Phys.*, **1982**, *76*, 5665.

836. Morrison, T. I.; Shenoy, G. K.; Niarchos, D. *J. Appl. Crystallogr.*, **1982**, *15*, 388.

837. Morrison, T. I.; Shenoy, G. K.; Nielsen, L. *Inorg. Chem.*, **1981**, *20*, 3565.

838. Motta, N.; DeCrescenzi, M.; Balzarotti, A. *Springer Ser. Chem. Phys.*, **1983**, *27*, 103.

839. Motta, N.; DeCrescenzi, M.; Balzarotti, A. *Phys. Rev. B*, **1983**, *27*, 4712.

840. Mueller, J. E.; Jepsen, O.; Wilkins, J. W. *Solid State Commun.*, **1982**, *42*, 365.

841. Muench, R.; Hochheimer, H. D.; Werner, A.; Materlik, G.; Jayaraman, A.; Rao, K. V. *Phys. Rev. Lett.*, **1983**, *50*, 1619.

842. Munoz, M. C.; Durham, P. J.; Gyorffy, B. L. *J. Phys. F.*, **1982**, *12*, 1497.

843. Muramatsu, S.; Sugiura, C. *J. Chem. Phys.*, **1982**, *76*, 2107.

844. Muranaka, T.; Kiyono, S.; Watanabe, T. *Jpn. J. Appl. Phys.*, **1981**, *20*, 1939.

845. Murata, T.; Lagarde, P.; Fontaine, A.; Raoux, M. *Springer Ser. Chem. Phys.*, **1983**, *27*, 271.

846. Murugesan, T.; Sarode, P. R.; Gopalakrishnan, J.; Rao, C. N. R. *J. Chem. Soc., Dalton Trans.*, **1980**, *5*, 837.

847. Nagarajan, R.; Sampathkumaran, E. V.; Gupta, L. C.; Vijayaraghavan, R.; Bhaktdarshan; Padalia, B. D. *Phys. Lett. A*, **1981**, *81*, 397.

848. Nagarajan, R.; Sampathkumaran, E. V.; Gupta, L. C.; Vijayaraghavan, R.; Prabhawalkar, V.; Bhaktdarshan; Padalia, B. D. *Phys. Lett. A*, **1981**, *84*, 275.

849. Namikawa, K.; Hosoya, S. *Jpn. J. Appl. Phys., Part 2*, **1982**, *21*, 687.

850. Natoli, C. R. *Springer Ser. Chem. Phys.*, **1983**, *27*, 43.

851. Natoli, C. R.; Misemer, D. K; Doniach, S.; Kutzler, F. W. *Phys. Rev. A*, **1980**, *22*, 1104.

852. Nemanich, R. J.; Connell, G. A. N.; Hayes, T. M.; Street, R. A. *Phys. Rev. B*, **1978**, *18*, 6900.

853. Nigam, A. N.; Kumar, A.; Srivastava, T. S.; Agarwala, U. C. *Indian J. Phys. A*, **1978**, *52*, 255.

854. Nigam, H. L. *Proc. Indian Sci. Congr. 68th*, **1981**, *No. IV*.

855. *Ninomiya, K.; Ishiguro, E.; Iwata, S.; Mikuni, A.; Sasaki, T. J. Phys. B*, **1981**, *14*, 1777.

856. Noguera, C.; Spanjaard, D. *J. Phys. F*, **1981a**, *11*, 1133.

857. Noguera, C.; Spanjaard, D. *Surf. Sci.*, **1981b**, *108*, 381.

858. Nomura, M.; Asakura, K.; Kaminaga, U.; Matsushita, T.; Kohra, K.; Kuroda, H. *Bull. Chem. Soc. Jpn.*, **1982**, *55*, 3911.

859. Noorman, P. E.; Schrijver, J. *Physica*, **1967**, *36*, 547.

860. Norman, D. *Daresbury Lab. Rep. DL/SCI/R17*, **1981**, 28.

861. Norman, D. *AIP Conf. Proc.*, **1982**, *94*, 745.

862. Norman, D.; Stöhr, J.; Jaeger, R.; Durham, P. J.; Pendry, J. B. *Phys. Rev. Lett.*, **1983**, *51*, 2052.

863. Norman, D.; Brennan, S.; Jaeger, R.; Stöhr, J. *Surf. Sci.*, **1981**, *105*, L297.

864. Norman, D.; Durham, P. J. *Proc. SPIE-Int. Soc. Opt. Eng.*, **1984**, *447*, 102.

865. Norman, D.; Durham, P. J.; Pendry, J. B.; Stöhr, J.; Jaeger, R. *Springer Ser. Chem. Phys.,* **1983**, *27*, 146.

866. Nukui, A.; Chiba, T. *Seramikkusu,* **1979**, *14*, 609.

867. Ohno, Y.; Hirama, K.; Nakai, S.; Sugiura, C.; Okada, S. *Synth. Met.,* **1983**, *6*, 149.

868. Ohno, Y.; Hirama, K.; Nakai, S.; Sugiura, C.; Okada, S. *Phys. Rev. B,* **1983**, *27*, 3811.

869. Ohno, Y.; Watanabe, H.; Kawata, A.; Nakai, S.; Sugiura, C. *Phys. Rev. B,* **1982**, *25*, 815.

870. Ohno, Y.; Hirama, K.; Nakai, S.; Sugiura, C.; Okada, S. *J. Phys. C,* **1983**, *16*, 6695.

871. Ohno, Y.; Kaneda, K.; Okada, S.; Hirama, K. *J. Solid State Chem.,* **1984**, *54*, 170.

872. Ohta, T. *Nippon Kinzoku Gakkai Kaiho,* **1981**, *20*, 570.

873. Oikawa, T.; Hosoi, J.; Inoue, M.; Harada, Y. *JEOL News, (Ser.) Electron Opt. Instrum. E,* **1982**, *20*, 8.

874. Okada, M. *Shokuhin no Bussei,* **1980**, *6*, 131.

875. Okada, M.; Seisanbu, S. *New Food Ind.,* **1980**, *22*, 47.

876. Okamoto, T.; Fukushima, Y. *J. Non-Cryst. Solids,* **1984**, *61-62*, 379.

877. Olson, C. G.; Lynch, D. W. *Solid State Commun.,* **1980**, *36*, 513.

878. Ovsyannikova, I. A.; Batsanov, S. S.; Nasonova, L. I.; Batsanova, L. R.; Nekrasova, E. A. *Bull. Acad. Sci. USSR, Phys. Ser.,* **1967**, *31*, 936.

879. Oyanagi, H.; Tanaka, K.; Hosoya, S.; Minomura, S. *J. Phys., Colloq.,* **1981**, *C4*, 221.

880. Oyanagi, H.; Tsuji, K.; Hosoya, S.; Minomura, S.; Fukamachi, T. *J. Non-Cryst. Solids,* **1980**, *35*, 555.

881. Oyanagi, H.; Matsushita, T.; Ito, M.; Kuroda, H. *Natl. Lab. High Energy Phys., KEK,* **1984**, *83*, 27.

882. Padalia, B. D.; Hatwar, T. K.; Ghatikar, M. N. *J. Phys. C,* **1983**, *16*, 1537.

883. Padalia, B. D.; Koul, P. N.; Ghatikar, M. N. *Proc. Nucl. Phys. Solid State Phys. Symp.,* **1983**, *23*, 79.

884. Pan, H. K.; Knapp, G. S.; Cooper, S. L. *Colloid Polym. Sci.,* **1984**, *262*, 734.

885. Pan, H. K.; Yarusso, D. J.; Knapp, G. S.; Cooper, S. L. *Proc.- Electrochem. Soc.,* **1983**, *83-3*, 15.

886. Pan, H. K.; Yarusso, D. J.; Knapp, G. S.; Cooper, S. L. *J. Polym. Sci., Polym. Phys. Ed.*, **1983**, *21*, 1389.

887. Pan, H. K.; Yarusso, D. J.; Knapp, G. S.; Pineri, M.; Meagher, A.; Coey, J. M. D.; Cooper, S. L. *J. Chem. Phys.*, **1983**, *79*, 4736.

888. Pande, C. S.; Viswanathan, R. *Solid State Commun.*, **1978**, *26*, 893.

889. Pantos, E. *Nucl. Instrum. Methods Phys. Res.*, **1983**, *208*, 449.

890. Pantos, E.; Firth, D. *Springer Ser. Chem. Phys.*, **1983**, *27*, 110.

891. Pardee, W. J.; Robertson, W. M.; James, M. R. *Scr. Metall.*, **1980**, *14*, 1333.

892. Parham, T. G.; Merrill, R. P. *J. Catal.*, **1984**, *85*, 295.

893. Park, J. W.; Chen, H. *Phys. Chem. Glasses*, **1982**, *23*, 107.

894. Park, R. L. *Appl. Surf. Sci.*, **1980**, *4*, 250.

895. Park, R. L. *Appl. Surf. Sci.*, **1982**, *13*, 231.

896. Park, R. L.; Houston, J. E. *J. Vac. Sci. Technol.*, **1974**, *11*, 1.

897. Parthasarathy, R.; Prasad, R. V.; Sarode, P. R.; Rao, K. J. *Proc. Indian Acad. Sci., Chem. Sci.*, **1982**, *91*, 201.

898. Parthasarathy, R.; Rao, K. J.; Rao, C. N. R. *J. Phys. C*, **1982**, *15*, 3649.

899. Parthasarathy, R.; Sarode, P. R.; Rao, K. J. *J. Mater. Sci.*, **1981**, *16*, 3222.

900. Parthasarathy, R.; Sarode, P. R.; Rao, K. J.; Rao, C. N. R. *Proc. Indian Natl. Sci. Acad. A*, **1982**, *48*, 119.

901. Pascal, J. L.; Potier, J.; Jones, D. J.; Roziere, J.; Michalowicz, A. *Inorg. Chem.*, **1984**, *23*, 2068.

902. Paschina, G.; Piccaluga, G.; Pinna, G.; Magini, M. *Chem. Phys. Lett.*, **1983**, *98*, 157.

903. Pauling, P. *"The Nature of the Chemical Bond,"* 3rd edition, Cornell University Press, Ithaca, New York, **1960**, p. 98.

904. Pavlychev, A. A.; Vinogradov, A. S.; Zimkina, T. M.; Onopko, D. E.; Titov, S. A. *Opt. Spektrosk.*, **1982**, *52*, 506.

905. Peisach, J.; Powers, L.; Blumberg, W. E.; Chance, B. *Biophys. J.*, **1982**, *38*, 277.

906. Pendry, J. B. *Daresbury Lab. Rep. DL/SCI/R17*, **1981**, 5.

907. Pendry, J. B. *Springer Ser. Chem. Phys.*, **1983**, *27*, 4.

908. Pendry, J. B.; Gurman, S. J. in *"The Structure of Non-Crystalline Materials,"* P. H. Gaskell; Davis, E. A., ed., Taylor and Francis, London, **1977**, p. 61.

909. Penn, D. R. *Phys. Rev. B,* **1976,** *13,* 5248.

910. Penner-Hahn, J. E. *Report SSRL-84/03, Order No. DE84011318;* NTIS: *Energy Res. Abstr.,* **1984,** *9,* No. 29347.

911. Penner-Hahn, J. E.; McMurry, T. J.; Renner, M.; Latos-Grazynsky, L.; Eble, K. S.; Davis, E. M.; Balch, A. L.; Groves, J. T.; Dawson, J. H.; Hodgson, K. O. *J. Biol. Chem.,* **1983,** *258,* 12761.

912. Perutz, M. F.; Hasnain, S. S.; Duke, P. J.; Sessler, J. L.; Hahn, J. E. *Nature* (London), **1982,** *295,* 535.

913. Petiau, J.; Calas, G. *J. Phys., Colloq.,* **1982,** *C9,* 47.

914. Petiau, J.; Calas, G. *Springer Ser. Chem. Phys.,* **1983,** *27,* 144.

915. Petiau, J.; Calas, G.; Bondot, P.; Lapeyre, C.; Levitz, P.; Loupias, G. *Daresbury Lab. Rep. DL/SCI/R17,* **1981,** 127.

916. Pettifer, R. F. in *"Trends Phys.,"* Woolfson, M. M., ed., Adam Hilger, Bristol, England, **1979,** p. 522.

917. Pettifer, R. F. in *"Charact. Catal.,"* Thomas, J. M.; Lambert, R. M., ed., Wiley, Chichester, U.K., **1980,** p. 264.

918. Pettifer, R. F. *Daresbury Lab. Rep. DL/SCI/R17,* **1981a,** 57.

919. Pettifer, R. F. in *"Inn.-Shell X-Ray Phys. At. Solids,"* Fabian, D. J., Kleinpoppen, H., Watson, L. M., ed., Plenum, N.Y., **1981b,** p. 653.

920. Pettifer, R. F.; Cox, A. D. *Springer Ser. Chem. Phys.,* **1983,** *27,* 66.

921. Pettifer, R. F.; McMillan, P. W. *Phil. Mag.,* **1977,** *35,* 871.

922. Pettifer, R. F.; McMillan, P. W.; Gurman, S. J. in *"Struct. Non-Cryst. Mater.,"* Gaskell, P. H.; Davis, E. A., ed., **1977,** p. 63.

923. Peuckert, M.; Keim, W.; Storp, S.; Weber, R. S. *J. Mol. Catal.,* **1983,** *20,* 115.

924. Phillips, J. C. *J. Phys. E,* **1981,** *14,* 1425.

925. Phillips, J. C.; Bordas, J.; Foote, A. M.; Koch, M. H. J.; Moody, M. F. *Biochemistry,* **1982,** *21,* 830.

926. Piacentini, M. *Springer Ser. Chem. Phys.,* **1983,** *27,* 193.

927. Pilar, F. L. *"Elementary Quantum Chemistry,"* McGraw-Hill, New York, **1968;** pp 163, 318.

928. Platzman, P. M.; Wolff, P. A. *"Waves and Interactions in Solid State Plasmas,"* Academic, New York, **1973.**

929. Poncet, J. L.; Guilard, R.; Friant, P.; Goulon, J. *Polyhedron,* **1983,** *2,* 417.

930. Powell, C. J. *Surface Sci.,* **1974,** *44,* 29.

268

931. Powers, L. *Biochim. Biophys. Acta*, **1982**, *683*, 1.

932. Powers, L.; Blumberg, W. E.; Chance, B.; Barlow, C. H.; Leigh, J. S., Jr.; Smith, J.; Yonetani, T.; Vik, S.; Peisach, J. in *"Frontiers of Biological Energetics,"* Vol. 2, Dutton, P. L.; Leigh, J. S.; Scarpa, A., ed., Academic Press, Inc., **1978**, p. 863.

933. Powers, L.; Blumberg, W. E.; Chance, B.; Barlow, C. H.; Leigh, J. S., Jr.; Smith, J.; Yonetani, T.; Vik, S.; Peisach, J. *Biochim. Biophys. Acta*, **1979**, *546*, 520.

934. Powers, L.; Blumberg, W. E.; Chance, B.; Barlow, C. H.; Leigh, J. S., Jr.; Smith, J.; Yonetani, T.; Vik, S.; Peisach, J. in *"Cytochrome Oxidase,"* King, T. E., ed., Elsevier/North-Holland Biomedical Press, **1979**, p. 189.

935. Powers, L.; Chance, B.; Ching, Y.; Angiobillo, P. *Biophys. J.*, **1981**, *34*, 465.

936. Powers, L.; Chance, B.; Ching, Y.; Muhoberac, B.; Weintraub, S. T.; Wharton, D. C. *FEBS Lett.*, **1982**, *138*, 245.

937. Powers, L.; Eisenberger, P.; Stamatoff, J. *Ann. N. Y. Acad. Sci.*, **1978**, *307*, 113.

938. Powers, L.; Sessler, J. L.; Woolery, G. L.; Chance, B. *Biochemistry*, **1984**, *23*, 5519.

939. Prabhawalkar, V.; Padalia, B. D. *Phys. Status Solidi B*, **1982**, *110*, 659.

940. Prabhawalkar, V.; Padalia, B. D. *Curr. Sci.*, **1983**, *52*, 799.

941. Prasad, J.; Krishna, V.; Nigam, H. L. *J. Chem. Soc., Dalton Trans.*, **1976**, *23*, 241.

942. Prasad, J.; Nigam, H. L.; Agarwala, U. *J. Phys. C*, **1976**, *9*, 4349.

943. Prasad, S. K.; Singhal, S. P.; Herman, H.; Del Cueto, J. A.; Shevchik, N. J. *Scr. Metall.*, **1979**, *13*, 549.

944. Purdum, H.; Montano, P. A.; Shenoy, G. K.; Morrison, T. *Phys. Rev. B*, **1982**, *25*, 4412.

945. Puschmann, A.; Haase, J. *Surf. Sci.*, **1984** *144*, 559.

946. Rabe, P. *Daresbury Lab. Rep. DL/SCI/R17*, **1981**, 76.

947. Rabe, P. *Springer Ser. Chem. Phys.*, **1983**, *27*, 73.

948. Rabe P.; Haensel, R. *Festkoerperprobleme*, **1980**, *20*, 43.

949. Rabe, P.; Tolkiehn, G.; Werner, A. *J. Phys. C*, **1979a**, *12*, L545.

950. Rabe, P.; Tolkiehn, G.; Werner, A. *J. Phys. C*, **1979b**, *12*, 899.

951. Rabe, P.; Tolkiehn, G.; Werner, A. *J. Phys. C*, **1979c**, *12*, 1173.

952. Rabe, P.; Tolkiehn, G.; Werner, A. *J. Phys. C*, **1980a**, *13*, 1857.

953. Rabe, P.; Tolkiehn, G.; Werner, A. *Nucl. Instrum. Methods,* **1980b,** *171,* 329.

954. Rabe, P.; Tolkiehn, G.; Werner, A.; Haensel, R. *Z. Naturforsch. A,* **1979,** *34,* 1528.

955. Rao, B. J.; Chetal, A. R. *J. Phys. C,* **1982,** *15,* 6281.

956. Rao, B. J.; Chetal, A. R. *J. Phys. D.,* **1982,** *15,* L195.

957. Rao, C. N. R.; Sarode, P. R.; Parthasarathy, R.; Rao, K. J. *Philos. Mag. B,* **1980,** *41,* 581.

958. Rao, Y. L.; Chourasia, A. R.; Mande, C. *J. Non-Cryst. Solids,* **1981,** *46,* 13.

959. Rao, K. J.; Wong, J.; Rao, B. G. *Phys. Chem. Glasses,* **1984,** *25,* 57.

960. Rao, K. J.; Wong, J.; Weber, M. J. *J. Chem. Phys.,* **1983,** *78,* 6228.

961. Raoux, D.; Flank, A. M.; Sadoc, A. *Springer Ser. Chem. Phys.,* **1983,** *27,* 232.

962. Raoux, D.; Fontaine, A.; Lagarde, P.; Sadoc, A. *Phys. Rev. B,* **1981,** *24,* 5547.

963. Raoux, D.; Petiau, J.; Bondot, P.; Calas, G.; Fontaine, A.; Lagarde, P.; Levitz, P.; Loupias, G.; Sadoc, A. *Rev. Phys. Appl.,* **1980,** *15,* 1079.

964. Raoux, D.; Sadoc, A.; Lagarde, P.; Fontaine, A. *Daresbury Lab. Rep. DL/SCI/R17,* **1981,** 124.

965. Raoux, D.; Sadoc, J. F.; Lagarde, P.; Sadoc, A.; Fontaine, A. *J. Phys., Colloq.,* **1980,** *C8,* 207.

966. Ravot, D.; Godart, C.; Achard, J. C.; Lagarde, P. in *"Valence Fluctuations Solids,"* St. Barbara Inst. Theor. Phys. Conf., Falicov, L. M., Hanke, W., Maple, M. B., ed., **1981,** p. 423.

967. Reed, J.; Eisenberger, P. *J. Chem. Soc., Chem. Commun.,* **1977,** 628.

968. Reed, J.; Eisenberger, P.; Hastings, J. *Inorg. Chem.,* **1978,** *17,* 481.

969. Reed, J.; Eisenberger, P.; Teo, B. K.; Kincaid, B. M. *J. Am. Chem. Soc.,* **1977,** *99,* 5217.

970. Reed, J.; Eisenberger, P.; Teo, B. K.; Kincaid, B. M. *J. Am. Chem. Soc.,* **1978,** *100,* 2375.

971. Reedijk, J. *Chem. Weekbl. Mag.,* **1978,** 741,739.

972. Rehn, V. *Nav. Res. Rev.,* **1983,** *35,* 36.

973. Rehr, J. J.; Chou, S. H. *Springer Ser. Chem. Phys.,* **1983,** *27,* 22.

974. Rehr, J. J.; Stern, E. A. *Phys. Rev. B,* **1976,** *14,* 4413.

975. Rehr, J. J.; Stern, E. A.; Martin, R. L.; Davidson, E. R. *Phys. Rev. B,* **1978**, *17*, 560.

976. Rennert, P.; Vasvari, B. *J. Phys. F,* **1983**, *13*, 89.

977. Renouprez, A.; Fouilloux, P.; Moraweck, B. *Stud. Surf. Sci. Catal.,* **1980**, 421.

978. Richard, P.; Poncet, J. L.; Barbe, J. M.; Guilard, R.; Goulon, J.; Rinaldi, D.; Cartier, A.; Tola, P. *J. Chem. Soc., Dalton Trans.,* **1982**, *8*, 1451.

979. Ritsko, J. I.; Schnatterly, S. E.; Gibbons, P. G. *Phys. Rev. Lett.,* **1974**, *32*, 671.

980. Robertson, A. S. *Lawrence Berkeley Lab. Report LBL-9840, Energy Res. Abstr.,* **1980**, *5*, No. 3427.

981. Röhler, J.; Kappler, J. P.; Krill, G. *Nucl. Instrum. Methods Phys. Res.,* **1983**, *208*, 647.

982. Röhler, J.; Krill, G.; Kappler, J. P.; Ravet, M. F. *Springer Ser. Chem. Phys.,* **1983**, *27*, 213.

983. Rosenberg, R. A.; LaRoe, P. R.; Rehn, V.; Stöhr, J.; Jaeger R.; Parks, C. C. *Phys. Rev. B,* **1983**, *28*, 3026.

984. Roe, A. L.; Schneider, D. J.; Mayer, R. J.; Pyrz, J. W.; Widom, J.; Que, L., Jr. *J. Am. Chem. Soc.,* **1984**, *106*, 1676.

985. Ross, I.; Binstead, N.; Blackburn, N. J.; Bremner, I.; Diakun, G. P.; Hasnain, S. S.; Knowles, P. F.; Vasak, M.; Garner, C. D. *Springer Ser. Chem. Phys.,* **1983**, *27*, 337.

986. Rothberg, G. M.; Choudhary, K. M.; Den Boer, M. L.; Williams, G. P.; Hecht, M. H.; Lindau, I. *Phys. Rev. Lett.,* **1984**, *53*, 1183.

987. Sadoc, A. *J. Non-Cryst. Solids,* **1984**, *61-62*, 403.

988. Sadoc, A.; Calvayrac, Y.; Quivy, A.; Harmelin, M.; Flank, A. M. *J. Non-Cryst. Solids,* **1984**, *65*, 109.

989. Sadoc., A.; Flank, A. M.; Raoux, D.; Lagarde, P. *J. Phys., Colloq.,* **1982**, *C9*, 43.

990. Sadoc, A.; Fontaine, A.; Lagarde, P.; Raoux, D. *J. Am. Chem. Soc.,* **1981**, *103*, 6287.

991. Sadoc, A.; Raoux, D.; Lagarde, P.; Fontaine, A. *J. Non-Cryst. Solids,* **1982**, *50*, 331.

992. Sahasrabudhe, V. *Solid State Commun.,* **1983**, *46*, 697.

993. Sahastabudhe, V.; Vaingankar, A. S. *Solid State Commun.,* **1982**, *43*, 299.

994. Sahasrabudhe, V. S.; Vaingankar, A. S. *Proc. Nucl. Phys. Solid State Phys. Symp.*, **1983**, *23*, 61.

995. Sakka, S.; Kamiya, K.; Hayashi, M. *Bull. Inst. Chem. Res.*, **1981**, *59*, 172.

996. Salem, S. I.; Chang, C. N.; Lee, P. L.; Severson, V. *J. Phys. C*, **1978**, *11*, 4085.

997. Salem, S. I.; Chang, C. N.; Nash, T. J. *Phys. Rev. B*, **1978**, *18*, 5168.

998. Salem, S. I.; Dev, B.; Lee, P. L. *Phys. Rev. A*, **1980**, *22*, 2679.

999. Salem, S. I.; Kumar, A.; Schiessel, K. G.; Lee, P. L. *Phys. Rev. A*, **1982**, *26*, 3334.

1000. Sampathkumaran, E. V.; Gupta, L. C.; Vijayaraghavan, R.; Hatwar, T. K.; Ghatikar, M. N.; Padalia, B. D. *Mater. Res. Bull.*, **1980**, *15*, 939.

1001. Sandstrom, D. R. *J. Chem. Phys.*, **1979**, *71*, 2381.

1002. Sandstrom, D. R. *Nuovo Cimento D*, **1984**, *3*, 825.

1003. Sandstrom, D. R.; Dodgen, H. W.; Lytle, F. W. *J. Chem. Phys.*, **1977**, *67*, 473.

1004. Sandstrom, D. R.; Filby, R.; Lytle, F. W.; Greegor, R. B. in *Fuel*, **1982**, *61*, p. 195.

1005. Sandstrom, D. R.; Lytle, F. W. *Ann. Rev. Phys. Chem.*, **1979**, *30*, 215.

1006. Sandstrom, D. R.; Lytle, F. W.; Wei, P. S. P.; Greegor, R. B.; Wong, J.; Schultz, P. *J. Non-Cryst. Solids*, **1980**, *41*, 201.

1007. Sandstrom, D. R.; Stults, B. R.; Greegor, R. B. in *"EXAFS Spectroscopy: Techniques and Applications,"* Teo, B. K., Joy, D. C., ed., Plenum, New York, **1981**, p. 139.

1008. Sankar, G.; Sarode, P. R.; Srinivasan, A.; Rao, C. N. R.; Vasudevan, S.; Thomas, J. M. *Proc.-Indian Acad. Sci., Chem. Sci.*, **1984**, *93*, 321.

1009. Sano, M.; Maruo, T.; Yamatera, H. *Chem. Phys. Lett.*, **1983**, *101*, 211.

1010. Sano, Mitsuru; Maruo, T.; Yamatera, H. *Bull. Chem. Soc. Jpn.*, **1983**, *56*, 3287.

1011. Sano, M.; Taniguchi, K.; Yamatera, H. *Chem. Lett.*, **1980**, *10*, 1285.

1012. Sano, M.; Maruo, T.; Masuda, Y.; Yamatera, H. *Inorg. Chem.*, **1984**, *23*, 4466.

1013. Sano, M.; Maruo, T.; Yamatera, H. *Bull. Chem. Soc. Jpn.*, **1984**, *57*, 2757.

1014. Sarode, P. R.; Rao, K. J.; Hegde, M. S.; Rao, C. N. R. *J. Phys. C*, **1979**, *12*, 4119.

1015. Sarode, P. R.; Sankar, G.; Rao, C. N. R. *Proc.-Indian Acad. Sci.*, **1983**, *92*, 527.

1016. Sarode, P. R.; Sarma, D. D.; Rao, C. N. R.; Sampathkumaran, E. V.; Gupta, L. C.; Vijayaraghavan, R. *Mater. Res. Bull.*, **1981**, *16*, 175.

1017. Sato, Y.; Iwasawa, Y.; Kuroda, H. *Chem. Lett.*, **1982**, *7*, 1101.

1018. Satpathy, S.; Dow, J. D.; Bowen, M. A. *Phys. Rev. B*, **1983**, *28*, 4255.

1019. Saxena, K. N.; Saxena, C. P.; Anikhindi, R. G.; Kaveeshwar, A. S. *Phys. Lett. A*, **1980**, *78*, 325.

1020. Saxena, S. G.; Chauhan, H. S.; Chandra, S.; Garg, K. B. *Springer Ser. Chem. Phys.*, **1983**, *27*, 171.

1021. Saxena, S. G.; Garg, K. B. *Proc. Nucl. Phys. Solid State Phys. Symp. C*, **1982**, *24*, 147.

1022. Sayers, D. E. in *"Amorphous Liq. Semicond.,"* Spear, W. E., ed., Univ. Edinburgh, Edinburgh, Scotland, **1977**, p. 61.

1023. Sayers, D. E.; Heald, S. M.; Pick, M. A.; Budnick, J. I.; Stern, E. A.; Wong, J. *Nucl. Instrum. Methods Phys. Res.*, **1983**, *208*, 631.

1024. Sayers, D. E.; Lytle, F. W.; Stern, E. A. *Adv. X-Ray Anal.*, **1970**, *13*, 248.

1025. Sayers, D. E.; Stern, E. A.; Herriott, J. R. *J. Chem. Phys.*, **1976**, *64*, 427.

1026. Sayers, D. E.; Stern, E. A.; Lytle, F. W. *Phys. Rev. Lett.*, **1971**, *27*, 1204.

1027. Sayers, D. E.; Stern, E. A.; Lytle, F. W. *Program Ext. Abstr. - Int. Conf. Phys. X-Ray Spectra*, **1976**, 84.

1028. Sayers, D. E.; Theil, E. C.; Rennick, F. J. *J. Biol. Chem.*, **1983**, *258*, 14076.

1029. Schaich, W. *Phys. Rev. B*, **1973**, *8*, 4078.

1030. Schaich, W. L. *Phys. Rev. B*, **1976**, *14*, 4420.

1031. Schiraiwa, T.; Ishimura, T.; Sawada, M. *J. Phys. Soc. Jpn.*. **1958**, *138*, 848.

1032. Schmueckle, F.; Lamparter, P.; Steeb, S. *Z. Naturforsch., A*, **1982**, *37*, 572.

1033. Schneider, D. J.; Roe, A. L.; Mayer, R. J.; Que, L., Jr. *J. Biol. Chem.*, **1984**, *259*, 9699.

1034. Scott, R. A. *NATO Adv. Study. Inst. Ser. C*, **1982**, *89*, 475.

1035. Scott, R. A.; Hahn, J. E.; Doniach, S.; Freeman, H. C.; Hodgson, K. O. *J. Am. Chem. Soc.*, **1982**, *104*, 5364.

1036. Scott, R. A.; Wallin, S. A.; Czechowski, M.; Der Vartanian, D. V.; LeGall, J.; Peck, H. D., Jr.; Moura, I. *J. Am. Chem. Soc.*, **1984**, *106*, 6864.

1037. Seah, M. P.; Dench, W. A. *Surf. Inter. Anal.*, **1979**, *1*, 2.

1038. Sevillano, E.; Meuth, H.; Rehr, J. J. *Phys. Rev. B*, **1979**, *20*, 4908.

1039. Sevillano, E.; Meuth, H.; Rehr, J. J. *Phys. Rev. B*, **1979**, *20*, 4908.

1040. Sham, T. K. *Springer Ser. Chem. Phys.*, **1983**, *27*, 165.

1041. Sham, T. K.; Brunschwig, B. S. *J. Am. Chem. Soc.*, **1981**, *103*, 1590.

1042. Sham, T. K.; Brunschwig, B. S. *Springer Ser. Chem. Phys.*, **1983**, *27*, 168.

1043. Sham, T. K.; Hastings, J. B.; Perlman, M. L. *J. Am. Chem. Soc.*, **1980**, *102*, 5904.

1044. Sham, T. K.; Hastings, J. B.; Perlman, M. L. *Chem. Phys. Lett.*, **1981**, *83*, 391.

1045. Sham, T. K.; Holroyd, R. A. *J. Chem. Phys.*, **1984**, *80*, 1026.

1046. Sharma, B. K.; Chandra, S.; Garg, K. B. *Proc. Nucl. Phys. Solid State Phys. Symp.*, **1983**, *23*, 67.

1047. Shaw, C. F., III; Schaeffer, N. A.; Elder, R. C.; Eidsness, M. K.; Trooster, J. M.; Calis, G. H. M. *J. Am. Chem. Soc.*, **1984**, *106*, 3511.

1048. Shevchik, N. J.; Fischer, D. A. *Rev. Sci. Instrum.*, **1979**, *50*, 577.

1049. Shimomura, O.; Kawamura, T.; Fukamachi, T.; Hosoya, S.; Hunter, S.; Bienestock, A. in "*High Pressure Sci. Technol.*," 7th Proc. Int. AIRAPT Conf., Vodar, B., Marteau, P., ed., **1980**, *1*, p. 534.

1050. Shore, B. W.; Menzel, D. H. "*Principles of Atomic Spectra,*" John Wiley & Sons, New York, **1968**.

1051. Short, D. R.; Khalid, S. M.; Katzer, J. R.; Kelley, M. J. *J. Catal.*, **1981**, *72*, 288.

1052. Short, D. R.; Mansour, A. N.; Cook, J. W., Jr.; Sayers, D. E.; Katzer, J. R. *J. Catal.*, **1983**, *82*, 299.

1053. Shrivastava, B. D.; Jain, R. K. *Indian J. Pure Appl. Phys.*, **1981**, *19*, 762.

1054. Shrivastava, B. D.; Landge, P. R. *Nuovo Cimento B*, **1979**, *49*, 118.

1055. Shulman, R. G. *Trends Biochem. Sci.*, **1978**, *3*, N282.

1056. Shulman, R. G.; Eisenberger, P.; Blumberg, W. E.; Stombaugh, N. A. *Proc. Nat. Acad. Sci. U.S.A.*, **1975**, *72*, 4002.

1057. Shulman, R. G.; Eisenberger, P.; Kincaid, B. M. *Ann. Rev. Biophys. Bioeng.*, **1978**, *7*, 559.

274

1058. Shulman, R. G.; Eisenberger, P.; Teo, B. K.; Kincaid, B. M.; Brown, G. S. *J. Mol. Biol.*, **1978**, *124*, 305.

1059. Shulman, R. G.; Sugano, S. *Phys. Rev.*, **1963**, *130*, 506.

1060. Shulman, R. G.; Yafet Y.; Eisenberger, P.; Blumberg, W. E. *Proc. Natl. Acad. Sci. U.S.A.*, **1976**, *73*, 1384.

1061. Siddons, D. P.; Hart, M. *Springer Ser. Chem. Phys.*, **1983**, *27*, 373.

Sinfelt, J. H. *J. Catal.*, **1973**, *29*, 308.   62.

1063. Sinfelt, J. H.; Via, G. H.; Lytle, F. W. *J. Chem. Phys.*, **1978**, *68*, 209.

1064. Sinfelt, J. H.; Via, G. H.; Lytle, F. W. *J. Chem. Phys.*, **1980**, *72*, 4832.

1065. Sinfelt, J. H.; Via, G. H.; Lytle, F. W. *J. Chem. Phys.*, **1982**, *76*, 2779.

1066. Sinfelt, J. H.; Via, G. H.; Lytle, F. W.; Greegor, R. B. *J. Chem. Phys.*, **1981**, *75*, 5527.

1067. Sinfelt, J. H.; Via, G. H.; Lytle, F. W. *Catal. Rev.*, **1984**, *26*, 81.

1068. Singh, H.; Garg, K. B. *Proc. Nucl. Phys. Solid State Phys. Symp. C*, **1982**, *24*, 151.

1069. Singh, K. K.; Sarode, P. R.; Ganguly, P. *J. Chem. Soc., Dalton Trans.*, **1983**, 1895.

1070. Slater, J. C. *Phys. Rev.*, **1951**, *81*, 385; *82*, 538.

1071. Slater, J. C. *"Quantum Theory of Molecules and Solids,"* McGraw Hill, New York, **1974**.

1072. Slater, J. C.; Mann, J. B.; Wilson, T. M.; Wood, J. H. *Phys. Rev.*, **1969**, *184*, 672.

1073. Slater, J. C.; Wilson, T. M.; Wood, J. H. *Phys. Rev.*, **1969**, *179*, 28.

1074. Slater, J. C.; Wood, J. H. *Inter. J. Quantum Chem.*, **1971**, *4*, 3.

1075. Smith, D. A.; Heeg, M. J.; Heineman, W. R.; Elder, R. C. *J. Am. Chem. Soc.*, **1984**, *106*, 3053.

1076. Sokolenko, V. I.; Zhurakovskii, E. A.; Sokolenko, A. I. *Dopov. Akad. Nauk Ukrsr. A, USSR*, **1979**, *3*, 209.

1077. Somorjai, G. A. *"Chemistry in Two Dimensions: Surfaces,"* Cornell University, Ithaca, **1981**, p. 41.

1078. Somorjai, G. A.; Farrell, H. H. *Advan. Chem. Phys.*, **1971**, *20*, 215.

1079. Spicer, W. E.; Lindau, I.; Helms, C. R. *Res. Develop.*, **1977**, *28*, 20.

1080. Spieker, P.; Ando, M.; Kamiya, N. *Nucl. Instrum. Methods Phys. Res. A*, **1984**, *222*, 196.

1081. Spira, D. J.; Co. M. S.; Solomon, E. I.; Hodgson, K. O. *Biochem. Biophys. Res. Commun.*, **1983**, *112*, 746.

1082. Spiro, C. L.; Wong, J.; Lytle, F. W.; Greegor, R. B.; Maylotte, D. H.; Lamson, S. H. *Science,* **1984**, *226*, 48.

1083. Spiro, T. G.; Brown, J. M.; Larrabee, J. A.; Powers, L.; Kincaid, B. in *"Oxidases Relat. Redox Syst.,"* King, T. E.; Mason, H. S.; Morrison, M., ed., Pergamon, New York, **1982**, p. 291.

1084. Spiro, T. G.; Wollery, G. L.; Brown, J. M.; Powers, L.; Winkler, M. E.; Solomon, E. I. in *"Copper Coord. Chem.: Biochem. Inorg. Perspect.,"* Karlin, K. D.; Zubieta, J., ed., Adenine Press: Guilderland, N.Y., **1983**, p. 23.

1085. Srivastava, K. S.; Harsh, O. K.; Kumar, V. *Phys. Status Solidi B,* **1979**, *91*, K169.

1086. Srivastava, K. S.; Kumar, V. *J. Phys. Chem. Solids,* **1981**, *42*, 275.

1087. Srivastava, K. S.; Kumar, V.; Harsh, O. K. *Indian J. Pure Appl. Phys.,* **1981**, *19*, 398.

1088. Srivastava, K. S.; Shrivastava, R. L.; Harsh, O. K.; Kumar, V. *J. Phys. Chem. Solids,* **1979**, *40*, 489.

1089. Srivastava, U. C. *Indian J. Pure Appl. Phys.,* **1978**, *16*, 114.

1090. Srivastava, U. C. *Indian J. Pure Appl. Phys.,* **1980**, *18*, 258.

1091. Srivastava, U. C.; Nigam, H. L. *Coord. Chem. Rev.,* **1972-1973**, *9*, 275.

1092. Stearns, D. G. *Philos. Mag. B,* **1984**, *49*, 541.

1093. Stearns, D. G.; Stearns, M. B. *Phys. Rev. B,* **1983**, *27*, 3842.

1094. Stearns, M. B. *Phys. Rev. B,* **1982**, *25*, 2382.

1095. Stern, E. A. *Phys. Rev. B,* **1974**, *10*, 3027.

1096. Stern, E. A. *Sci. Amer.,* **1976**, 96.

1097. Stern, E. A. *J. Vac. Sci. Technol.,* **1977**, *14*, 461.

1098. Stern, E. A. *Contemp. Phys.,* **1978**, *19*, 289.

1099. Stern, E. A. *AIP Conf. Proc.,* **1980a**, *61*, 197.

1100. Stern, E. A. *AIP Conf. Proc.,* **1980b**, *64*, 39.

1101. Stern, E. A. ed., *"Laboratory EXAFS Facilities-1980,"* AIP Conf. Proc., **1980c**, *64*, 165.

1102. Stern, E. A. in *"Anal. Electron Microsc.,"* Geiss, R. H., ed., San Francisco Press, San Francisco, Calif., **1981a**, p. 225.

1103. Stern, E. A. *Daresbury Lab* (Rep.), **1981b**, DL/SCI/R17, 40.

276

1104. Stern, E. A. *Optik*, **1982a**, *61*, 45.

1105. Stern, E. A. *Phys. Rev. Lett.*, **1982b**, *49*, 1353.

1106. Stern, E. A.; Bouldin, C. E.; Von Roedern, B.; Azoulay, J. *Phys. Rev. B*, **1983**, *27*, 6557.

1107. Stern, E. A.; Bunker, B. A.; Heald, S. M. *Phys. Rev. B.*, **1980**, *21*, 5521.

1108. Stern, E. A.; Bunker, B. A.; Heald, S. M. in *"EXAFS Spectroscopy: Techniques and Applications,"* Teo, B. K., Joy, D. C., ed., Plenum, New York, **1981**, p. 59.

1109. Stern, E. A.; Elam, W. T.; Bunker, B. A.; Lu, K. Q.; Heald, S. M. *Nucl. Instrum. Methods Phys. Res.*, **1982**, *195*, 345.

1110. Stern, E. A.; Heald, S. M. *Rev. Sci. Instrum.*, **1979**, *50*, 1579.

1111. Stern, E. A.; Heald, S. M. *Nucl. Instrum. Methods*, **1980**, *172*, 397.

1112. Stern, E. A.; Heald, S. M.; Bunker, B. *Phys. Rev. Lett.*, **1979**, *42*, 1372.

1113. Stern, E. A.; Lu, K. Q. *Nucl. Instrum. Methods Phys. Res.*, **1982**, *195*, 415.

1114. Stern, E. A.; Rinaldi, S.; Callen, E.; Heald, S.; Bunker, B. *J. Magn. Magn. Mater*, **1978**, *7*, 188.

1115. Stern, E. A.; Sayers, D. E.; Dash, J. G.; Shechter, H.; Bunker, B. *Phys. Rev. Lett.*, **1977**, *38*, 767.

1116. Stern, E. A.; Sayers, D. E.; Lytle, F. W. *Phys. Rev. B*, **1975**, *11*, 4836.

1117. Stöhr, J. *J. Vac. Sci. Technol.*, **1979**, *16*, 37.

1118. Stöhr, J. in *"Emission and Scattering Techniques,"* Day, P., ed., Reidel Publishing Co., **1981a**, 213.

1119. Stöhr, J. *NATO Adv. Study Inst. Ser. C*, **1981b**, *73*, 213.

1120. Stöhr, J.; Bauer, R. S.; McMenamin, J. C.; Johansson, L. I.; Brennan, S. *J. Vac. Sci. Technol.*, **1979**, *16*, 1195.

1121. Stöhr, J.; Denley, D.; Perfetti, P. *Phys. Rev. B*, **1978**, *18*, 4132.

1122. Stöhr, J.; Gland, J. L; Eberhardt, W.; Outka, D.; Madix, R. J.; Sette, F.; Koestner, R. J.; Doebler, U. *Phys. Rev. Lett.*, **1983**, *51*, 2414.

1123. Stöhr, J.; Jaeger, R. *J. Vac. Sci. Technol.*, **1982a**, *21*, 619.

1124. Stöhr, J.; Jaeger, R. *Phys. Rev. B*, **1982b**, *26*, 4111.

1125. Stöhr, J.; Jaeger, R. *Phys. Rev. B*, **1983**, *27*, 5146.

1126. Stöhr, J.; Jaeger, R.; Brennan, S. *Surf. Sci.*, **1982**, *117*, 503.

1127. Stöhr, J.; Jaeger, R.; Feldhaus, J.; Brennan, S.; Norman, D.; Apai, G. *Appl. Opt.*, **1980**, *19*, 3911.

1128. Stöhr, J.; Jaeger, R.; Kendelewicz, T. *Phys. Rev. Lett.*, **1982**, *49*, 142.

1129. Stöhr, J.; Jaeger, R.; Rossi, G.; Kendelewicz, T.; Lindau, I. *Surf. Sci.*, **1983**, *134*, 813.

1130. Stöhr, J.; Johansson, L. I.; Brennan, S.; Hecht, M.; Miller, J. N. *Phys Rev. B*, **1980**, *22*, 4052.

1131. Stöhr, J.; Johansson, L. I.; Lindau, I.; Pianetta, P. *J. Vac. Sci. Technol*, **1979a**, *16*, 1221.

1132. Stöhr, J.; Johansson, L.; Lindau, I.; Pianetta, P. *Phys. Rev. B*, **1979b**, *20*, 664.

1133. Stöhr, J.; Noguera, C.; Kendelewicz, T. *Phys. Rev. B*, **1984**, *30*, 5571.

1134. Stöhr, J.; Sette, F.; Johnson, A. L. *Phys. Rev. Lett.*, **1984**, *53*, 1684.

1135. Stoller, Ch.; Wölfli, W.; Bonani, M.; Stöckli, M.; Suter, M. *Phys. Lett. A*, **1976**, *58*, 18.

1136. Stout, C. D.; Ghosh, D.; Pattabhi, V.; Robbins, A. H. *J. Bio. Chem.*, **1980**, *255*, 1797.

1137. Stout, G. H.; Jensen, L. H. *"X-Ray Structure Determination,"* Macmillan, New York, **1968**.

1138. Stucky, G. D.; Iton, L.; Morrison, T.; Shenoy, G.; Suib, S.; Zerger, R. P. *J. Mol. Catal.*, **1984**, *27*, 71.

1139. Stuhrmann, H. B. *Quarter. Rev. Biophy.* **1978**, *11*, 71.

1140. Stults, B. R.; Friedman, R. M.; Koenig, K.; Knowles, W.; Greegor, R. B.; Lytle, F. W. *J. Am. Chem. Soc.*, **1981**, *103*, 3235.

1141. Stutius, W.; Boyce, J. B.; Mikkelsen, J. C., Jr. *Solid State Commun.*, **1979**, *31*, 539.

1142. Sugiura, C. *J. Chem. Phys.*, **1981**, *74*, 215.

1143. Sugiura, C. *J. Chem. Phys.*, **1982**, *77*, 681.

1144. Suib, S. L.; Zerger, R. P.; Stucky, G. D.; Morrison, T. I.; Shenoy, G. K. *J. Chem. Phys.*, **1984**, *80*, 2203.

1145. Sukhorukov, V. L.; Demekhina, L. A.; Yavna, V. A.; Demekhin, V. F. *Fiz. Tverd. Tela (Leningrad)*, **1979**, *21*, 2976.

1146. Tanabe, S.; Ida, T.; Tsuiki, H.; Ueno, A.; Kotera, Y.; Tohji, K.; Udagawa, Y. *Chem. Lett.*, **1984**, 1271.

1147. Tanabe, S.; Ueno, A.; Tohji, K.; Udagawa, Y. *Chem. Lett.*, **1983**, 1089.

1148. Tang, C.; Georgopoulos, P.; Cohen, J. B. *J. Am. Ceram. Soc.*, **1982**, *65*, 625.

1149. Taniguchi, K.; Yamaki, N.; Ikeda, S. *Jpn. J. Appl. Phys., Part 1*, **1984**, *23*, 909.

1150. Taniguchi, K.; Oka, K.; Yamaki, N.; Ikeda, S. *Adv. X-Ray Anal*, **1981**, *24*, 177.

1151. Taylor, J. M.; McMillan, P. W. in *"Struct. Non-Cryst. Mater.,"* Proc. 2nd. Int. Conf., Gaskell, P. H.; Parker, J. M.; Davis, E. A., ed., Taylor & Francis: London, **1983**, p. 589.

1152. Teo, B. K. *Acc. Chem. Res.*, **1980**, *13*, 412.

1153. Teo, B. K. *J. Am. Chem. Soc.*, **1981a**, *103*, 3990.

1154. Teo, B. K. in *"EXAFS Spectroscopy: Techniques and Applications,"* Teo, B. K., Joy, D. C., ed., Plenum, New York, **1981b**, p. 13.

1155. Teo, B. K. *Springer Ser. Chem. Phys.*, **1983**, *27*, 11.

1156. Teo, B. K. in *"New Frontiers in Organometallic and Inorganic Chemistry,"* Huang, Y., Yamamoto, A., Teo, B. K., ed., Science Press, Beijing, China, **1984**, p. 341.

1157. Teo, B. K.; Antonio, M. R.; Averill, B. A. *J. Am. Chem. Soc.*, **1983**, *105*, 3751.

1158. Teo, B. K.; Antonio, M. R.; Coucouvanis, D.; Simhon, E. D.; Stremple, P. P. *J. Am. Chem. Soc.*, **1983**, *105*, 5767.

1159. Teo, B. K.; Antonio, M. R.; Tieckelmann, R. H.; Silvis, H. C.; Averill, B. A. *J. Am. Chem. Soc.*, **1982**, *104*, 6126.

1160. Teo, B. K.; Averill, B. A. *Biochem. Biophys. Res. Commun.*, **1979**, *88*, 1454.

1161. Teo, B. K.; Chen, H. S.; Wang, R.; Antonio, M. R. *J. Non-Cryst. Solids*, **1983**, *58*, 249.

1162. Teo, B. K.; Eisenberger, P.; Kincaid, B. M. *J. Am. Chem. Soc.*, **1978**, *100*, 1735.

1163. Teo, B. K.; Eisenberger, P.; Reed, J.; Barton, J. K.; Lippard, S. J. *J. Am. Chem. Soc.*, **1978**, *100*, 3225.

1164. Teo, B. K.; Joy, D. C., ed., *"EXAFS Spectroscopy: Techniques and Applications,"* Plenum, New York, **1981**.

1165. Teo, B. K.; Kijima, K.; Bau, R. *J. Am. Chem. Soc.*, **1978**, *100*, 621.

1166. Teo, B. K.; Lee, P. A. *J. Am. Chem. Soc.*, **1979**, *101*, 2815.

1167. Teo, B. K.; Lee, P. A.; Simons, A. L.; Eisenberger, P.; Kincaid, B. M. *J. Am. Chem. Soc.*, **1977**, *99*, 3854.

1168. Teo, B. K.; Shulman, R. G. in *"Iron-Sulfur Biochemistry,"* Vol. 4 of *"Metal Ions in Biology,"* Spiro, T. G., ed., John Wiley & Sons, New York, **1982**, p. 345.

1169. Teo, B. K.; Shulman, R. G.; Brown, G. S.; Meixner, A. E. *J. Am. Chem. Soc.,* **1979**, *101*, 5624.

1170. Terauchi, H.; Iida, S.; Tanabe, K.; Maeda, H.; Hida, M.; Kamijo, N.; Takashige, M.; Nakamura, T. *J. Phys. Soc. Jpn.,* **1984**, *53*, 1598.

1171. Terauchi, H.; Iida, S.; Tanabe, K.; Kikukawa, K.; Maeda, H.; Hida, M.; Kamijo, N. *J. Phys. Soc. Jpn.,* **1983**, *52*, 3700.

1172. Terauchi, H.; Iida, S.; Tanabe, K.; Kikukawa, K.; Maeda, H.; Hida, M.; Kamijo, N.; Yamada, Y. *J. Phys. Soc. Jpn.,* **1983**, *52*, 4041.

1173. Terauchi, H.; Maeda, H.; Tanabe, K.; Kamijo, N.; Hida, M.; Takashige, M.; Nakamura, T.; Ozawa, H.; Uno, R. *Ferroelectrics,* **1981**, *37*, 599.

1174. Terauchi, H.; Tanabe, K.; Maeda, H.; Hida, M.; Kamijo, N.; Takashige, M.; Nakamura, T.; Ozawa, H.; Uno, R. *J. Phys. Soc. Jpn.,* **1981**, *50*, 3977.

1175. Theil, E. C.; Sayers, D. E.; Brown, M. A. *J. Biol. Chem.,* **1979**, *254*, 8132.

1176. Theye, M. L. *Rev. Roum. Phys.,* **1981**, *26*, 869.

1177. Theye, M. L.; Gheorghiu, A.; Launois, H. *J. Phys. C,* **1980**, *13*, 6569.

1178. Thoai, D. B. Tran.; Ekardt, W. *Solid State Commun.,* **1981**, *40* 269.

1179. Thulke, W.; Haensel, R.; Rabe, P. *Phys. Status Solidi A,* **1983**, *78*, 539.

1180. Thulke, W.; Haensel, R.; Rabe, P. *Rev. Sci Instrum.,* **1983**, *54*, 277.

1181. Thulke, W.; Haensel, R.; Rabe, P. *Springer Ser. Chem. Phys.,* **1983**, *27*, 409.

1182. Thulke, W.; Rabe, P. *J. Phys. C.,* **1983**, *16*, L955.

1183. Tohji, K.; Udagawa, Y. *Jpn. J. Appl. Phys., Part 1,* **1983**, *22*, 882.

1184. Tohji, K.; Udagawa, Y.; Kawasaki, T.; Masuda, K. *Rev. Sci. Instrum.,* **1983**, *54*, 1482.

1185. Tohji, K.; Udagawa, Y.; Tanabe, S.; Ida, T.; Ueno, A. *J. Am. Chem. Soc.,* **1984**, *106*, 5172.

1186. Tohji, K.; Udagawa, Y.; Tanabe, S.; Ueno, A. *J. Am. Chem. Soc.,* **1984**, *106*, 612.

1187. Tokumoto, M.; Oyanagi, H.; Ishiguro, T.; Shirakawa, H.; Nemoto, H.; Matsushita, T.; Ito, M.; Kuroda, H.; Kohra, K. *Solid State Commun.,* **1983**, *48*, 861.

280

1188. Tolkiehn, G.; Rabe, P.; Werner, A. in *"Inn.-Shell X-Ray Phys. At. Solids,"* Fabian, D. J.; Kleinpoppen, H.; Watson, L. M., ed., Plenum, New York, N.Y., **1981**, p. 675.

1189. Topsoe, H.; Candia, R.; Topsoe, N. Y.; Clausen, B. S. *Bull. Soc. Chim. Belg.,* **1984**, *93*, 783.

1190. Torensma, R.; Phillips, J. C. *Biochem. J.,* **1983**, *209*, 373.

1191. Tranquada, J. M.; Ingalls, R. *Phys. Rev. B,* **1983**, *28*, 3520.

1192. Tullius, T. D.; Conradson, S. D.; Berg, J. M.; Hodgson, K. O. in *"Molybdenum Chem. Biol. Significance,"* Newton, W. E.; Otsuka, S., ed., Plenum, New York, N.Y., **1980**, p. 139.

1193. Tullius, T.; Frank, P.; Hodgson, K. O. *Proc. Natl. Acad. Sci., USA,* **1978**, *75*, 4069.

1194. Tullius, T. D.; Gillum, W. O.; Carlson, R. M. K.; Hodgson, K. O. *J. Am. Chem. Soc.,* **1980**, *102*, 5670.

1195. Tullius, T. D.; Kurtz, D. M., Jr.; Conradson, S. D.; Hodgson, K. O. *J. Am. Chem. Soc.,* **1979**, *101*, 2776.

1196. Vaingankar, A. S.; Khasbardar, B. V.; Patil, R. N. *J. Phys. F,* **1979**, *9*, 2301.

1197. Vaingankar, A. S.; Patil, S. A.; Sahasrabudhe, V. S. *Trans. Indian Inst. Met.,* **1981**, *34*, 387.

1198. Vanderheyden, J. L.; Ketring, A. R.; Libson, K.; Heeg, M. J.; Roecker, L.; Motz, P.; Whittle, R.; Elder, R. C.; Deutsch, E. *Inorg. Chem.,* **1984** *23*, 3184.

1199. Van't Blik, H. F. J.; Van Zon, J. B. A. D.; Koningsberger, D. C.; Prins, R. *J. Mol. Catal.,* **1984**, *25*, 379.

1200. Van't Blik, H. F. J.; Van Zon, J. B. A. D.; Huizinga, T.; Vis, J. C.; Koningsberger, D. C.; Prins, R. *J. Phys. Chem.,* **1983**, *87*, 2264.

1201. Van Zon, J. B. A. D.; Koningsberger, D. C.; Van't Blik, H. F. J.; Prins, R.; Sayers, D. E. *J. Chem. Phys.,* **1984**, *80*, 3914.

1202. Vasin, V. V. *Springer Ser. Chem. Phys.,* **1983**, *27*, 114.

1203. Vedrine, J. in *"Chem. Phys. Aspects Catal. Oxid.,"* Portefaix, J. L.; Figueras, F., ed., CNRS, Paris, Fr., **1980**, p. 367.

1204. Vedrinskii, R. V.; Gegusin, I. I.; Datsyuk, V. N.; Novakovich, A. A.; Kraizman, V. L. *Phys. Status Solidi B,* **1982**, *111*, 433.

1205. Verdaguer, M.; Julve, M.; Michalowicz, A.; Kahn, O. *Inorg. Chem.,* **1983**, *22*, 2624.

1206. Verdaguer, M.; Michalowicz, A.; Girerd, J. J.; Berding, N. A.; Kahn, O. *Inorg. Chem.*, **1980**, *19*, 3271.

1207. Vergand, F.; Fargues, D.; Belin, E.; Bonnelle, C. *J. Phys. F*, **1981**, *11*, 1887.

1208. Vernon, S. P.; Stearns, M. B. *Phys. Rev. B*, **1984**, *29*, 6968.

1209. Via, G. H.; Meitzner, G.; Lytle, F. W.; Sinfelt, J. H. *J. Chem. Phys.*, **1983**, *79*, 1527.

1210. Via, G. H.; Sinfelt, J. H.; Lytle, F. W. *J. Chem. Phys.*, **1979**, *71*, 690.

1211. Via, G. H.; Sinfelt, J. H.; Lytle, F. W. in *"EXAFS Spectroscopy: Techniques and Applications,"* Teo, B. K., Joy, D. C., ed., **1981**, p. 159.

1212. Via, G. H.; Sinfelt, J. H.; Lytle, F. W.; Greegor, R. B. *Prepr. — Am. Chem. Soc., Div. Pet. Chem.*, **1983**, *28*, 460.

1213. Victoreen, J. A. *J. Appl. Phys.*, **1948**, *19*, 855.

1214. Vijayavargiya, V. P.; Gupta, S. N.; Padalia, B. D. *Phys. Status Solidi B*, **1977**, *80*, 83.

1215. Vinogradov, A. S.; Dukhnyakov, A. Yu.; Ipatov, V. M.; Onopko, D. E.; Pavlychev, A. A.; Titov, S. A. *Fiz. Tverd. Tela (Leningrad)*, **1982**, *24*, 1417.

1216. Vlaic, G.; Bart, J. C. J. *Recl. J. R. Neth. Chem. Soc.*, **1982**, *101*, 171.

1217. Vlaic, G.; Bart, J. C. J.; Cavigiolo, W.; Michalowicz, A. *Springer Ser. Chem. Phys.*, **1983**, *27*, 307.

1218. Vlaic, G.; Bart, J. C. J.; Cavigiolo, W.; Mobilio, S. *Chem. Phys. Lett.*, **1980**, *76*, 453.

1219. Vlaic, G.; Bart, J. C. J.; Cavigiolo, W.; Mobilio, S.; Navarra, G. *Daresbury Lab. Rep. DL/SCI/R17*, **1981a**, 133.

1220. Vlaic, G.; Bart, J. C. J.; Cavigiolo, W.; Mobilio, S.; Navarra, G. *Z. Naturforsch. A*, **1981b**, *36*, 1192.

1221. Vlaic, G.; Bart, J. C. J.; Cavigiolo, W.; Mobilio, S.; Navarra, G. *Chem. Phys.*, **1982**, *64*, 115.

1222. Vulli, M.; Starke, K. *J. Microsc. Spectrosc. Electron.*, **1978**, *3*, 45.

1223. Walter, B. *Fortschr. Röntgenstr.*, **1927**, *35*, 929, 1308.

1224. Waseda, Y. *"The Structure of Non-Crystalline Materials,"* McGraw Hill, New York, **1980**.

1225. Waychunas, G. A. *J. Mater. Sci.*, **1983**, *18*, 195.

1226. Waychunas, G. A.; Rossman, G. R. *Phys. Chem. Miner.*, **1983**, *9*, 212.

1227. Watanabe, T.; Ishizuka, H.; Kuramoto, Y.; Horie, C. *J. Phys. Soc. Jpn,* **1980**, *49*, 299.

1228. Watson, R. E.; Perlman, M. L. *Science,* **1978**, *199*, 1295.

1229. Weber, W. M. *Phys. Status Solidi B,* **1979**, *91*, 667.

1230. Weber, W. M. *Phys. Lett. A.,* **1980**, *78*, 51.

1231. Weber, W.; Peisl, J. in *"Proceedings of the Yamada Conference V on Point Defects and Defect Interactions in Metals,"* Takamura, J. I.; Doyama, M.; Kiritani, M., ed., Univ. Tokyo Press, Tokyo, Japan, **1982**, p. 368.

1232. Weber, W.; Peisl, J. *Phys. Rev. B,* **1983**, *28*, 806.

1233. Wendin, G. *Springer Ser. Chem. Phys.,* **1983**, *27*, 29.

1234. Werner, A.; Hochheimer, H. D.; Lengeler, B. *Rev. Sci. Instrum.,* **1982**, *53*, 1467.

1235. Werner, A.; Hochheimer, H. D.; Lengeler, B. *Solid State Commun.,* **1983**, *45*, 1035.

1236. Wesner, D.; Krummacher, S.; Carr, R.; Sham, T. K.; Strongin. M.; Eberhardt, W.; Weng, S. L.; Williams, G.; Howells, M.; et. al. *Phys. Rev. B,* **1983**, *28*, 2152.

1237. White, J. M. *Science,* **1982**, *218*, 429.

1238. Wilkinson, D. H. *"Ionization Chambers and Counters,"* Cambridge University Press, Cambridge, U.K., **1950**.

1239. Williams, A. *Rev. Sci. Instrum.,* **1983**, *54*, 193.

1240. Williams, A. R.; Johnson, W. L. *Patent US4446568A,* **1984**.

1241. Williams, R. S.; Denley, D.; Shirley, D. A.; Stöhr, J. *J. Am. Chem. Soc.,* **1980**, *102*, 5717.

1242. Winick, H.; Bienenstock, A. *Ann. Rev. Nucl. Part. Sci.,* **1978**, *28*, 33.

1243. Winick, H.; Doniach, S. *"Synchrotron Radiation Research,"* Plenum, New York, **1980**.

1244. Wolff, T. E.; Berg, J. M.; Hodgson, K. O.; Frankel, R. B.; Holm, R. H. *J. Am. Chem. Soc.,* **1979**, *101*, 4140.

1245. Wolff, T. E.; Berg, J. M.; Warrick, C.; Hodgson, K. O.; Holm, R. H.; Frankel, R. B. *J. Amer. Chem. Soc.,* **1978**, *100*, 4630.

1246. Wölfli, W.; Stoller, Ch.; Bonani, G.; Suter, M.; Stöckli, M. *Phys. Rev. Lett.,* **1975**, *35*, 656.

1247. Wong, J. in *"Glassy Metals. Part 1. Ionic Structure, Electronic Transport, and Crystallization,"* Guntherodt, H. J.; Beck, H., ed., Springer-Verlag, Berlin, **1981**, p. 45.

1248. Wong, J. *Proc. Soc. Photo-Opt. Instrum. Eng.,* **1980b**, *204*, 44.

1249. Wong, J. in *"Topics in Applied Physics,"* Guntherodt, G. J., Beck, H., ed., Springer, New York, **1981**, *46*, p. 45.

1250. Wong, J. *Springer Ser. Chem. Phys.,* **1983**, *27*, 280.

1251. Wong, J.; Liebermann, H. H. *Phys. Rev. B,* **1984**, *29*, 651.

1252. Wong, J.; Lytle, F. W. *J. Appl. Phys.,* **1980a**, *51*, 280.

1253. Wong, J.; Lytle, F. W. *J. Non-Cryst. Solids,* **1980b**, *37*, 273.

1254. Wong, J.; Lytle, F. W.; Greegor, R. B.; Liebermann, H. H.; Walter, J. L.; Luborsky, F. E. in *"Rapidly Quenched Met.,"* *Vol. 2*, Cantor, B., ed., Met. Soc., London, England, **1978**, p. 345.

1255. Wong, J.; Lytle, F. W.; Liebermann, H. H.; Tanner, L. E. in *"Proc. - Conf. Met. Glasses: Sci. Technol.,"* Hargitai, C., Bakonyi, I., Kemeny, T., ed., **1981**, *1*, p. 383.

1256. Wong, J.; Lytle, F. W.; Messmer, R. P.; Maylotte, D. H. *Springer Ser. Chem. Phys.,* **1983**, *27*, 130.

1257. Wong, J.; Maylotte, D. H.; Lytle, F. W.; Greegor, R. B.; St. Peters, R. L. *Springer Ser. Chem. Phys.,* **1983**, *27*, 206.

1258. Wong, J.; Maylotte, D. H.; St. Peters, R. L.; Lytle, F. W.; Greegor, R. B. in *"Process Mineral.,"* Hagni, R. D., ed., Metall. Soc. AIME, Warrendale, Pa., **1982**, p. 335.

1259. Wong, J.; Rao, K. J. *Solid State Commun.,* **1983**, *45*, 853.

1260. Woodruff, D. P. *Vide Couches Minces,* **1983**, *38*, 189.

1261. Woodruff, D. P. *Chem. Ind. (London),* **1981a**, 171.

1262. Woodruff, D. P. *Vacuum,* **1981b**, *31*, 399.

1263. Woodruff, D. P.; Jones, R. G. *Daresbury Lab. Rep. DL/SCI/R17,* **1981**, 101.

1264. Woolery, G. L.; Powers, L.; Winkler, M.; Solomon, E. I.; Lerch, K.; Spiro, T. G. *Biochim. Biophys. Acta,* **1984**, *788*, 155.

1265. Woolery, G. L.; Powers, L.; Winkler, M.; Solomon, E. I.; Spiro, T. G. *J. Am. Chem. Soc.,* **1984**, *106*, 86.

1266. Woolery, G. L.; Powers, L.; Peisach, J.; Spiro, T. G. *Biochemistry,* **1984**, *23*, 3428.

1267. Yaakobi, B.; Deckman, H.; Bourke, P.; Letzring, S.; Soures, J. M. *Appl. Phys. Lett.*, **1980**, *37*, 767.

1268. Yachandra, V.; Powers, L.; Spiro, T. G. *J. Am. Chem. Soc.*, **1983**, *105*, 6596.

1269. Yamaguchi, T.; Lindqvist, O.; Claeson, T.; Boyce, J. B. *Chem. Phys. Lett.*, **1982**, *93*, 528.

1270. Yamaguchi, T.; Lindqvist, O.; Boyce, J. B.; Claeson, T. *Acta Chem. Scand. A*, **1984**, 38, 423.

1271. Yarusso, D. J.; Cooper, S. L.; Knapp, G. S.; Georgopoulos, P. *J. Polym. Sci.*, **1980**, *18*, 557.

1272. Yarusso, D. J.; Knapp, G. S.; Goergopoulos, P.; Cooper, S. L. *Polym. Prepr., Am. Chem. Soc.*, **1980**, *21*, 78.

1273. Yates, J. T. *Chem. & Eng. News*, **1974**, 19.

1274. Yigarashi, M.; Fujikawa, T. *J. Electron Spectrosc. Relat. Phenom.*, **1984**, *33*, 347.

1275. Yokoyama, T.; Yamazaki, K.; Kosugi, N.; Kuroda, H.; Ichikawa, M.; Fukushima, T. *J. Chem. Soc., Chem. Commun.*, **1984**, *15*, 962.

1276. Zunger, A.; Jaffe, J. E. *Phys. Rev. Lett.*, **1983**, *51*, 662.

APPENDIX I

The Periodic Table

PERIODIC TABLE OF THE ELEMENTS

The Atomic Weights shown on this chart are based on atomic mass of C¹² – 12. The Weights in parentheses ( ) are mass numbers of the most stable or best-known isotopes.

KEY

ATOMIC NUMBER →	42	95.94 ← ATOMIC WEIGHT
	Mo ← SYMBOL
ELECTRON CONFIGURATION →	Molybdenum ← NAME
	2 8 18
	13 1

INERT GASES

| GROUP I-A | II-A | III-B | IV-B | V-B | VI-B | VII-B | VIII | | | I-B | II-B | III-A | IV-A | V-A | VI-A | VII-A | |

APPENDIX II

X-Ray Absorption Edges

and

Characteristic X-Ray Emission Lines

Reproduced from Woldseth, R., *"X-Ray Energy Spectrometry,"* Kevex, California, **1973**, Appendix D, which was taken from Johnson, G. G.; White, E. W., *"X-Ray Emission Wavelengths and KeV Tables for Nondiffractive Analysis,"* ASTM Data Series DS 46, **1970**.

X-ray emission line and absorption-edge energies (keV) as a function of atomic number Z.

M SERIES (columns M_{IV}: $M\beta$; M_V: $M\alpha$; M_ζ…) — no values tabulated for $Z = 1$ to 34.

L SERIES

Z	Element	$L\ell$	$L\alpha_{1,2}$	$L\eta$	$L\beta_1$	$L\beta_3$	L_{III} (ab)	L_{II} (ab)	L_{I} (ab)
19	K	0.260		0.262					
20	Ca	0.303	0.341	0.306	0.345		0.346	0.350	0.400
21	Sc	0.348	0.395	0.353	0.400		0.403	0.407	0.463
22	Ti	0.395	0.452	0.401	0.458		0.454	0.460	0.530
23	V	0.446	0.511	0.453	0.519	0.585	0.513	0.520	0.604
24	Cr	0.500	0.573	0.510	0.583	0.654	0.574	0.583	0.682
25	Mn	0.556	0.637	0.567	0.649	0.721	0.641	0.652	0.754
26	Fe	0.615	0.705	0.628	0.718	0.792	0.709	0.721	0.842
27	Co	0.678	0.776	0.694	0.791	0.866	0.779	0.794	0.929
28	Ni	0.743	0.851	0.762	0.869	0.941	0.855	0.872	1.012
29	Cu	0.811	0.930	0.832	0.950	1.023	0.932	0.952	1.100
30	Zn	0.884	1.012	0.906	1.034	1.107	1.021	1.044	1.196
31	Ga	0.957	1.098	0.984	1.125	1.197	1.117	1.134	1.300
32	Ge	1.036	1.188	1.068	1.218	1.294	1.218	1.249	1.420
33	As	1.120	1.282	1.155	1.317	1.388	1.325	1.360	1.530
34	Se	1.204	1.379	1.244	1.419	1.490	1.436	1.477	1.653

K SERIES

Z	Element	$K\alpha_2$	$K\alpha_1$	$K\beta_1$	$K\beta_2$	K (ab)
1	H					0.0136
2	He					0.025
3	Li					0.055
4	Be					0.112
5	B					0.192
6	C					0.283
7	N					0.399
8	O					0.531
9	F					0.687
10	Ne					0.867
11	Na		1.041	1.067		1.072
12	Mg		1.253	1.295		1.305
13	Al	1.486	1.486	1.553		1.559
14	Si	1.739	1.740	1.829		1.838
15	P	2.012	2.013	2.136		2.142
16	S	2.306	2.307	2.464		2.472
17	Cl	2.620	2.622			2.822
18	Ar	2.955	2.957	3.190		3.202
19	K	3.310	3.313	3.589		3.607
20	Ca	3.687	3.691	4.012		4.038
21	Sc	4.085	4.090	4.460		4.496
22	Ti	4.504	4.510	4.931		4.965
23	V	4.944	4.951	5.426		5.465
24	Cr	5.405	5.414	5.946		5.989
25	Mn	5.887	5.898	6.489		6.540
26	Fe	6.390	6.403	7.057		7.112
27	Co	6.914	6.929	7.648		7.709
28	Ni	7.460	7.477	8.263		8.333
29	Cu	8.026	8.046	8.904		8.979
30	Zn	8.614	8.637	9.570	9.656	9.659
31	Ga	9.223	9.250	10.263	10.365	10.368
32	Ge	9.854	9.885	10.980	11.099	11.104
33	As	10.506	10.542	11.724	11.862	11.868
34	Se	11.179	11.220	12.494	12.650	12.658

K - SERIES (energies in eV)

Z	Element	K(ab)	Kβ₂	Kβ₁	Kβ₃	Kα₁	Kα₂
35	Br	13 474	13 467	13 291	13 282	11 922	11 876
36	Kr	14 322	14 312	14 110	14 102	12 648	12 596
37	Rb	15 201	15 183	14 959	14 949	13 393	13 333
38	Sr	16 105	16 082	15 833	15 822	14 163	14 095
39	Y	17 037	17 013	16 735	16 723	14 956	14 880
40	Zr	17 998	17 967	17 665	17 651	15 772	15 688
41	Nb	18 986	18 949	18 619	18 603	16 612	16 518
42	Mo	20 002	19 962	19 605	19 587	17 476	17 371
43	Tc	21 054	21 002	20 615	20 595	18 364	18 248
44	Ru	22 118	22 070	21 653	21 631	19 276	19 147
45	Rh	23 224	23 169	22 720	22 695	20 213	20 070
46	Pd	24 350	24 295	23 815	23 787	21 174	21 017
47	Ag	25 514	25 452	24 938	24 907	22 159	21 987
48	Cd	26 711	26 639	26 091	26 057	23 170	22 980
49	In	27 940	27 856	27 271	27 233	24 206	23 998
50	Sn	29 200	29 104	28 481	28 439	25 267	25 040
51	Sb	30 491	30 388	29 721	29 674	26 355	26 106
52	Te	31 813	31 698	30 990	30 933	27 468	27 199
53	I	33 169	33 036	32 289	32 234	28 607	28 312
54	Xe	34 582	34 408	33 619	33 556	29 774	29 453
55	Cs	35 959	35 815	34 981	34 913	30 968	30 620
56	Ba	37 441	37 251	36 372	36 298	32 188	31 812
57	La	38 925	38 723	37 795	37 714	33 436	33 028
58	Ce	40 449	40 226	39 251	39 163	34 714	34 273
59	Pr	41 998	41 767	40 741	40 646	36 020	35 544
60	Nd	43 571	43 327	42 264	42 159	37 355	36 841
61	Pm	45 207	44 929	43 818	43 705	38 718	38 165
62	Sm	46 835	46 566	45 405	45 281	40 111	39 516
63	Eu	48 515	48 248	47 030	46 896	41 535	40 895
64	Gd	50 240	49 952	48 648	48 547	42 989	42 302
65	Tb	51 996	51 715	50 374	50 221	44 474	43 737
66	Dy	53 789	53 500	52 110	51 949	45 991	45 200
67	Ho	55 615	55 315	53 868	53 702	47 539	46 692
68	Er	57 483	57 204	55 672	55 485	49 119	48 213

L - SERIES (energies in eV)

Z	Element	Lℓ	Lα₂	Lα₁	Lβ₂	L_III(ab)	Lη	Lβ₁	Lγ₁	L_II(ab)	Lβ₄	Lβ₃	Lγ₃	L_I(ab)
35	Br	1 293		1 480		1 550	1 339	1 526		1 596		1 596		1 794
36	Kr			1 586		1 675		1 636		1 756	1 697	1 706		1 920
37	Rb	1 482	1 692	1 694		1 806	1 542	1 752		1 866	1 817	1 826		2 067
38	Sr	1 582	1 804	1 806		1 940	1 649	1 871		2 007	1 936	1 947		2 216
39	Y	1 685	1 920	1 922		2 079	1 761	1 995		2 145	2 060	2 072		2 369
40	Zr	1 792	2 040	2 042	2 219	2 223	1 876	2 124	2 302	2 307	2 187	2 201		2 547
41	Nb	1 902	2 163	2 166	2 367	2 371	1 996	2 257	2 461	2 465	2 319	2 334		2 698
42	Mo	2 015	2 289	2 293	2 518	2 520	2 120	2 394	2 623	2 625	2 455	2 473		2 866
43	Tc			2 424		2 677		2 536		2 795				3 054
44	Ru	2 252	2 554	2 558	2 835	2 837	2 382	2 683	2 964	2 966	2 741	2 763		3 236
45	Rh	2 376	2 692	2 696	3 001	3 003	2 519	2 834	3 143	3 146	2 890	2 915		3 419
46	Pd	2 503	2 838	2 838	3 171	3 173	2 660	2 990	3 328	3 330	3 045	3 072		3 617
47	Ag	2 633	2 978	2 984	3 347	3 351	2 806	3 150	3 519	3 524	3 203	3 234		3 806
48	Cd	2 767	3 126	3 133	3 528	3 537	2 956	3 316	3 716	3 727	3 367	3 401		4 019
49	In	2 904	3 279	3 286	3 713	3 730	3 112	3 487	3 920	3 938	3 535	3 572		4 237
50	Sn	3 044	3 435	3 443	3 904	3 929	3 272	3 662	4 130	4 156	3 708	3 750		4 465
51	Sb	3 188	3 595	3 604	4 100	4 132	3 436	3 843	4 347	4 381	3 886	3 932		4 698
52	Te	3 335	3 758	3 769	4 301	4 341	3 605	4 029	4 570	4 612	4 069	4 120		4 939
53	I	3 484	3 925	3 937	4 507	4 557	3 780	4 220	4 800	4 852	4 257	4 313		5 188
54	Xe			4 109		4 781				5 100				5 452
55	Cs	3 794	4 272	4 286	4 935	5 011	4 141	4 619	5 279	5 358	4 649	4 716		5 720
56	Ba	3 953	4 450	4 465	5 156	5 247	4 330	4 827	5 530	5 624	4 851	4 926		5 995
57	La	4 124	4 650	4 650	5 383	5 483	4 524	5 041	5 788	5 891	5 061	5 143	6 073	6 267
58	Ce	4 287	4 822	4 839	5 612	5 724	4 731	5 261	6 051	6 165	5 276	5 364	6 340	6 549
59	Pr	4 452	5 013	5 033	5 849	5 968	4 935	5 488	6 321	6 443	5 497	5 591	6 615	6 846
60	Nd	4 632	5 207	5 229	6 088	6 208	5 145	5 721	6 601	6 722	5 721	5 828	6 900	7 126
61	Pm		5 407	5 432	6 338	6 466		5 960	6 891	7 018		6 070		7 448
62	Sm	4 994	5 607	5 635	6 586	6 717	5 588	6 205	7 177	7 312	6 195	6 317	7 485	7 737
63	Eu	5 176	5 816	5 845	6 842	6 983	5 816	6 455	7 479	7 624	6 438	6 570	7 795	8 069
64	Gd	5 361	6 024	6 056	7 102	7 243	6 049	6 712	7 784	7 931	6 686	6 830	8 104	8 376
65	Tb	5 546	6 237	6 272	7 365	7 515	6 283	6 977	8 100	8 252	6 939	7 095	8 422	8 708
66	Dy	5 742	6 457	6 494	7 634	7 850	6 533	7 246	8 417	8 621	7 203	7 369	8 752	9 043
67	Ho	5 942	6 679	6 719	7 910	8 071	6 787	7 524	8 746	8 919	7 470	7 650	9 086	9 395
68	Er	6 152	6 904	6 947	8 188	8 364	7 057	7 809	9 087	9 263	7 744	7 938	9 429	9 776

M - SERIES (energies in keV)

Z	Element	M_IV(ab)	Mβ
57	La	0.851	0.854
58	Ce	0.902	0.902
59	Pr		0.949
60	Nd	1.004	0.996
62	Sm	1.108	1.100
63	Eu		1.153
64	Gd	1.221	1.209
65	Tb	1.280	1.266
66	Dy		1.325
67	Ho	1.390	1.383
68	Er		1.443

Z	Element	K(ab)	Kβ₃	Kβ₁	Kβ₂	Kα₁	Kα₂	L_I(ab)	Lγ₃	Lβ₃	Lβ₄	L_II(ab)	Lγ₁	Lβ₁	Lη	L_III(ab)	Lβ₂	Lα₁	Lα₂	Lℓ	M_IV(ab)	Mβ	M_V(ab)	Mζ₁	Mζ₂
69	Tm	59 390	57 293	57 506	59 085	50 733	49 764	10 116	9 778	8 229	8 024	9 618	9 424	8 100	7 308	8 648	8 467	7 179	7 132	6 341	1 515	1 503	—	—	—
70	Yb	61 332	59 141	59 356	60 974	52 380	51 345	10 486	10 141	8 535	8 312	9 978	9 778	8 400	7 579	8 943	8 757	7 414	7 366	6 544	1 578	1 567	—	—	—
71	Lu	63 304	61 037	61 272	62 956	54 061	52 956	10 867	10 509	8 845	8 605	10 345	10 142	8 708	7 856	9 241	9 047	7 654	7 604	6 752	—	1 631	—	—	—
72	Hf	65 351	62 969	63 222	64 969	55 781	54 602	11 264	10 889	9 162	8 904	10 739	10 514	9 021	8 138	9 561	9 346	7 898	7 843	6 958	1 718	1 697	—	—	—
73	Ta	67 414	64 938	65 212	67 001	57 523	56 267	11 680	11 276	9 486	9 211	11 139	10 893	9 342	8 427	9 881	9 650	8 145	8 086	7 172	1 793	1 765	—	—	—
74	W	69 524	66 940	67 233	69 089	59 308	57 972	12 098	11 672	9 817	9 524	11 542	11 284	9 671	8 723	10 204	9 960	8 396	8 334	7 386	1 871	1 835	1 809	1 775	1 773
75	Re	71 662	68 983	69 298	71 219	61 130	59 708	12 522	12 080	10 158	9 845	11 955	11 683	10 008	9 026	10 531	10 274	8 651	8 585	7 602	—	1 906	—	—	—
76	Os	73 860	71 065	71 401	73 390	62 990	61 476	12 965	12 498	10 509	10 174	12 383	12 093	10 354	9 335	10 869	10 597	8 910	8 840	7 821	—	1 978	—	—	—
77	Ir	76 112	73 190	73 548	75 606	64 885	63 276	13 424	12 922	10 866	10 509	12 824	12 510	10 706	9 649	11 215	10 919	9 174	9 098	8 040	2 116	2 053	2 041	1 980	1 975
78	Pt	78 395	75 355	75 735	77 864	66 821	65 112	13 892	13 359	11 233	10 852	13 273	12 943	11 069	9 973	11 564	11 249	9 441	9 360	8 267	2 202	2 127	2 122	2 050	2 046
79	Au	80 723	77 567	77 991	80 172	68 792	66 978	14 353	13 807	11 608	11 203	13 733	13 379	11 440	10 307	11 918	11 583	9 712	9 626	8 493	2 291	2 204	2 206	2 123	2 118
80	Hg	83 103	79 809	80 240	82 530	70 807	68 883	14 846	14 262	11 993	11 561	14 209	13 828	11 821	10 649	12 284	11 922	9 987	9 896	8 720	2 385	2 282	2 295	—	—
81	Tl	85 528	82 104	82 562	84 933	72 859	70 820	15 344	14 734	12 344	11 929	14 698	14 289	12 211	10 992	12 657	12 270	10 267	10 171	8 952	2 485	2 362	2 389	2 270	2 265
82	Pb	88 006	84 436	84 922	87 351	74 956	72 792	15 860	15 215	12 791	12 304	15 194	14 762	12 612	11 347	13 035	12 621	10 550	10 448	9 183	2 586	2 442	2 484	2 345	2 339
83	Bi	90 527	86 819	87 328	89 846	77 095	74 802	16 385	15 708	13 204	12 689	15 711	15 245	13 021	11 710	13 418	12 978	10 837	10 729	9 419	2 687	2 525	2 579	2 422	2 416
84	Po	93 112	89 231	89 781	92 383	79 279	76 851	16 935	—	13 635	13 043	16 244	15 741	13 445	—	13 817	13 338	11 129	11 014	9 662	—	—	—	—	—
85	At	95 740	91 707	92 287	94 974	81 499	78 930	17 490	—	14 065	—	16 784	16 249	13 874	—	14 215	—	11 425	11 303	—	—	—	—	—	—
86	Rn	98 418	94 230	94 850	97 622	83 768	81 051	18 058	—	14 509	—	17 337	16 764	14 313	—	14 618	—	11 725	11 596	—	—	—	—	—	—
87	Fr	101 147	96 791	97 460	100 307	86 089	83 217	18 638	18 354	14 973	14 745	17 904	17 300	14 758	13 661	15 028	14 448	12 029	11 893	10 620	—	—	—	—	—
88	Ra	103 927	99 415	100 113	103 051	88 454	85 419	19 233	—	15 442	—	18 484	17 845	15 233	—	15 442	14 839	12 338	12 194	—	—	—	—	—	—
89	Ac	106 759	102 084	102 829	105 849	90 868	87 660	19 842	—	15 929	—	19 078	18 405	15 710	14 507	15 865	—	12 650	12 499	—	—	—	—	—	—
90	Th	109 649	104 813	105 591	108 699	93 334	89 938	20 470	19 503	16 423	15 640	19 692	18 979	16 199	14 944	16 300	15 621	12 967	12 807	11 117	3 491	3 145	3 332	2 996	2 986
91	Pa	112 581	107 576	108 409	111 605	95 852	92 271	21 102	20 094	16 927	16 101	20 311	19 565	16 699	15 397	16 731	16 022	13 288	13 120	11 364	—	3 239	—	3 082	3 072
92	U	115 603	110 387	111 281	114 587	98 422	94 649	21 756	20 709	17 452	16 573	20 947	20 164	17 217	15 874	17 167	16 425	13 612	13 437	11 616	3 728	3 336	3 552	3 170	3 159
93	Np	118 619	113 725	116 943	120 350	100 781	96 844	22 417	21 336	17 986	17 054	21 596	20 781	17 747	16 330	17 614	16 837	13 942	13 757	11 847	—	—	—	—	—
94	Pu	121 760	116 943	120 350	123 960	103 300	99 164	23 095	21 979	18 537	17 553	22 263	21 414	18 291	—	18 053	17 252	14 276	14 082	12 122	—	—	—	—	—
95	Am	124 876	120 350	123 960	126 490	105 949	101 607	23 793	—	19 103	18 060	22 944	22 061	18 849	—	18 526	17 673	14 615	14 409	12 381	—	—	—	—	—
96	Cm	128 088	122 733	126 490	—	108 737	104 168	24 503	—	—	—	23 640	22 703	19 399	—	18 990	18 096	14 953	14 740	—	—	—	—	—	—
97	Bk	131 357	126 490	130 484	—	111 676	106 862	25 230	—	—	—	24 352	23 389	19 961	—	19 461	18 529	15 304	15 080	—	—	—	—	—	—
98	Cf	134 683	127 794	133 290	—	114 778	109 699	25 971	—	—	—	25 080	24 070	20 557	—	19 938	18 983	15 652	15 418	—	—	—	—	—	—

APPENDIX III

Victoreen's C and D Values for True Absorption

C_1, D_1 are for X-ray energies $E > E_K$; C_2, D_2 for $E_K > E > E_{L_I}$; C_3, D_3 for $E_{L_I} > E > E_{L_{II}}$; C_4, D_4 for $E_{L_{II}} > E > L_{III}$; C_5, D_5 for $E_{L_{III}} > E > E_{M_I}$. The energies of the absorption edges are given in Appendix II. (Adapted from "International Tables for X-Ray Crystallography," Vol. III, pp. 171-173 which was based on Victoreen, J. A. *J. Appl. Phys.*, **1984**, *19*, 855; **1949**, *20*, 1141; **1943**, *14*, 95.)

	Z	C_1	D_1	C_2	D_2
H	1	0.0127	0.466×10^{-5}		
He	2	0.0514	7.52×10^{-5}		
Li	3	0.150	0.494×10^{-3}		
Be	4	0.365	2.13×10^{-3}		
B	5	0.609	0.00451		
C	6	1.22	0.0142		
N	7	2.05	0.0317		
O	8	3.18	0.0654		
F	9	4.60	0.112		
Ne	10	6.51	0.206		
Na	11	8.67	0.330		
Mg	12	11.3	0.539		
Al	13	14.4	0.803		
Si	14	18.2	1.10		
P	15	22.6	1.55		
S	16	27.6	2.18		
Cl	17	33.4	3.03		
A	18	40.0	4.18		
K	19	47.4	5.59		
Ca	20	55.8	7.56		
Sc	21	65.2	9.81		
Ti	22	75.6	12.3	5.15	0.153
V	23	86.9	15.1	6.14	0.203
Cr	24	99.0	18.2	7.24	0.268
Mn	25	112	22.3	8.51	0.344

	Z	C_1	D_1	C_2	D_2
Fe	26	126	27.2	9.95	0.433
Co	27	141	33.2	11.6	0.535
Ni	28	158	40.1	13.4	0.651
Cu	29	176	48.3	15.6	0.779
Zn	30	195	57.5	17.8	0.937
Ga	31	216	68.6	20.2	1.13
Ge	32	238	81.1	22.7	1.37
As	33	262	95.4	25.3	1.67
Se	34	287	112	28.0	2.02
Br	35	314	130	30.9	2.43
Kr	36	343	151	33.9	2.92
Rb	37	374	174	37.1	3.48
Sr	38	406	200	40.5	4.14
Y	39	441	229	44.1	4.88
Zr	40	477	261	47.9	5.72
Nb	41	515	296	51.9	6.67
Mo	42	555	336	56.2	7.73
Tc	43	597	379	60.7	8.91
Ru	44	641	427	65.5	10.1
Rh	45	686	479	70.5	11.4
Pd	46	734	537	75.8	12.8
Ag	47	784	599	81.4	14.3
Cd	48	835	667	87.4	15.9
In	49	889	741	93.6	17.7
Sn	50	944	821	100	19.7
Sb	51	1000	908	107	21.8
Te	52	1060	1000	114	24.3
I	53	1120	1100	122	27.1
Xe	54	1180	1210	130	30.2
Cs	55	1250	1330	139	33.6

	Z	C_1	D_1	C_2	D_2
Ba	56	1310	1460	147	37.3
La	57	1380	1590	157	41.3
Ce	58	1450	1740	166	45.6
Pr	59	*1418*	*1831*	176	50.2
Nd	60	*1489*	*1993*	187	55.1
Pm	61	*1569*	*2175*	198	60.3
Sm	62	*1645*	*2362*	210	65.8
Eu	63	*1744*	*2591*	222	71.6
Gd	64	*1804*	*2773*	234	77.7
Tb	65	*1910*	*3035*	247	84.1
Dy	66	*1996*	*3278*	261	90.8
Ho	67	*2103*	*3572*	275	98.0
Er	68	*2213*	*3881*	290	105
Tm	69	*2336*	*4229*	305	113
Yb	70	*2429*	*4538*	321	121
Lu	71	*2557*	*4928*	337	130
Hf	72	*2666*	*5299*	354	140
Ta	73	*2796*	*5727*	372	150
W	74	*2923*	*6170*	390	161
Re	75	*3064*	*6662*	409	173
Os	76	*3182*	*7126*	429	187
Ir	77	*3339*	*7697*	449	202
Pt	78	*3485*	*8925*	492	240
Au	79	*3656*	*8925*	492	240
Hg	80	*3800*	*9542*	514	265
Tl	81	*3946*	*10189*	540	295
Pb	82	*4115*	*10925*	570	333
Bi	83	*4312*	*11766*	605	382

	Z	C_3	D_3	C_4	D_4	C_5	D_5
Pd	46						
Ag	47						
Cd	48						
In	49						
Sn	50						
Sb	51						
Te	52						
I	53					12.4	0.761
Xe	54					14.7	1.06
Cs	55					17.0	1.38
Ba	56					19.2	1.72
La	57					21.5	2.07
Ce	58					23.8	2.44
Pr	59					26.1	2.83
Nd	60					28.4	3.24
Pm	61					30.7	3.67
Sm	62					33.1	4.12
Eu	63					35.4	4.61
Gd	64					37.8	5.11
Tb	65					40.3	5.64
Dy	66					42.8	6.20
Ho	67					45.4	6.79
Er	68					48.0	7.41
Tm	69					50.7	8.06
Yb	70					53.4	8.75
Lu	71					56.2	9.47
Hf	72					59.2	10.2
Ta	73	*325*	*128*	*188*	*56.6*	62.2	11.0
W	74	*342*	*138*	*199*	*62.0*	65.3	11.9
Re	75	*360*	*150*	*211*	*68.0*	68.4	12.7
Os	76	*375*	*161*	*222*	*73.9*	71.8	13.6
Ir	77	*395*	*175*	*235*	*81.0*	75.2	14.6
Pt	78	*413*	*188*	*248*	*88.2*	78.7	15.6
Au	79	*435*	*204*	*262*	*96.5*	82.3	16.6
Hg	80	*453*	*219*	*275*	*104*	86.1	17.7

	Z	C_3	D_3	C_4	D_4	C_5	D_5
Tl	81	*471*	*234*	*288*	*113*	90.0	18.9
Pb	82	*493*	*251*	*303*	*123*	94.1	20.0
Bi	83	*517*	*271*	*320*	*134*	98.3	21.3

Numbers given in italics are of low accuracy. C_1, D_1 for Pr through Bi as well as C_3, D_3 and C_4, D_4 for Ta through Bi are new (calculated based on Victoreen, J. A. *J. App. Phys.*, **1948**, *19*, 855; **1949**, *20*, 1141).

APPENDIX IV

Fluorescence Yields

Reproduced from Woldseth, R., "X-Ray Energy Spectrometry," Kevex, California, **1973**, Table 1.1: Ref. Bambynek, W., *et al*, *Rev. Mod. Phys.*, **1972**, *44*, 716.

Element	Z	K^a		L^b		M^c
Be	4	4.5	.04			
B	5	10.1	.04			
C	6	2.0	.03			
N	7	3.5	.03			
O	8	5.8	.03			
F	9	9.0	.03			
Ne	10	1.34	.02			
Na	11	1.92	.02			
Mg	12	2.65	.02			
Al	13	3.57	.02			
Si	14	4.70	.02			
P	15	6.04	.02			
S	16	7.61	.02			
Ci	17	9.42	.02			
Ar	18	1.15	.01			
K	19	1.38	.01			
Ca	20	1.63	.01			
Sc	21	1.90	.01			
Ti	22	2.19	.01			
V	23	2.50	.01	2.4	.03	
Cr	24	2.82	.01	3.0	.03	
Mn	25	3.14	.01			
Fe	26	3.47	.01			
Co	27	3.81	.01			
Ni	28	4.14	.01			
Cu	29	4.45	.01	5.6	.03	
Zn	30	4.79	.01			
Ga	31	5.10	.01	6.4	.03	
Ge	32	5.40	.01			
As	33	5.67	.01			
Se	34	5.96	.01			
Br	35	6.22	.01			

Element	Z	K^a		L^b		M^c
Kr	36	6.46	.01	1.0	.02	
Rb	37	6.69	.01	1.0	.02	
Sr	38	6.91	.01			
Y	39	7.11	.01	3.2	.02	
Zr	40	7.30	.01			
Nb	41	7.48	.01			
Mo	42	7.64	.01	6.7	.02	
Tc	43	7.79	.01			
Ru	44	7.93	.01			
Rh	45	8.07	.01			
Pd	46	8.19	.01			
Ag	47	8.30	.01	5.6	.02	
Cd	48	8.40	.01			
In	49	8.50	.01			
Sn	50	8.59	.01			
Sb	51	8.67	.01	1.2	.01	
Te	52	8.75	.01	1.2	.01	
I	53	8.82	.01			
Xe	54	8.89	.01	1.1	.01	
Cs	55	8.95	.01	8.9	.02	
Ba	56	9.01	.01	9.3	.02	
La	57	9.06	.01	1.0	.01	
ce	58	9.11	.01	1.6	.01	
Pr	59	9.15	.01	1.7	.01	
Nd	60	9.20	.01	1.7	.01	
Pm	61	9.24	.01			
Sm	62	9.28	.01	1.9	.01	
Eu	63	9.31	.01	1.7	.01	
Gd	64	9.34	.01	2.0	.01	
Tb	65	9.37	.01	2.0	.01	
Dy	66	9.40	.01	1.4	.01	
Ho	67	9.43	.01			
Er	68	9.45	.01			
Tm	69	9.48	.01			
Yb	70	9.50	.01			
Lu	71	9.52	.01	2.9	.01	
Hf	72	9.54	.01	2.6	.01	
Ta	73	9.56	.01	2.3	.01	
W	74	9.57	.01	3.0	.01	
Re	75	9.59	.01			
Os	76	9.61	.01	3.5	.01	
Ir	77	9.62	.01	3.0	.01	

Element	Z	K^a		L^b		M^c	
Pt	78	9.63	.01	3.3	.01		
Au	79	9.64	.01	3.9	.01		
Hg	80	9.66	.01	3.9	.01		
Tl	81			4.6	.01		
Pb	82	9.68	.01	3.8	.01	2.9	.02
Bi	83			4.1	.01	3.6	.02
U	92	9.76	.01	5.2	.01	6	.02

a. Empirical "best fit" values.

b. Average experimental.

c. Experimental.

APPENDIX V

Backscattering Amplitude,

Backscattering Phase,

and

Central Atom Phase

Functions

Tables I-VIII are reproduced from Teo, B. K.; Lee, P. A., *J. Am. Chem. Soc.,* **1979**, *101*, 2815.

Table 1. Backscattering Amplitudes $F(k)$ in Å vs. Photoelectron Wave Vector k in Å^{-1}, Calculated Using Clementi-Roetti Wave Functions

Z	CHEM	k=3.7795 / 8.5038	4.2519 / 9.4486	4.7243 / 10.3935	5.1967 / 11.3384	5.6692 / 12.2832	6.1416 / 13.2281	6.6140 / 14.1729	7.0865 / 15.1178	7.5589
6	C	0.5063	0.3566	0.2800	0.2430	0.2099	0.1719	0.1430	0.1325	0.1182
		0.0839	0.0712	0.0562	0.0481	0.0397	0.0349	0.0297	0.0267	
8	O	0.7041	0.5781	0.4629	0.3909	0.3248	0.2661	0.2205	0.1913	0.1700
		0.1256	0.1016	0.0814	0.0674	0.0563	0.0489	0.0416	0.0365	
9	F	0.7175	0.6275	0.5294	0.4503	0.3924	0.3273	0.2733	0.2349	0.2038
		0.1532	0.1204	0.0956	0.0792	0.0652	0.0566	0.0481	0.0423	
11	NA	0.6630	0.6007	0.5436	0.4669	0.4310	0.3905	0.3482	0.3023	0.2657
		0.2028	0.1610	0.1263	0.1014	0.0864	0.0715	0.0629	0.0569	
15	P	0.7829	0.7034	0.6517	0.6112	0.5533	0.4806	0.4471	0.4164	0.3671
		0.2938	0.2344	0.1874	0.1571	0.1295	0.1089	0.0939	0.0845	
16	S	0.8140	0.8173	0.7569	0.6523	0.5843	0.5505	0.5075	0.4508	0.3936
		0.3283	0.2608	0.2151	0.1731	0.1459	0.1191	0.1016	0.0854	
17	CL	0.8269	0.8413	0.8188	0.7130	0.6344	0.5755	0.5391	0.4938	0.4361
		0.3475	0.2864	0.2289	0.1901	0.1580	0.1330	0.1100	0.0954	
20	CA	0.6478	0.8064	0.8361	0.7698	0.7051	0.6457	0.5971	0.5404	0.4972
		0.3981	0.3317	0.2776	0.2160	0.1910	0.1630	0.1359	0.1146	
22	TI	0.6225	0.7179	0.7580	0.8075	0.8073	0.7404	0.6822	0.6198	0.5537
		0.4707	0.3844	0.3378	0.2653	0.2197	0.1905	0.1587	0.1391	
24	CR	0.4569	0.5482	0.6666	0.7161	0.7604	0.7668	0.7479	0.6865	0.6354
		0.5288	0.4438	0.3661	0.3089	0.2541	0.2195	0.1863	0.1623	
26	FE	0.3625	0.4285	0.5148	0.5849	0.6800	0.7274	0.7201	0.6942	0.6615
		0.5798	0.4949	0.4189	0.3474	0.2902	0.2451	0.2086	0.1768	
29	CU	0.2757	0.2577	0.3559	0.4352	0.5124	0.6240	0.6680	0.6723	0.6682
		0.6325	0.5587	0.4846	0.4113	0.3480	0.2951	0.2513	0.2178	
32	GE	0.3062	0.2491	0.2677	0.3006	0.3833	0.4995	0.5529	0.5724	0.5982
		0.6124	0.5788	0.5249	0.4674	0.3988	0.3458	0.2934	0.2543	
35	BR	0.3836	0.3144	0.2785	0.2894	0.3386	0.4210	0.5020	0.5485	0.5694
		0.6120	0.5832	0.5513	0.4975	0.4456	0.3826	0.3362	0.2899	

Table II. Backscattering Phase Shifts $\phi_b(k)$ in Radian vs. Photoelectron Wave Vector k in Å⁻¹. Calculated Using Clementi–Roetti Wave Functions.

Z	CHEM	k=3.7795 / 8.5038	4.2519 / 9.4486	4.7243 / 10.3935	5.1967 / 11.3384	5.6692 / 12.2832	6.1416 / 13.2281	6.6140 / 14.1729	7.0865 / 15.1178	7.5589
6	C	0.0481	-0.0992	-0.3146	-0.5489	-0.7327	-0.9119	-1.0949	-1.2448	-1.3145
		-1.6305	-1.8127	-2.0773	-2.2421	-2.4449	-2.5762	-2.7436	-2.8657	
8	O	0.5056	0.4555	0.3436	0.1858	0.0530	-0.0643	-0.1986	-0.3627	-0.5035
		-0.7327	-0.9763	-1.1969	-1.4081	-1.6035	-1.7800	-1.9252	-2.0925	
9	F	0.6436	0.5801	0.5072	0.3723	0.2630	0.1815	0.0750	-0.0616	-0.1872
		-0.4056	-0.6354	-0.8457	-1.0587	-1.2427	-1.4357	-1.5878	-1.7465	
11	NA	1.4871	1.4189	1.2842	1.1232	1.0379	0.9245	0.8037	0.7278	0.6167
		0.4132	0.1683	-0.0389	-0.2383	-0.3511	-0.4625	-0.8664	-1.0499	
15	P	3.5897	3.4657	3.3259	3.1419	3.0078	2.8527	2.7088	2.5551	2.4108
		2.1811	1.9250	1.7299	1.4634	1.2613	1.0318	0.8387	0.6943	
16	S	3.6618	3.5265	3.4586	3.3502	3.1962	3.0299	2.8992	2.7921	2.6714
		2.4337	2.1703	1.9651	1.7471	1.5259	1.3306	1.1155	0.9559	
17	CL	3.7781	3.6982	3.6428	3.5596	3.4464	3.3174	3.1796	3.0631	2.9656
		2.7275	2.4810	2.2481	2.0427	1.8078	1.6304	1.4153	1.2463	
20	CA	5.5111	5.1885	5.1244	5.0356	4.9437	4.7719	4.6259	4.4780	4.3182
		4.0505	3.7842	3.5296	3.1526	2.9924	2.8367	2.6625	2.4975	
22	TI	5.0578	5.1829	5.0969	5.0310	5.0209	4.9533	4.8189	4.7412	4.6626
		4.3738	4.1297	3.8988	3.6356	3.4189	3.1788	2.9330	2.6651	
24	CR	4.6789	5.0195	5.1303	5.0781	5.0492	5.0313	5.0002	4.9126	4.8156
		4.6585	4.4184	4.2252	3.9968	3.7829	3.5680	3.3656	3.1832	
26	FE	4.6376	5.1013	5.2163	5.1940	5.1656	5.1196	5.1716	5.1143	5.0787
		4.9115	4.7426	4.5374	4.3659	4.1507	3.9500	3.7421	3.5227	
29	CU	4.3259	4.9690	5.3797	5.4192	5.4286	5.4584	5.4798	5.4447	5.3991
		5.3159	5.1710	5.0326	4.8523	4.6898	4.4958	4.3136	4.1174	
32	GE	4.4485	4.9937	5.4287	5.6053	5.6784	5.7439	5.7959	5.7799	5.7404
		5.6405	5.5145	5.3850	5.2535	5.1026	4.9454	4.7666	4.6126	
35	BR	4.6324	5.0076	5.5637	5.9716	6.2120	6.3555	6.3982	6.3956	6.3713
		6.2729	6.1482	5.9926	5.8585	5.7174	5.5499	5.3929	5.2145	

Table III. Central Atom Phase Shifts $\phi_a^{l=1}(k)$ in Radian vs. Photoelectron Wave Vector k in Å⁻¹, Calculated Using Clementi–Roetti Wave Functions and the $(Z+1)$ Ion Approximation

Z	CHEM	k=3.7795 / 8.5038	4.2519 / 9.4486	4.7243 / 10.3935	5.1967 / 11.3384	5.6692 / 12.2832	6.1416 / 13.2281	6.6140 / 14.1729	7.0865 / 15.1178	7.5589
12	MG	-3.5617	-3.9685	-4.3733	-4.7701	-5.0669	-5.3543	-5.6428	-5.8338	-6.0636
		-6.4312	-6.7567	-7.0415	-7.2993	-7.5158	-7.7085	-7.8942	-8.0626	
14	SI	-2.3921	-2.9872	-3.4091	-3.7521	-4.0891	-4.4444	-4.7279	-4.9464	-5.1959
		-5.6061	-5.9793	-6.2763	-6.5604	-6.8066	-7.0308	-7.2375	-7.4235	
16	S	-1.5587	-2.1042	-2.5678	-2.9541	-3.2763	-3.6236	-3.9540	-4.2141	-4.4305
		-4.9052	-5.2781	-5.6170	-5.9076	-6.1779	-6.4301	-6.6363	-6.8490	
21	SC	1.1795	0.6230	0.1224	-0.3994	-0.7927	-1.1724	-1.5605	-1.8614	-2.1806
		-2.7211	-3.2072	-3.6470	-4.0255	-4.3628	-4.6688	-4.9459	-5.2051	
23	V	1.5350	0.9061	0.3468	-0.0757	-0.4409	-0.8628	-1.1871	-1.4872	-1.8112
		-2.3372	-2.8357	-3.2666	-3.6301	-3.9812	-4.2905	-4.5668	-4.8298	
26	FE	2.0856	1.4838	0.9136	0.4869	0.1332	-0.2651	-0.6151	-0.9053	-1.2124
		-1.7320	-2.2309	-2.6475	-3.0347	-3.3885	-3.7033	-3.9952	-4.2651	
29	CU	2.5785	1.9474	1.4064	1.0160	0.6236	0.2288	-0.0929	-0.3817	-0.6966
		-1.2194	-1.7082	-2.1233	-2.5115	-2.8684	-3.1826	-3.4880	-3.7629	
32	GE	3.2853	2.6446	2.1916	1.7551	1.3016	0.9493	0.6306	0.2722	-0.0379
		-0.5737	-1.0568	-1.5087	-1.8904	-2.2483	-2.5956	-2.8984	-3.1768	
35	BR	4.2011	3.5744	3.0770	2.6538	2.2224	1.8180	1.4768	1.1480	0.8028
		0.2540	-0.2749	-0.7220	-1.1320	-1.5120	-1.8640	-2.1812	-2.4812	
40	ZR	7.1080	6.4124	5.8533	5.2006	4.7033	4.2814	3.8093	3.4067	3.0375
		2.3621	1.7442	1.1970	0.7108	0.2704	-0.1353	-0.5099	-0.8646	
45	RH	6.4865	6.0246	5.5155	4.9999	4.5687	4.2071	3.8180	3.4570	3.1318
		2.5292	1.9858	1.4785	1.0327	0.6200	0.2359	-0.1138	-0.4461	
52	TE	9.1879	8.5796	8.0058	7.4207	6.9617	6.5146	6.0481	5.6693	5.2845
		4.6178	3.9937	3.4310	2.9335	2.4710	2.0384	1.6423	1.2821	

Table IV. Backscattering Amplitudes $F(k)$ in Å vs. Photoelectron Wave Vector k in Å⁻¹, Calculated Using Herman–Skillman Wave Functions

Z	CHEM	k=3.7795 / 8.5038	4.2519 / 9.4486	4.7243 / 10.3935	5.1967 / 11.3384	5.6692 / 12.2832	6.1416 / 13.2281	6.6140 / 14.1729	7.0865 / 15.1178	7.5589

Z	El																	
14	SI	0.7326	0.2745	0.7068	0.2215	0.6358	0.1771	0.5533	0.1466	0.5208	0.1183	0.4815	0.1004	0.4223	0.0836	0.3760	0.0727	0.3439
17	CL	0.7740	0.3441	0.8130	0.2857	0.7927	0.2288	0.7050	0.1905	0.6256	0.1597	0.5710	0.1338	0.5316	0.1111	0.4883	0.0960	0.4323
20	CA	0.6805	0.3976	0.7672	0.3310	0.8078	0.2771	0.7555	0.2213	0.7002	0.1915	0.6389	0.1618	0.5916	0.1350	0.5387	0.1135	0.4937
40	ZR	0.6463	0.5842	0.5510	0.5784	0.4291	0.5596	0.3454	0.5253	0.3190	0.4791	0.3513	0.4349	0.4217	0.3904	0.4807	0.3366	0.5192
42	MO	0.7500	0.5779	0.6977	0.5960	0.5944	0.5729	0.4233	0.5452	0.3582	0.5019	0.3596	0.4509	0.3894	0.4085	0.4570	0.3668	0.5172
44	RU	0.8645	0.5732	0.7911	0.5976	0.7219	0.5900	0.5553	0.5623	0.4243	0.5218	0.3835	0.4755	0.3894	0.4239	0.4305	0.3837	0.4903
46	PD	0.9115	0.5519	0.8554	0.5944	0.7922	0.5994	0.6655	0.5813	0.5083	0.5413	0.4197	0.4974	0.3822	0.4470	0.3984	0.4068	0.4530
47	AG	0.9143	0.5330	0.8729	0.5923	0.8121	0.6008	0.6993	0.5881	0.5558	0.5517	0.4354	0.5066	0.3745	0.4616	0.3895	0.4198	0.4358
50	SN	1.1146	0.5174	1.0135	0.5849	0.8812	0.6142	0.7798	0.6031	0.6296	0.5789	0.4939	0.5380	0.4107	0.4940	0.4014	0.4395	0.4261
53	I	1.1468	0.5065	1.0382	0.5883	0.9153	0.6233	0.8085	0.6313	0.6808	0.6063	0.5630	0.5680	0.4660	0.5190	0.4215	0.4749	0.4238
57	LA	0.6221	0.5067	0.7004	0.5901	0.7250	0.6420	0.7630	0.6381	0.6704	0.6242	0.5967	0.6089	0.5238	0.5548	0.4512	0.5164	0.4503
58	CE	0.6161	0.4573	0.6850	0.5524	0.6812	0.6158	0.7354	0.6300	0.6726	0.6188	0.5940	0.6083	0.5272	0.5663	0.4443	0.5188	0.4224
65	TB	0.3467	0.3606	0.4390	0.3836	0.4390	0.4589	0.5228	0.5367	0.5789	0.5707	0.5519	0.6010	0.5369	0.5999	0.4821	0.5772	0.4263
70	YB	0.2904	0.3677	0.3517	0.3169	0.3241	0.3367	0.3897	0.4202	0.4710	0.4868	0.4782	0.5316	0.4918	0.5826	0.4765	0.5791	0.4360
74	W	0.3196	0.3985	0.2523	0.3366	0.1939	0.3169	0.1996	0.3750	0.2903	0.4554	0.4000	0.5122	0.4404	0.5417	0.4525	0.5693	0.4603
76	OS	0.4614	0.4310	0.3804	0.3461	0.2549	0.3195	0.1401	0.3555	0.1989	0.4355	0.2965	0.4984	0.3971	0.5467	0.4521	0.5628	0.4538
78	PT	0.5711	0.4431	0.5106	0.3649	0.3803	0.3193	0.1694	0.3409	0.0819	0.4138	0.1982	0.4798	0.3370	0.5289	0.4127	0.5488	0.4430
80	HG	0.6538	0.4496	0.5718	0.3826	0.4872	0.3250	0.3073	0.3269	0.0806	0.3914	0.1015	0.4466	0.2527	0.5088	0.3666	0.5499	0.4265
82	PB	0.8557	0.4701	0.7215	0.4067	0.5926	0.3415	0.3856	0.3295	0.1469	0.3866	0.0945	0.4381	0.2350	0.4958	0.3655	0.5415	0.4384

Table V. Backscattering Phase Shifts $\phi_A(k)$ in Radian vs. Photoelectron Wave Vector k in Å^{-1}, Calculated Using Herman–Skillman Wave Functions

Z	CHEM	k=3.7795	4.2519	4.7243	5.1967	5.6692	6.1416	6.6140	7.0865	7.5589
		8.5038	9.4486	10.3935	11.3384	12.2832	13.2281	14.1729	15.1178	
14	SI	2.8451	2.7102	2.5684	2.4237	2.2930	2.1695	2.0314	1.8893	1.7979
		1.5790	1.3425	1.1418	0.9229	0.7272	0.5244	0.3251	0.1337	
17	CL	3.5339	3.4935	3.4518	3.3727	3.2504	3.1405	3.0158	2.9084	2.8179
		2.5775	2.3437	2.1201	1.9182	1.6835	1.5147	1.3126	1.1424	
20	CA	4.9368	4.9506	4.9006	4.8040	4.7095	4.5616	4.4313	4.2976	4.1491
		3.8851	3.6277	3.3828	3.0205	2.8617	2.7022	2.5257	2.3615	
40	ZR	5.5087	5.6976	5.9968	6.3248	6.7537	7.1845	7.3301	7.3862	7.3997
		7.3259	7.2062	7.0627	6.8848	6.7291	6.5750	6.4172	6.2162	
42	MO	5.3241	5.6417	5.9025	6.2277	6.5820	6.9913	7.3814	7.5168	7.5415
		7.5779	7.4839	7.3453	7.1972	7.0307	6.8815	6.7131	6.5743	
44	RU	5.1521	5.5560	5.8075	6.1610	6.4952	6.8481	7.2943	7.5823	7.6837
		7.7764	7.7103	7.6053	7.4610	7.3223	7.1668	7.0090	6.8538	
46	PD	4.9785	5.4527	5.7202	6.0619	6.3308	6.6559	7.1120	7.4861	7.6767
		7.8956	7.8671	7.8056	7.6815	7.5570	7.4254	7.2556	7.1135	
47	AG	4.9498	5.4256	5.7322	6.0104	6.2758	6.5895	7.0201	7.4073	7.6789
		7.9364	7.9624	7.8971	7.7941	7.6706	7.5470	7.3844	7.2409	
50	SN	6.2323	6.5485	6.7944	6.9256	7.0671	7.3134	7.5998	7.9268	8.2718
		8.5524	8.6258	8.5639	8.4687	8.3487	8.2004	8.0240	7.8928	
53	I	7.2328	7.4433	7.5482	7.6736	7.7480	7.9092	8.1375	8.3787	8.6663
		9.0611	9.1503	9.1388	9.0401	8.9299	8.8010	8.6309	8.4567	
57	LA	9.1054	9.1140	9.0879	9.0990	9.0917	9.1506	9.2783	9.4672	9.6694
		10.0405	10.1739	10.1463	10.0304	9.9063	9.7864	9.6034	9.4633	
58	CE	8.8314	8.8587	8.7713	8.8249	8.8322	8.8543	8.9636	9.1388	9.3441
		9.7813	9.9848	9.9920	9.9276	9.8101	9.7087	9.5561	9.4087	
65	TB	8.9896	9.0186	8.8981	8.8092	8.7686	8.7275	8.7431	8.7896	8.8404
		9.2244	9.6925	9.9728	10.0507	10.0328	9.9813	9.9134	9.8120	
70	YB	9.1561	9.1654	9.1020	8.9196	8.8057	8.7093	8.6750	8.6419	8.6522
		8.8262	9.2384	9.7232	9.9597	10.0334	10.0505	10.0326	9.9615	
74	W	10.8847	10.8005	10.3731	9.7269	9.3337	9.2044	9.1181	9.0581	9.0780
		9.1798	9.5185	9.9579	10.2952	10.4195	10.4622	10.4376	10.3847	
76	OS	10.9741	11.2710	11.3216	10.2771	9.5414	9.2519	9.1696	9.1667	9.1447
		9.2602	9.5576	9.9875	10.3635	10.5346	10.6037	10.5956	10.5386	

78	11.1468	11.5015	11.6393	11.5297	10.1376	9.3255	9.2371	9.2168	9.1978
PT	9.3455	9.6205	10.0458	10.4390	10.6673	10.7535	10.7584	10.7178	9.3491
80	11.5236	11.6950	11.8774	12.0838	11.8658	9.2967	9.2339	9.3059	9.7369
HG	9.4588	9.7259	10.1046	10.4877	10.7982	10.9296	10.9448	10.9055	
82	12.3853	12.5708	12.7450	12.9479	13.2543	8.9454	9.4699	9.6543	
PB	9.8628	10.1082	10.4315	10.8212	11.1250	11.2779	11.2993	11.2529	

Table VI. Central Atom Phase Shifts $\phi_a{}^{l=0}(k)$ in Radian vs. Photoelectron Wave Vector k in Å^{-1}, Calculated Using Herman–Skillman Wave Functions

Z	CHEM	k = 3.7795 / 8.5038	4.2519 / 9.4486	4.7243 / 10.3935	5.1967 / 11.3384	5.6692 / 12.2832	6.1416 / 13.2281	6.6140 / 14.1729	7.0865 / 15.1178	7.5589
11	NA+	-5.6231	-6.3933	-6.8848	-7.3767	-7.8876	-8.2287	-8.6431	-8.9622	-9.2652
		-9.8182	-10.3291	-10.7726	-11.1661	-11.5138	-11.8216	-12.1021	-12.3581	
12	MG++	-3.8448	-4.6324	-5.3798	-5.8169	-6.3347	-6.8851	-7.2518	-7.6238	-8.0287
		-8.6331	-9.1791	-9.7016	-10.1483	-10.5207	-10.8800	-11.2092	-11.5071	
14	SI	-5.1257	-5.7878	-6.3136	-6.7332	-7.1618	-7.5758	-7.9339	-8.2163	-8.5254
		-9.0594	-9.5388	-9.9391	-10.3355	-10.6638	-10.9752	-11.2682	-11.5251	
17	CL	-3.9386	-4.5306	-5.0641	-5.5371	-5.9210	-6.3182	-6.7155	-7.0543	-7.3263
		-7.9082	-8.3831	-8.8327	-9.2170	-9.5822	-9.9003	-10.2143	-10.4905	
19	K+	-0.8143	-1.6909	-2.2709	-2.9487	-3.4504	-3.9979	-4.4169	-4.8718	-5.2530
		-5.9536	-6.5740	-7.1175	-7.5960	-8.0316	-8.4189	-8.7911	-9.1260	
20	CA++	0.6406	-0.0014	-0.8699	-1.4697	-2.0743	-2.6683	-3.1234	-3.6604	-4.0268
		-4.8218	-5.5110	-6.1175	-6.6567	-7.1396	-7.5766	-7.9730	-8.3377	
20	CA	-1.6038	-2.3361	-2.8941	-3.4800	-3.9310	-4.4009	-4.8048	-5.1989	-5.5518
		-6.2042	-6.7698	-7.2768	-7.7304	-8.1389	-8.5121	-8.8554	-9.1691	
22	TI	-1.3496	-2.0813	-2.6117	-3.1165	-3.6305	-4.0214	-4.3987	-4.8077	-5.1324
		-5.7680	-6.3163	-6.8133	-7.2728	-7.6814	-8.0480	-8.3895	-8.7136	
26	FE	-0.5981	-1.2789	-1.8973	-2.3640	-2.7732	-3.2195	-3.6079	-3.9363	-4.2863
		-4.9006	-5.4475	-5.9420	-6.3928	-6.7960	-7.1690	-7.5194	-7.8358	
28	NI	-0.2909	-0.9459	-1.5472	-2.0204	-2.4204	-2.8508	-3.2421	-3.5710	-3.9074
		-4.5158	-5.0653	-5.5374	-5.9978	-6.4019	-6.7785	-7.1283	-7.4465	
32	GE	0.8703	0.2092	-0.2978	-0.8170	-1.3128	-1.7093	-2.0806	-2.4920	-2.8277
		-3.4527	-4.0051	-4.5191	-4.9725	-5.3879	-5.7904	-6.1490	-6.4830	
35	BR	1.7659	1.1103	0.5794	0.0977	-0.4215	-0.8458	-1.2233	-1.6066	-1.9886
		-2.6082	-3.2051	-3.7131	-4.1773	-4.6187	-5.0200	-5.3906	-5.7394	
40	ZR	3.8309	3.1251	2.5669	1.9563	1.4417	1.0104	0.5316	0.1307	-0.2532
		-0.9471	-1.5886	-2.1614	-2.6682	-3.1331	-3.5692	-3.9683	-4.3512	
42	MO	4.2333	3.4817	2.8584	2.3604	1.8328	1.3441	0.9345	0.5199	0.1211
		-0.5622	-1.1870	-1.7663	-2.2750	-2.7419	-3.1318	-3.5866	-3.9524	
44	RU	4.5455	3.8317	3.1836	2.6907	2.1956	1.6875	1.2868	0.9049	0.5003
		-0.1749	-0.8243	-1.3863	-1.8955	-2.3768	-2.8095	-3.2053	-3.5863	
46	PD	4.8624	4.1300	3.5221	3.0478	2.5299	2.0388	1.6580	1.2552	0.8468
		0.1802	-0.4482	-1.0313	-1.5420	-2.0109	-2.4527	-2.8626	-3.2327	

50	SN	6.1712	5.5288	4.8544	4.2728	3.8081	3.2835	2.8175	2.4217	1.9980
		1.3042	0.6382	0.0332	-0.5017	-0.9909	-1.4469	-1.8849	-2.2776	
53	I	6.9861	6.3488	5.6996	5.0999	4.5908	4.0966	3.6020	3.1848	2.7917
		2.0548	1.3868	0.7883	0.2369	-0.2833	-0.7488	-1.1859	-1.5902	
57	LA	8.9217	8.2145	7.4576	6.8800	6.2577	5.7083	5.2118	4.7188	4.2983
		3.5052	2.7852	2.1321	1.5268	0.9808	0.4739	-0.0014	-0.4375	
58	CE	8.8166	8.1511	7.4003	6.8131	6.2281	5.6672	5.1862	4.6962	4.2806
		3.4880	2.7814	2.1388	1.5480	1.0024	0.5109	0.0419	-0.4007	
64	GD	9.4593	8.8153	8.0828	7.4980	6.9600	6.3999	5.9272	5.4477	5.0296
		4.2472	3.5518	2.9159	2.3326	1.8002	1.3049	0.8393	0.4053	
70	YB	9.7131	9.0941	8.4302	7.8181	7.2945	6.7735	6.3181	5.8661	5.4388
		4.6757	4.0225	3.4051	2.8361	2.3099	1.8207	1.3664	0.9411	
74	W	10.4268	9.6477	9.0202	8.4815	7.9303	7.3854	6.9506	6.5046	6.0647
		5.3324	4.6571	4.0207	3.4588	2.9322	2.4358	1.9818	1.5598	
76	OS	10.6930	9.9356	9.2906	8.7545	8.2274	7.6945	7.2307	6.8067	6.3706
		5.6367	4.9411	4.3220	3.7535	3.2160	2.7260	2.2752	1.8352	
78	PT	10.9689	10.2018	9.5890	9.0822	8.5108	7.9961	7.5602	7.0946	6.6577
		5.9321	5.2571	4.6173	4.0381	3.5196	3.0234	2.5544	2.1299	
80	HG	11.2355	10.5896	9.9143	9.3646	8.8521	8.2996	7.8570	7.4171	6.9807
		6.2308	5.5580	4.9246	4.3319	3.8022	3.3126	2.8496	2.4138	
82	PB	11.9467	11.2913	10.5835	10.0159	9.4675	8.9075	8.4542	7.9969	7.5528
		6.7782	6.0730	5.4323	4.8243	4.2836	3.7834	3.3072	2.8597	

Table VII. Central Atom Phase Shifts $\phi_a^{l=1}(k)$ in Radian vs. Photoelectron Wave Vector k in Å^{-1}. Calculated Using Herman–Skillman Wave Functions

Z	CHEM	k=3.7795	4.2519	4.7243	5.1967	5.6692	6.1416	6.6140	7.0865	7.5589
		8.5038	9.4486	10.3935	11.3384	12.2832	13.2281	14.1729	15.1178	
6	C	-7.0744	-7.2817	-7.4487	-7.6994	-7.9676	-8.1821	-8.3345	-8.4618	-8.6018
		-8.8888	-9.0707	-9.2798	-9.4242	-9.5691	-9.6967	-9.8045	-9.9151	
8	O	-6.5189	-6.7480	-6.9315	-7.1700	-7.4127	-7.6278	-7.7959	-7.9196	-8.0470
		-8.3582	-8.5693	-8.7617	-8.9497	-9.0839	-9.2360	-9.3545	-9.4670	
9	F	-6.2972	-6.5199	-6.7071	-6.9492	-7.2004	-7.4174	-7.5870	-7.7204	-7.8349
		-8.1329	-8.3485	-8.5462	-8.7425	-8.8804	-9.0351	-9.1566	-9.2720	
11	NA+	-3.2155	-3.8481	-4.2669	-4.6247	-5.0507	-5.2996	-5.6235	-5.8562	-6.0813
		-6.4784	-6.8412	-7.1488	-7.4189	-7.6600	-7.8716	-8.0583	-8.2261	
12	MG++	-1.4390	-2.0405	-2.7238	-3.0884	-3.4938	-3.9705	-4.2603	-4.5530	-4.8926
		-5.3568	-5.7625	-6.1516	-6.4918	-6.7651	-7.0197	-7.2475	-7.4645	
14	SI	-2.8651	-3.4002	-3.8494	-4.1870	-4.5117	-4.8393	-5.1257	-5.3292	-5.5617
		-5.9614	-6.3161	-6.5963	-6.8764	-7.1045	-7.3137	-7.5183	-7.6926	
17	CL	-1.7939	-2.2287	-2.6765	-3.0897	-3.3912	-3.6954	-4.0199	-4.2977	-4.5038
		-4.9602	-5.3250	-5.6685	-5.9513	-6.2241	-6.4524	-6.6771	-6.8757	
19	K+	1.3571	0.6147	0.0905	-0.4883	-0.9257	-1.4047	-1.7551	-2.1531	-2.4673
		-3.0579	-3.5781	-4.0250	-4.4108	-4.7602	-5.0656	-5.3588	-5.6179	
20	CA++	2.9048	2.2965	1.5766	0.9778	0.4902	-0.0731	-0.4507	-0.9338	-1.2464
		-1.9394	-2.5310	-3.0439	-3.4957	-3.8939	-4.2513	-4.5735	-4.8644	
20	CA	0.5733	-0.0952	-0.5599	-1.0552	-1.4490	-1.8348	-2.1860	-2.5083	-2.8093
		-3.3537	-3.8148	-4.2257	-4.5900	-4.9121	-5.2042	-5.4720	-5.7105	
22	TI	0.7928	0.1399	-0.3505	-0.7510	-1.1891	-1.5297	-1.8307	-2.1774	-2.4535
		-2.9774	-3.4271	-3.8348	-4.2071	-4.5351	-4.8252	-5.0924	-5.3461	
26	FE	1.3837	0.8536	0.3077	-0.1080	-0.4466	-0.8114	-1.1523	-1.4268	-1.7096
		-2.2304	-2.6766	-3.0871	-3.4565	-3.7820	-4.0841	-4.3658	-4.6186	
28	NI	1.6350	1.1435	0.6110	0.1706	-0.1425	-0.4901	-0.8352	-1.1129	-1.3810
		-1.8995	-2.3468	-2.7365	-3.1171	-3.4458	-3.7537	-4.0366	-4.2916	
32	GE	2.8531	2.2248	1.7630	1.3573	0.9298	0.5690	0.2700	-0.0767	-0.3758
		-0.8942	-1.3665	-1.7991	-2.1741	-2.5210	-2.8564	-3.1522	-3.4270	
35	BR	3.7269	3.1353	2.6303	2.2422	1.8157	1.4317	1.0971	0.7827	0.4517
		-0.0860	-0.5993	-1.0274	-1.4246	-1.7951	-2.1336	-2.4457	-2.7370	
40	ZR	5.8079	5.1256	4.6373	4.1281	3.6490	3.2773	2.8561	2.4865	2.1632
		1.5447	0.9799	0.4818	0.0414	-0.3642	-0.7440	-1.0869	-1.4163	

No.	Symbol									
42	MO	6.1471	5.5279	4.9250	4.4785	4.0464	3.6009	3.2270	2.8816	2.5219
		1.9237	1.3660	0.8583	0.4160	0.0058	-0.3779	-0.7269	-1.0414	
44	RU	6.4206	5.8617	5.2665	4.7926	4.3806	3.9385	3.5686	3.2431	2.8949
		2.2888	1.7143	1.2250	0.7741	0.3513	-0.0242	-0.3677	-0.6993	
46	PD	6.7635	6.1581	5.5692	5.1463	4.7175	4.2611	3.9233	3.5903	3.2188
		2.6376	2.0726	1.5596	1.1130	0.6981	0.3114	-0.0450	-0.3665	
47	AG	7.4130	7.0555	6.6507	6.3647	6.0790	5.7689	5.5393	5.3076	5.0642
		4.6584	4.2777	3.9214	3.6012	3.3222	3.0718	2.8187	2.6021	
50	SN	8.0817	7.4989	6.9398	6.3880	5.9590	5.5205	5.0864	4.7366	4.3678
		3.7451	3.1433	2.6045	2.1309	1.6957	1.2903	0.9009	0.5570	
53	I	8.9247	8.3161	7.7791	7.2392	6.7528	6.3289	5.8870	5.4941	5.1532
		4.4825	3.8899	3.3529	2.8544	2.3887	1.9751	1.5843	1.2241	
55	CS+	10.3098	9.6092	8.8908	8.3605	7.9268	7.4458	6.9389	6.5101	6.1637
		5.4374	4.8397	4.2264	3.7234	3.2053	2.7855	2.3499	1.9675	
57	LA	10.8567	10.2554	9.5608	9.0124	8.4708	7.9414	7.5065	7.0439	6.6639
		5.9371	5.2814	4.6881	4.1370	3.6430	3.1842	2.7531	2.3596	
58	CE	10.7507	10.1561	9.4991	8.9259	8.4226	7.8903	7.4572	7.0182	6.6274
		5.9071	5.2682	4.6874	4.1510	3.6564	3.2123	2.7865	2.3858	
64	GD	11.3615	10.7710	10.1419	9.5739	9.1174	8.6007	8.1617	7.7395	7.3430
		6.6353	6.0082	5.4315	4.9007	4.4168	3.9662	3.5430	3.1502	
70	YB	11.5744	10.9738	10.4332	9.8476	9.3850	8.9363	8.5022	8.1120	7.7131
		7.0261	6.4381	5.8762	5.3578	4.8803	4.4379	4.0281	3.6442	
74	W	12.2519	11.6107	10.9990	10.4924	10.0924	9.5418	9.1289	8.7431	8.3425
		7.6748	7.0550	6.4830	5.9731	5.4897	5.0400	4.6301	4.2456	
76	OS	12.5102	11.8931	11.2782	10.7721	10.3217	9.8559	9.4097	9.0322	8.6483
		7.9683	7.3377	6.7810	6.2573	5.7681	5.3254	4.9137	4.5123	
78	PT	12.8279	12.1642	11.5505	11.0983	10.6215	10.1311	9.7348	9.3312	8.9200
		8.2696	7.6463	7.0656	6.5410	6.0655	5.6126	5.1879	4.8013	
80	HG	13.0831	12.4816	11.9145	11.3844	10.9332	10.4476	10.0268	9.6428	9.2403
		8.5622	7.9474	7.3650	6.8229	6.3431	5.8984	5.4754	5.0767	
82	PB	13.7859	13.2106	12.6005	12.0451	11.5681	11.0660	10.6298	10.2276	9.8130
		9.1126	8.4636	7.8730	7.3162	6.8247	6.3675	5.9295	5.5188	

Table VIII. Central Atom Phase Shifts $\phi_a^{l=2}(k)$ in Radian vs. Photoelectron Wave Vector k in Å^{-1}, Calculated Using Herman–Skillman Wave Functions

Z	CHEM	k = 3.7795 / 8.5038	4.2519 / 9.4486	4.7243 / 10.3935	5.1967 / 11.3384	5.6692 / 12.2832	6.1416 / 13.2281	6.6140 / 14.1729	7.0865 / 15.1178	7.5589
11	NA+	-0.3995	-0.8132	-0.9636	-1.1910	-1.4274	-1.5356	-1.7250	-1.8052	-1.9704
		-2.1820	-2.3796	-2.5390	-2.6900	-2.8270	-2.9523	-3.0639	-3.1649	
12	MG++	1.3366	0.9309	0.4564	0.3056	-0.0218	-0.3142	-0.4408	-0.6764	-0.8554
		-1.1806	-1.4292	-1.6413	-1.8642	-2.0579	-2.2275	-2.3535	-2.4818	
14	SI	-0.7863	-1.0065	-1.2323	-1.3593	-1.4973	-1.6609	-1.7749	-1.8344	-1.9419
		-2.1014	-2.2746	-2.4059	-2.5426	-2.6554	-2.7539	-2.8504	-2.9472	
17	CL	-0.2270	-0.2936	-0.4897	-0.7129	-0.8225	-0.9409	-1.1002	-1.2162	-1.2796
		-1.5121	-1.6571	-1.8275	-1.9597	-2.0875	-2.2033	-2.3084	-2.4182	
19	K+	2.9156	2.5286	2.1377	1.8014	1.5062	1.2084	1.0116	0.7518	0.5703
		0.2063	-0.1042	-0.3633	-0.5900	-0.8017	-0.9859	-1.1670	-1.3169	
20	CA++	4.7443	4.1594	3.7548	3.2475	2.9594	2.5239	2.2975	1.9436	1.7547
		1.2801	0.8908	0.5610	0.2602	0.0020	-0.2311	-0.4373	-0.6244	
20	CA	2.0448	1.6636	1.3692	1.1028	0.8626	0.6594	0.4565	0.2822	0.1137
		-0.1949	-0.4470	-0.6673	-0.8742	-1.0508	-1.2141	-1.3598	-1.4922	
22	TI	2.0353	1.7127	1.3722	1.1717	0.9422	0.7419	0.6008	0.4051	0.2559
		-0.0259	-0.2628	-0.4696	-0.6788	-0.8663	-1.0169	-1.1526	-1.2932	
26	FE	2.2020	1.9926	1.7277	1.4388	1.2717	1.1124	0.9207	0.7721	0.6362
		0.3515	0.1303	-0.0759	-0.2748	-0.4459	-0.6032	-0.7529	-0.8865	
28	NI	2.2964	2.0904	1.8684	1.5675	1.4034	1.2655	1.0768	0.9185	0.8013
		0.5233	0.3003	0.1068	-0.0928	-0.2627	-0.4257	-0.5750	-0.7118	
32	GE	3.4808	3.1522	2.7816	2.5499	2.3530	2.1165	1.9519	1.7749	1.5910
		1.3132	1.0420	0.8017	0.6026	0.4131	0.2241	0.0594	-0.0915	
35	BR	4.2896	4.0442	3.6549	3.3775	3.1748	2.9537	2.7140	2.5398	2.3480
		2.0083	1.7061	1.4580	1.2218	1.0069	0.8098	0.6283	0.4587	
40	ZR	6.3946	6.0007	5.6098	5.3205	4.9774	4.7086	4.4491	4.1646	3.9606
		3.5365	3.1589	2.8311	2.5430	2.2758	2.0247	1.8051	1.5931	
42	MO	6.6133	6.3255	5.9501	5.5844	5.3187	5.0429	4.7477	4.5307	4.2950
		3.8789	3.4912	3.1549	2.8612	2.5850	2.3293	2.1041	1.9009	
44	RU	6.8945	6.5710	6.2631	5.8888	5.5835	5.3282	5.0698	4.8339	4.6216
		4.1896	3.8014	3.4727	3.1599	2.8769	2.6276	2.4024	2.1809	

46	PD	7.1714	6.8840	6.5244	6.1577	5.9111	5.6099	5.3509	5.1543	4.9069
		4.5062	4.1031	3.7560	3.4523	3.1646	2.9073	2.6681	2.4569	
50	SN	8.7603	8.2233	7.8591	7.4918	7.1216	6.8426	6.5317	6.2581	6.0202
		5.5650	5.1250	4.7401	4.4031	4.0928	3.8059	3.5254	3.2909	
53	I	9.6422	9.1056	8.6831	8.3551	7.9584	7.6306	7.3421	7.0258	6.7741
		6.2836	5.8591	5.4556	5.0838	4.7453	4.4476	4.1595	3.9002	
57	LA	11.6200	11.0909	10.6303	10.1350	9.7425	9.3073	8.9637	8.6106	8.2968
		7.7315	7.2254	6.7693	6.3422	5.9646	5.6144	5.2813	4.9833	
58	CE	11.4894	10.9460	10.5122	10.0288	9.6359	9.2402	8.8656	8.5562	8.2289
		7.6787	7.1910	6.7468	6.3326	5.9523	5.6157	5.2869	4.9811	
64	GD	11.9957	11.4600	11.0329	10.5967	10.2245	9.8649	9.4819	9.1841	8.8624
		8.3220	7.8415	7.3944	6.9822	6.6076	6.2605	5.9345	5.6351	
70	YB	12.0252	11.5305	11.1428	10.7483	10.3436	10.0665	9.7133	9.4184	9.1284
		8.5982	8.1509	7.7170	7.3189	6.9550	6.6196	6.3096	6.0174	
74	W	12.5569	12.1504	11.7813	11.3239	10.9935	10.6783	10.3303	10.0274	9.7497
		9.2114	8.7382	8.3088	7.9078	7.5280	7.1859	6.8713	6.5722	
76	OS	12.8359	12.4158	12.0424	11.6109	11.2547	10.9740	10.6219	10.3022	10.0360
		9.4885	9.0202	8.5885	8.1705	7.7954	7.4539	7.1305	6.8183	
78	PT	13.1248	12.7441	12.3083	11.8942	11.5823	11.2487	10.9034	10.6135	10.3072
		9.7937	9.3002	8.8599	8.4476	8.0702	7.7199	7.3916	7.0878	
80	HG	13.5611	13.0062	12.5998	12.2395	11.8503	11.5302	11.2065	10.9026	10.6126
		10.0773	9.5912	9.1331	8.7092	8.3387	7.9918	7.6568	7.3423	
82	PB	14.3165	13.7561	13.3507	12.9237	12.5148	12.1799	11.8216	11.5009	11.1963
		10.6319	10.1103	9.6426	9.2027	8.8174	8.4533	8.1012	7.7740	

APPENDIX VI

Tables of

Scattering Amplitude, $F(\beta, k)$ in Å,

and Phase, $\theta(\beta, k)$ in Radians, Functions

for Some Elements

There are 37 sets of functions per element, covering scattering angles, β, from 0° to 180°, in steps of 5°. Each set contains 20 points each of amplitude and phase functions for $k = 0.9449$, 1.8898, 2.8346, 3.7795, 4.2519, 4.7243, 5.1967, 5.6692, 6.1416, 6.6140, 7.0865, 7.5589, 8.5038, 9.4486, 10.3935, 11.3384, 12.2832, 13.2281, 14.1729, 15.1178 Å^{-1}. (Calculated using Herman-Skillman Wave functions, Teo, B. K., **1982**.)

```
0  20
  1.5596   1.2945   1.4648   1.6966   1.7322   1.7675   1.7815   1.7727
  1.7777   1.7725   1.7668   1.7563   1.7785   1.7713   1.7577   1.7468
  1.7369   1.7274   1.7210   1.7135
  2.6157   2.0474   1.3092   1.1779   1.1293   1.0922   1.0614   1.0437
  1.0209   0.9998   0.9777   0.9572   0.9210   0.8875   0.8553   0.8234
  0.7935   0.7678   0.7429   0.7180
5  20
  1.5566   1.2923   1.4544   1.6769   1.7085   1.7390   1.7487   1.7360
  1.7361   1.7262   1.7158   1.7005   1.7100   1.6919   1.6677   1.6457
  1.6242   1.6025   1.5831   1.5624
  2.6153   2.0484   1.3138   1.1822   1.1335   1.0967   1.0661   1.0488
  1.0263   1.0054   0.9835   0.9633   0.9272   0.8939   0.8616   0.8297
  0.7995   0.7732   0.7475   0.7218
10  20
  1.5475   1.2857   1.4237   1.6189   1.6391   1.6561   1.6535   1.6298
  1.6166   1.5941   1.5708   1.5431   1.5196   1.4745   1.4249   1.3775
  1.3307   1.2834   1.2383   1.1929
  2.6139   2.0515   1.3277   1.1953   1.1465   1.1106   1.0808   1.0644
  1.0431   1.0232   1.0022   0.9826   0.9476   0.9154   0.8837   0.8522
  0.8221   0.7950   0.7681   0.7418
15  20
  1.5325   1.2746   1.3743   1.5264   1.5289   1.5255   1.5049   1.4656
  1.4340   1.3944   1.3542   1.3111   1.2474   1.1730   1.0989   1.0299
  0.9642   0.9011   0.8426   0.7879
  2.6116   2.0564   1.3512   1.2179   1.1692   1.1351   1.1068   1.0924
  1.0736   1.0561   1.0373   1.0195   0.9883   0.9599   0.9317   0.9040
  0.8777   0.8540   0.8307   0.8098
20  20
  1.5115   1.2590   1.3087   1.4049   1.3856   1.3579   1.3167   1.2609
  1.2099   1.1538   1.0984   1.0422   0.9472   0.8563   0.7736   0.7017
  0.6388   0.5831   0.5354   0.4958
  2.6083   2.0630   1.3849   1.2514   1.2033   1.1724   1.1471   1.1364
  1.1225   1.1098   1.0958   1.0825   1.0618   1.0447   1.0281   1.0131
  1.0007   0.9914   0.9843   0.9813
25  20
  1.4848   1.2388   1.2302   1.2619   1.2190   1.1665   1.1058   1.0362
  0.9698   0.9026   0.8385   0.7771   0.6719   0.5864   0.5172   0.4640
  0.4234   0.3918   0.3674   0.3493
  2.6039   2.0708   1.4295   1.2976   1.2513   1.2262   1.2065   1.2027
  1.1979   1.1947   1.1909   1.1875   1.1913   1.1995   1.2083   1.2174
  1.2270   1.2362   1.2446   1.2502
30  20
  1.4526   1.2140   1.1425   1.1058   1.0400   0.9657   0.8901   0.8128
  0.7391   0.6700   0.6073   0.5511   0.4616   0.4021   0.3614   0.3351
  0.3180   0.3052   0.2935   0.2816
  2.5984   2.0796   1.4860   1.3597   1.3177   1.3025   1.2933   1.3017
  1.3140   1.3290   1.3447   1.3605   1.4092   1.4541   1.4882   1.5072
  1.5145   1.5142   1.5104   1.5019
35  20
  1.4151   1.1844   1.0499   0.9456   0.8602   0.7703   0.6872   0.6109
  0.5404   0.4804   0.4302   0.3890   0.3338   0.3061   0.2897   0.2786
  0.2697   0.2592   0.2455   0.2286
  2.5916   2.0889   1.5552   1.4418   1.4089   1.4114   1.4211   1.4509
  1.4935   1.5396   1.5870   1.6304   1.7229   1.7735   1.7895   1.7785
  1.7584   1.7402   1.7282   1.7206
```

```
40   20
  1.3725   1.1499   0.9564   0.7901   0.6908   0.5943   0.5133   0.4475
  0.3916   0.3503   0.3199   0.2975   0.2741   0.2641   0.2559   0.2457
  0.2330   0.2177   0.1997   0.1814
  2.5834   2.0979   1.6377   1.5499   1.5349   1.5680   1.6113   1.6757
  1.7642   1.8490   1.9252   1.9814   2.0525   2.0495   2.0205   1.9835
  1.9582   1.9495   1.9517   1.9591
45   20
  1.3252   1.1103   0.8658   0.6478   0.5425   0.4503   0.3822   0.3349
  0.3016   0.2810   0.2676   0.2570   0.2461   0.2384   0.2281   0.2135
  0.1969   0.1801   0.1638   0.1499
  2.5737   2.1063   1.7338   1.6915   1.7093   1.7934   1.8891   1.9958
  2.1262   2.2238   2.2891   2.3174   2.3062   2.2478   2.2022   2.1757
  2.1748   2.1896   2.1988   2.1920
50   20
  1.2735   1.0657   0.7808   0.5260   0.4244   0.3485   0.3019   0.2749
  0.2609   0.2520   0.2446   0.2365   0.2239   0.2127   0.1992   0.1831
  0.1676   0.1539   0.1412   0.1305
  2.5621   2.1131   1.8426   1.8748   1.9471   2.1028   2.2549   2.3804
  2.5051   2.5661   2.5846   2.5691   2.4908   2.4205   2.3989   2.4063
  2.4266   2.4351   2.4103   2.3550
55   20
  1.2179   1.0158   0.7033   0.4300   0.3427   0.2916   0.2667   0.2513
  0.2445   0.2361   0.2271   0.2173   0.2004   0.1872   0.1746   0.1611
  0.1484   0.1365   0.1251   0.1152
  2.5484   2.1175   1.9619   2.1042   2.2509   2.4727   2.6426   2.7402
  2.8159   2.8251   2.7995   2.7545   2.6610   2.6229   2.6432   2.6683
  2.6661   2.6256   2.5531   2.4722
60   20
  1.1589   0.9608   0.6335   0.3612   0.2964   0.2685   0.2555   0.2407
  0.2313   0.2187   0.2066   0.1958   0.1782   0.1678   0.1589   0.1473
  0.1343   0.1220   0.1112   0.1015
  2.5323   2.1185   2.0882   2.3727   2.5911   2.8314   2.9705   3.0234
  3.0500   3.0222   2.9726   2.9201   2.8614   2.8724   2.9089   2.9053
  2.8467   2.7573   2.6722   2.6127
65   20
  1.0969   0.9007   0.5702   0.3159   0.2746   0.2598   0.2483   0.2286
  0.2137   0.1975   0.1842   0.1747   0.1624   0.1572   0.1496   0.1359
  0.1211   0.1090   0.0994   0.0908
  2.5133   2.1149   2.2169   2.6583   2.9150   3.1279   3.2236   3.2445
  3.2411   3.1976   3.1450   3.1041   3.1041   3.1359   3.1405   3.0823
  2.9774   2.8773   2.8182   2.7888
70   20
  1.0325   0.8358   0.5111   0.2858   0.2625   0.2507   0.2355   0.2111
  0.1923   0.1755   0.1641   0.1586   0.1549   0.1518   0.1405   0.1237
  0.1089   0.0987   0.0908   0.0829
  2.4909   2.1049   2.3433   2.9356   3.1927   3.3661   3.4295   3.4373
  3.4246   3.3851   3.3463   3.3250   3.3621   3.3668   3.3151   3.2122
  3.0977   3.0225   2.9874   2.9492
75   20
  0.9663   0.7666   0.4531   0.2615   0.2493   0.2349   0.2159   0.1902
  0.1710   0.1573   0.1506   0.1502   0.1531   0.1460   0.1290   0.1115
  0.0995   0.0917   0.0842   0.0757
  2.4644   2.0865   2.4634   3.1927   3.4305   3.5732   3.6201   3.6321
  3.6279   3.6057   3.5847   3.5724   3.5967   3.5504   3.4492   3.3288
  3.2365   3.1856   3.1381   3.0584
```

O

```
 80  20
    0.8991    0.6940    0.3933    0.2366    0.2304    0.2126    0.1924    0.1700
    0.1546    0.1469    0.1459    0.1491    0.1523    0.1375    0.1168    0.1020
    0.0935    0.0866    0.0777    0.0685
    2.4330    2.0564    2.5746    3.4372    3.6516    3.7788    3.8233    3.8516
    3.8653    3.8583    3.8405    3.8159    3.7929    3.7009    3.5714    3.4590
    3.3913    3.3342    3.2484    3.1343
 85  20
    0.8316    0.6190    0.3289    0.2087    0.2064    0.1874    0.1698    0.1551
    0.1466    0.1456    0.1486    0.1523    0.1493    0.1275    0.1070    0.0966
    0.0899    0.0814    0.0710    0.0625
    2.3956    2.0102    2.6761    3.6910    3.8848    4.0117    4.0629    4.1062
    4.1288    4.1165    4.0812    4.0320    3.9583    3.8385    3.7035    3.6067
    3.5391    3.4513    3.3318    3.2163
 90  20
    0.7647    0.5436    0.2583    0.1798    0.1815    0.1650    0.1538    0.1490
    0.1477    0.1514    0.1553    0.1562    0.1438    0.1185    0.1016    0.0944
    0.0865    0.0755    0.0653    0.0585
    2.3509    1.9411    2.7688    3.9886    4.1616    4.2979    4.3504    4.3834
    4.3889    4.3498    4.2884    4.2170    4.1065    3.9766    3.8487    3.7546
    3.6640    3.5433    3.4147    3.3222
 95  20
    0.6994    0.4706    0.1804    0.1568    0.1628    0.1526    0.1487    0.1524
    0.1559    0.1611    0.1628    0.1587    0.1372    0.1125    0.1001    0.0934
    0.0823    0.0698    0.0614    0.0562
    2.2973    1.8390    2.8563    4.3697    4.5083    4.6432    4.6683    4.6518
    4.6179    4.5461    4.4632    4.3787    4.2478    4.1170    3.9927    3.8851
    3.7660    3.6266    3.5114    3.4357
100  20
    0.6367    0.4045    0.0952    0.1508    0.1593    0.1553    0.1559    0.1634
    0.1681    0.1716    0.1690    0.1592    0.1310    0.1098    0.1005    0.0915
    0.0772    0.0654    0.0592    0.0542
    2.2330    1.6898    2.9551    4.8342    4.9097    5.0055    4.9700    4.8839
    4.8074    4.7101    4.6150    4.5264    4.3867    4.2527    4.1210    3.9934
    3.8534    3.7136    3.6146    3.5324
105  20
    0.5778    0.3524    0.0069    0.1691    0.1751    0.1728    0.1728    0.1790
    0.1816    0.1811    0.1732    0.1583    0.1265    0.1094    0.1005    0.0877
    0.0723    0.0627    0.0576    0.0515
    2.1559    1.4787    4.1125    5.2833    5.2894    5.3182    5.2186    5.0729
    4.9638    4.8517    4.7527    4.6663    4.5221    4.3763    4.2305    4.0849
    3.9352    3.8060    3.7091    3.6040
110  20
    0.5242    0.3231    0.0956    0.2080    0.2066    0.2002    0.1954    0.1967
    0.1949    0.1891    0.1758    0.1569    0.1239    0.1100    0.0988    0.0826
    0.0685    0.0612    0.0556    0.0483
    2.0636    1.2050   -0.2081   -0.6599   -0.6986   -0.7300   -0.8746   -1.0575
   -1.1862   -1.3041   -1.4014   -1.4826   -1.6332   -1.7981   -1.9570   -2.1156
   -2.2666   -2.3864   -2.4965   -2.6256
115  20
    0.4773    0.3239    0.1980    0.2584    0.2469    0.2323    0.2208    0.2150
    0.2071    0.1954    0.1774    0.1557    0.1234    0.1106    0.0957    0.0775
    0.0660    0.0602    0.0530    0.0452
    1.9548    0.9039   -0.1287   -0.4268   -0.4890   -0.5618   -0.7313   -0.9310
   -1.0682   -1.1858   -1.2789   -1.3548   -1.5175   -1.7025   -1.8694   -2.0358
   -2.1857   -2.3049   -2.4343   -2.5775
```

```
120  20
     0.4386     0.3542     0.3032     0.3136     0.2908     0.2660     0.2470     0.2330
     0.2180     0.2003     0.1785     0.1555     0.1243     0.1108     0.0922     0.0737
     0.0649     0.0590     0.0502     0.0429
     1.8292     0.6332    -0.0779    -0.2655    -0.3414    -0.4409    -0.6210    -0.8228
    -0.9606    -1.0744    -1.1631    -1.2368    -1.4164    -1.6182    -1.7871    -1.9576
    -2.1093    -2.2358    -2.3819    -2.5249
125  20
     0.4093     0.4060     0.4093     0.3697     0.3349     0.2994     0.2730     0.2503
     0.2278     0.2044     0.1797     0.1563     0.1262     0.1107     0.0893     0.0719
     0.0647     0.0573     0.0477     0.0416
     1.6891     0.4247    -0.0378    -0.1493    -0.2348    -0.3519    -0.5334    -0.7276
    -0.8606    -0.9696    -1.0559    -1.1316    -1.3300    -1.5438    -1.7113    -1.8842
    -2.0412    -2.1783    -2.3337    -2.4656
130  20
     0.3901     0.4699     0.5141     0.4241     0.3773     0.3314     0.2981     0.2667
     0.2366     0.2080     0.1812     0.1580     0.1284     0.1102     0.0874     0.0716
     0.0646     0.0554     0.0460     0.0413
     1.5403     0.2753    -0.0044    -0.0622    -0.1557    -0.2846    -0.4615    -0.6423
    -0.7676    -0.8719    -0.9591    -1.0410    -1.2565    -1.4779    -1.6442    -1.8192
    -1.9830    -2.1302    -2.2860    -2.4052
135  20
     0.3807     0.5390     0.6159     0.4752     0.4170     0.3615     0.3220     0.2821
     0.2445     0.2112     0.1829     0.1604     0.1305     0.1092     0.0862     0.0721
     0.0642     0.0535     0.0450     0.0412
     1.3910     0.1702     0.0241     0.0051    -0.0956    -0.2324    -0.4013    -0.5655
    -0.6816    -0.7826    -0.8737    -0.9649    -1.1927    -1.4185    -1.5857    -1.7631
    -1.9337    -2.0889    -2.2382    -2.3513
140  20
     0.3798     0.6084     0.7127     0.5222     0.4531     0.3892     0.3444     0.2965
     0.2518     0.2143     0.1849     0.1631     0.1323     0.1078     0.0852     0.0723
     0.0631     0.0518     0.0445     0.0408
     1.2501     0.0957     0.0484     0.0583    -0.0490    -0.1913    -0.3501    -0.4964
    -0.6032    -0.7026    -0.8002    -0.9020    -1.1358    -1.3633    -1.5335    -1.7133
    -1.8914    -2.0517    -2.1914    -2.3079
145  20
     0.3855     0.6750     0.8027     0.5643     0.4854     0.4143     0.3650     0.3096
     0.2583     0.2172     0.1870     0.1659     0.1336     0.1059     0.0841     0.0719
     0.0614     0.0505     0.0441     0.0398
     1.1244     0.0422     0.0690     0.1007    -0.0126    -0.1585    -0.3063    -0.4351
    -0.5331    -0.6324    -0.7382    -0.8503    -1.0841    -1.3107    -1.4850    -1.6670
    -1.8540    -2.0171    -2.1475    -2.2749
150  20
     0.3955     0.7363     0.8842     0.6011     0.5136     0.4367     0.3835     0.3215
     0.2642     0.2198     0.1890     0.1686     0.1346     0.1039     0.0829     0.0709
     0.0596     0.0496     0.0437     0.0384
     1.0177     0.0033     0.0863     0.1348     0.0160    -0.1323    -0.2691    -0.3815
    -0.4719    -0.5724    -0.6868    -0.8082    -1.0373    -1.2614    -1.4403    -1.6234
    -1.8207    -1.9855    -2.1074    -2.2505
155  20
     0.4075     0.7908     0.9558     0.6323     0.5376     0.4560     0.3997     0.3319
     0.2694     0.2221     0.1908     0.1710     0.1353     0.1021     0.0822     0.0700
     0.0579     0.0491     0.0433     0.0371
     0.9308    -0.0251     0.1007     0.1618     0.0383    -0.1114    -0.2380    -0.3359
    -0.4199    -0.5225    -0.6454    -0.7746    -0.9966    -1.2184    -1.4032    -1.5850
    -1.7925    -1.9582    -2.0724    -2.2322
```

```
160  20
   0.4198     0.8370     1.0162     0.6580     0.5572     0.4721     0.4134     0.3406
   0.2737     0.2241     0.1924     0.1731     0.1359     0.1008     0.0822     0.0698
   0.0568     0.0489     0.0430     0.0362
   0.8629    -0.0457     0.1122     0.1829     0.0554    -0.0951    -0.2127    -0.2985
  -0.3774    -0.4825    -0.6131    -0.7488    -0.9633    -1.1855    -1.3783    -1.5560
  -1.7713    -1.9367    -2.0444    -2.2179
165  20
   0.4307     0.8739     1.0643     0.6779     0.5725     0.4849     0.4243     0.3476
   0.2772     0.2257     0.1937     0.1748     0.1364     0.1002     0.0831     0.0705
   0.0564     0.0489     0.0429     0.0360
   0.8124    -0.0602     0.1210     0.1988     0.0680    -0.0829    -0.1932    -0.2693
  -0.3444    -0.4519    -0.5891    -0.7301    -0.9386    -1.1649    -1.3673    -1.5389
  -1.7583    -1.9218    -2.0247    -2.2069
170  20
   0.4392     0.9008     1.0993     0.6922     0.5835     0.4941     0.4322     0.3527
   0.2797     0.2268     0.1946     0.1760     0.1368     0.1001     0.0846     0.0718
   0.0566     0.0490     0.0431     0.0364
   0.7776    -0.0699     0.1272     0.2098     0.0767    -0.0743    -0.1792    -0.2484
  -0.3210    -0.4305    -0.5728    -0.7177    -0.9222    -1.1553    -1.3671    -1.5322
  -1.7520    -1.9132    -2.0128    -2.1990
175  20
   0.4446     0.9172     1.1205     0.7007     0.5901     0.4997     0.4370     0.3558
   0.2812     0.2275     0.1952     0.1767     0.1371     0.1002     0.0860     0.0731
   0.0569     0.0491     0.0432     0.0368
   0.7572    -0.0755     0.1309     0.2163     0.0818    -0.0693    -0.1709    -0.2358
  -0.3070    -0.4178    -0.5633    -0.7108    -0.9131    -1.1525    -1.3709    -1.5314
  -1.7499    -1.9094    -2.0066    -2.1943
180  20
   0.4465     0.9227     1.1276     0.7036     0.5922     0.5016     0.4386     0.3568
   0.2817     0.2277     0.1954     0.1770     0.1372     0.1003     0.0865     0.0736
   0.0571     0.0491     0.0433     0.0371
   0.7505    -0.0773     0.1322     0.2185     0.0834    -0.0676    -0.1681    -0.2316
  -0.3023    -0.4135    -0.5602    -0.7086    -0.9102    -1.1522    -1.3729    -1.5318
  -1.7494    -1.9085    -2.0046    -2.1927
```

Cl

```
 0  20
    1.9463    4.2966    3.5996    3.3807    3.3341    3.3309    3.3410    3.2904
    3.3628    3.3783    3.3855    3.3902    3.3760    3.4107    3.5040    3.5257
    3.5283    3.5360    3.5759    3.5655
    2.4263    1.5715    1.4508    1.2889    1.2209    1.1743    1.1377    1.1184
    1.0810    1.0601    1.0408    1.0209    0.9898    0.9528    0.9220    0.8920
    0.8676    0.8424    0.8135    0.7926
 5  20
    1.9433    4.2625    3.5557    3.3128    3.2525    3.2346    3.2277    3.1650
    3.2160    3.2144    3.2046    3.1921    3.1471    3.1411    3.1759    3.1528
    3.1137    3.0742    3.0596    3.0042
    2.4265    1.5731    1.4548    1.2945    1.2271    1.1811    1.1452    1.1260
    1.0900    1.0699    1.0514    1.0324    1.0020    0.9670    0.9372    0.9079
    0.8833    0.8591    0.8312    0.8106
10  20
    1.9344    4.1612    3.4267    3.1161    3.0178    2.9593    2.9069    2.8128
    2.8081    2.7635    2.7120    2.6584    2.5433    2.4492    2.3636    2.2586
    2.1502    2.0411    1.9475    1.8430
    2.4269    1.5782    1.4668    1.3117    1.2464    1.2026    1.1689    1.1507
    1.1195    1.1022    1.0869    1.0714    1.0450    1.0183    0.9952    0.9722
    0.9518    0.9356    0.9181    0.9057
15  20
    1.9195    3.9957    3.2204    2.8100    2.6578    2.5437    2.4318    2.2994
    2.2269    2.1340    2.0392    1.9460    1.7729    1.6169    1.4596    1.3300
    1.2171    1.1192    1.0369    0.9711
    2.4275    1.5868    1.4872    1.3420    1.2810    1.2419    1.2137    1.1985
    1.1779    1.1680    1.1613    1.1552    1.1436    1.1414    1.1458    1.1496
    1.1558    1.1704    1.1944    1.2124
20  20
    1.8985    3.7709    2.9487    2.4239    2.2141    2.0440    1.8775    1.7162
    1.5906    1.4683    1.3538    1.2486    1.0761    0.9396    0.8219    0.7532
    0.7069    0.6761    0.6542    0.6363
    2.4282    1.5993    1.5164    1.3879    1.3354    1.3061    1.2896    1.2832
    1.2857    1.2943    1.3091    1.3276    1.3601    1.4187    1.4925    1.5482
    1.5955    1.6311    1.6778    1.6917
25  20
    1.8713    3.4934    2.6270    1.9933    1.7355    1.5240    1.3266    1.1597
    1.0196    0.9051    0.8113    0.7360    0.6356    0.5848    0.5567    0.5455
    0.5360    0.5213    0.5034    0.4788
    2.4289    1.6160    1.5552    1.4541    1.4181    1.4089    1.4187    1.4355
    1.4893    1.5430    1.6079    1.6802    1.7989    1.9176    2.0083    2.0425
    2.0613    2.0713    2.0982    2.1133
30  20
    1.8377    3.1713    2.2723    1.5552    1.2712    1.0465    0.8558    0.7164
    0.6159    0.5523    0.5152    0.4951    0.4762    0.4711    0.4638    0.4504
    0.4312    0.4101    0.3847    0.3637
    2.4294    1.6378    1.6048    1.5484    1.5449    1.5790    1.6480    1.7227
    1.8811    2.0140    2.1391    2.2472    2.3625    2.4059    2.4100    2.4116
    2.4315    2.4765    2.5267    2.5702
35  20
    1.7975    2.8141    1.9018    1.1440    0.8660    0.6672    0.5307    0.4544
    0.4342    0.4275    0.4279    0.4291    0.4221    0.4052    0.3846    0.3648
    0.3456    0.3336    0.3126    0.2968
    2.4294    1.6659    1.6670    1.6851    1.7478    1.8762    2.0659    2.2396
    2.4842    2.6189    2.6976    2.7429    2.7555    2.7351    2.7378    2.7858
    2.8514    2.9149    2.9458    2.9441
```

Cl

```
40  20
   1.7507     2.4320     1.5316     0.7891     0.5589     0.4344     0.3876     0.3741
   0.3942     0.3949     0.3889     0.3805     0.3601     0.3365     0.3211     0.3100
   0.2960     0.2781     0.2552     0.2375
   2.4288     1.7020     1.7447     1.8910     2.0880     2.3860     2.6857     2.8606
   3.0124     3.0582     3.0628     3.0566     3.0368     3.0503     3.1279     3.2078
   3.2535     3.2584     3.2420     3.2179
45  20
   1.6972     2.0364     1.1757     0.5148     0.3816     0.3570     0.3625     0.3603
   0.3651     0.3504     0.3320     0.3166     0.2975     0.2890     0.2872     0.2718
   0.2490     0.2221     0.2068     0.1965
   2.4271     1.7497     1.8430     2.2192     2.6304     3.0242     3.2382     3.3096
   3.3600     3.3570     3.3392     3.3317     3.3567     3.4324     3.5270     3.5632
   3.5550     3.5389     3.5347     3.5449
50  20
   1.6368     1.6393     0.8456     0.3434     0.3278     0.3523     0.3494     0.3323
   0.3142     0.2923     0.2748     0.2656     0.2637     0.2674     0.2563     0.2271
   0.2027     0.1868     0.1825     0.1741
   2.4241     1.8150     1.9735     2.7467     3.2597     3.5302     3.6246     3.6317
   3.6519     3.6510     3.6548     3.6769     3.7462     3.8107     3.8546     3.8572
   3.8461     3.8621     3.8709     3.8706
55  20
   1.5695     1.2537     0.5511     0.2837     0.3280     0.3440     0.3170     0.2886
   0.2633     0.2502     0.2465     0.2495     0.2560     0.2478     0.2170     0.1895
   0.1778     0.1739     0.1652     0.1507
   2.4192     1.9103     2.1668     3.4312     3.7683     3.9067     3.9546     3.9472
   3.9880     4.0118     4.0372     4.0651     4.1067     4.1278     4.1516     4.1645
   4.1736     4.1763     4.1668     4.1439
60  20
   1.4956     0.8957     0.3068     0.2961     0.3294     0.3206     0.2809     0.2541
   0.2388     0.2410     0.2492     0.2553     0.2478     0.2190     0.1849     0.1723
   0.1680     0.1595     0.1436     0.1332
   2.4118     2.0640     2.5344     4.0325     4.1751     4.2561     4.3115     4.3155
   4.3766     4.3886     4.3882     4.3888     4.3941     4.4182     4.4644     4.4730
   4.4679     4.4429     4.4402     4.4069
65  20
   1.4152     0.5910     0.1745     0.3298     0.3269     0.2995     0.2638     0.2469
   0.2437     0.2519     0.2588     0.2560     0.2261     0.1901     0.1686     0.1640
   0.1545     0.1398     0.1290     0.1256
   2.4014     2.3489     3.4919     4.4886     4.5540     4.6291     4.6992     4.6946
   4.7217     4.6855     4.6508     4.6383     4.6533     4.7190     4.7633     4.7379
   4.7286     4.7160     4.7098     4.6464
70  20
   1.3287     0.3991     0.2515     0.3680     0.3328     0.2977     0.2716     0.2610
   0.2589     0.2606     0.2560     0.2401     0.1970     0.1699     0.1593     0.1511
   0.1369     0.1263     0.1224     0.1173
   2.3870     2.9210     4.5020     4.8557     4.9220     5.0020     5.0420     4.9943
   4.9671     4.9014     4.8621     4.8671     4.9320     5.0177     5.0185     4.9946
   5.0084     4.9991     4.9466     4.8788
75  20
   1.2367     0.4036     0.3902     0.4095     0.3537     0.3148     0.2909     0.2782
   0.2672     0.2553     0.2371     0.2125     0.1718     0.1566     0.1474     0.1355
   0.1247     0.1199     0.1153     0.1069
   2.3674     3.6716     4.9014     5.1601     5.2474     5.3114     5.2866     5.1961
   5.1408     5.0817     5.0711     5.1198     5.2393     5.2845     5.2680     5.2902
   5.3013     5.2512     5.1792     5.1421
```

Cl

```
 80  20
     1.1399   0.5337   0.5159   0.4519   0.3833   0.3374   0.3063   0.2855
     0.2613   0.2356   0.2087   0.1843   0.1548   0.1442   0.1341   0.1248
     0.1188   0.1147   0.1077   0.1008
     2.3411   4.1330   5.1189   5.4067   5.5007   5.5307   5.4450   5.3329
     5.2839   5.2636   5.3078   5.4100   5.5414   5.5298   5.5582   5.6064
     5.5605   5.4878   5.4407   5.3965
 85  20
     1.0396   0.6733   0.6175   0.4895   0.4100   0.3525   0.3094   0.2777
     0.2410   0.2065   0.1795   0.1617   0.1416   0.1310   0.1247   0.1193
     0.1146   0.1096   0.1032   0.0979
     2.3059   4.3747   5.2760   5.5990   5.6809   5.6740   5.5478   5.4369
     5.4240   5.4677   5.5778   5.7131   5.8101   5.7938   5.8831   5.8866
     5.7905   5.7385   5.6930   5.5957
 90  20
     0.9372   0.7800   0.6923   0.5148   0.4241   0.3533   0.2973   0.2544
     0.2094   0.1739   0.1534   0.1426   0.1276   0.1200   0.1205   0.1152
     0.1106   0.1063   0.1006   0.0949
     2.2588   4.5282   5.4050   5.7441   5.8018   5.7630   5.6185   5.5291
     5.5781   5.7033   5.8665   5.9969   6.0631   6.1056   6.1935   6.1236
     6.0269   5.9849   5.8946   5.7710
 95  20
     0.8348   0.8413   0.7394   0.5221   0.4203   0.3374   0.2700   0.2175
     0.1706   0.1412   0.1291   0.1225   0.1131   0.1150   0.1182   0.1116
     0.1084   0.1041   0.0979   0.0918
     2.1956   4.6483   5.5170   5.8500   5.8767   5.8132   5.6708   5.6225
     5.7596   5.9763   6.1651   6.2703   6.3510   6.4529   6.4653   6.3535
     6.2755   6.1971   6.0667   5.9560
100  20
     0.7350   0.8549   0.7582   0.5080   0.3971   0.3056   0.2293   0.1700
     0.1280   0.1102   0.1055   0.1018   0.1034   0.1165   0.1164   0.1107
     0.1089   0.1023   0.0959   0.0898
     2.1102   4.7605   5.6175   5.9227   5.9135   5.8317   5.7102   5.7274
     5.9919   6.3092   6.5034   6.5944   6.7180   6.7923   6.7154   6.5983
     6.5114   6.3803   6.2455   6.1393
105  20
     0.6419   0.8235   0.7485   0.4716   0.3559   0.2599   0.1772   0.1151
     0.0857   0.0841   0.0869   0.0885   0.1056   0.1222   0.1171   0.1142
     0.1107   0.1018   0.0958   0.0890
     1.9946   4.8828   5.7095   5.9649   5.9132   5.8162   5.7326   5.8617
     6.3475   6.7707   6.9556   7.0461   7.1349   7.0944   6.9624   6.8430
     6.7171   6.5571   6.4323   6.3014
110  20
     0.5609   0.7552   0.7106   0.4147   0.2999   0.2033   0.1163   0.0560
     0.0547   0.0733   0.0841   0.0938   0.1204   0.1302   0.1224   0.1210
     0.1132   0.1038   0.0974   0.0889
     1.8392  -1.2485  -0.4875  -0.3089  -0.4165  -0.5360  -0.5726  -0.1681
     0.7922   1.1523   1.2531   1.2858   1.2200   1.0736   0.9184   0.7783
     0.6124   0.4508   0.3220   0.1618
115  20
     0.4992   0.6653   0.6460   0.3409   0.2344   0.1418   0.0514   0.0208
     0.0646   0.0889   0.1033   0.1178   0.1420   0.1400   0.1322   0.1287
     0.1164   0.1080   0.0993   0.0897
     1.6372  -1.0389  -0.4036  -0.3438  -0.5398  -0.7272  -0.8293   1.7202
     1.8278   1.8077   1.7640   1.6939   1.4993   1.3025   1.1322   0.9618
     0.7737   0.6161   0.4690   0.2976
```

Cl

```
120  20
    0.4644    0.5817    0.5573    0.2568    0.1707    0.0930    0.0410    0.0792
    0.1080    0.1244    0.1371    0.1512    0.1645    0.1519    0.1445    0.1354
    0.1205    0.1123    0.1006    0.0911
    1.3942   -0.7331   -0.3172   -0.4578   -0.8234   -1.2685   -2.6431   -3.6880
   -3.9260   -4.0601   -4.1860   -4.3199   -4.5698   -4.7816   -4.9702   -5.1658
   -5.3600   -5.5227   -5.6881   -5.8556
125  20
    0.4612    0.5495    0.4482    0.1759    0.1362    0.1052    0.1127    0.1458
    0.1614    0.1691    0.1775    0.1873    0.1856    0.1660    0.1569    0.1408
    0.1254    0.1153    0.1014    0.0922
    1.1367   -0.3176   -0.2182   -0.7552   -1.4067   -2.1488   -3.0571   -3.5058
   -3.6715   -3.8143   -3.9717   -4.1299   -4.3933   -4.6113   -4.8183   -5.0284
   -5.2221   -5.3982   -5.5734   -5.7361
130  20
    0.4875    0.6094    0.3243    0.1356    0.1660    0.1738    0.1916    0.2129
    0.2179    0.2177    0.2204    0.2234    0.2056    0.1820    0.1682    0.1459
    0.1307    0.1168    0.1022    0.0926
    0.9010    0.1166   -0.0810   -1.4334   -2.0536   -2.5802   -3.0994   -3.3955
   -3.5178   -3.6567   -3.8224   -3.9827   -4.2418   -4.4691   -4.6881   -4.9040
   -5.0979   -5.2881   -5.4649   -5.6300
135  20
    0.5352    0.7564    0.1958    0.1793    0.24'5    0.2596    0.2714    0.2792
    0.2746    0.2673    0.2636    0.2585    0.2257    0.1991    0.1782    0.1515
    0.1360    0.1178    0.1035    0.0925
    0.7091    0.4501    0.1880   -2.1207   -2.4048   -2.7375   -3.0868   -3.3120
   -3.4133   -3.5490   -3.7129   -3.8653   -4.1145   -4.3529   -4.5755   -4.7929
   -4.9898   -5.1890   -5.3636   -5.5375
140  20
    0.5951    0.9554    0.1040    0.2691    0.3328    0.3496    0.3495    0.3431
    0.3296    0.3158    0.3055    0.2920    0.2466    0.2162    0.1874    0.1584
    0.1411    0.1195    0.1056    0.0925
    0.5633    0.6701    1.0537    3.8471    3.7108    3.4809    3.2235    3.0382
    2.9448    2.8103    2.6520    2.5097    2.2694    2.0236    1.8033    1.5849
    1.3832    1.1815    1.0088    0.8242
145  20
    0.6596    1.1758    0.1638    0.3687    0.4269    0.4384    0.4239    0.4032
    0.3811    0.3614    0.3445    0.3231    0.2678    0.2323    0.1960    0.1661
    0.1458    0.1224    0.1083    0.0933
    0.4555    0.8132    2.1743    3.7066    3.6223    3.4533    3.2540    3.0928
    2.9999    2.8652    2.7132    2.5797    2.3443    2.0984    1.8819    1.6619
    1.4549    1.2546    1.0821    0.8896
150  20
    0.7230    1.3972    0.2896    0.4654    0.5176    0.5226    0.4926    0.4581
    0.4280    0.4030    0.3799    0.3509    0.2880    0.2466    0.2042    0.1738
    0.1500    0.1261    0.1110    0.0948
    0.3765    0.9090    2.5124    3.6342    3.5710    3.4428    3.2833    3.1373
    3.0412    2.9059    2.7601    2.6335    2.3993    2.1593    1.9448    1.7232
    1.5110    1.3142    1.1393    0.9437
155  20
    0.7815    1.6053    0.4157    0.5537    0.6008    0.5992    0.5540    0.5068
    0.4693    0.4397    0.4108    0.3748    0.3060    0.2587    0.2115    0.1804
    0.1535    0.1298    0.1130    0.0965
    0.3188    0.9751    2.6459    3.5925    3.5395    3.4407    3.3096    3.1731
    3.0726    2.9375    2.7984    2.6777    2.4425    2.2101    1.9950    1.7734
    1.5561    1.3636    1.1840    0.9897
```

Cl

```
160  20
    0.8325   1.7893   0.5288   0.6299   0.6735   0.6656   0.6065   0.5481
    0.5044   0.4710   0.4371   0.3948   0.3205   0.2685   0.2177   0.1854
    0.1562   0.1325   0.1139   0.0975
    0.2771   1.0212   2.7169   3.5670   3.5196   3.4425   3.3319   3.2014
    3.0970   2.9633   2.8318   2.7177   2.4805   2.2536   2.0360   1.8168
    1.5944   1.4072   1.2216   1.0319
165  20
    0.8738   1.9409   0.6226   0.6917   0.7329   0.7198   0.6487   0.5813
    0.5328   0.4966   0.4585   0.4106   0.3313   0.2760   0.2226   0.1884
    0.1580   0.1339   0.1136   0.0977
    0.2476   1.0531   2.7597   3.5509   3.5069   3.4457   3.3496   3.2229
    3.1158   2.9846   2.8615   2.7551   2.5161   2.2902   2.0699   1.8554
    1.6282   1.4471   1.2552   1.0719
170  20
    0.9043   2.0538   0.6925   0.7370   0.7769   0.7598   0.6797   0.6057
    0.5537   0.5156   0.4747   0.4224   0.3386   0.2813   0.2260   0.1899
    0.1590   0.1342   0.1126   0.0970
    0.2281   1.0741   2.7858   3.5410   3.4991   3.4489   3.3624   3.2382
    3.1294   3.0013   2.8861   2.7874   2.5474   2.3188   2.0965   1.8879
    1.6568   1.4816   1.2842   1.1072
175  20
    0.9229   2.1233   0.7357   0.7647   0.8039   0.7844   0.6987   0.6205
    0.5666   0.5275   0.4848   0.4298   0.3427   0.2845   0.2281   0.1904
    0.1596   0.1340   0.1115   0.0963
    0.2169   1.0860   2.8001   3.5356   3.4949   3.4511   3.3701   3.2473
    3.1377   3.0121   2.9026   2.8099   2.5694   2.3373   2.1141   1.9103
    1.6765   1.5057   1.3050   1.1330
180  20
    0.9292   2.1468   0.7504   0.7741   0.8130   0.7927   0.7050   0.6255
    0.5710   0.5316   0.4883   0.4323   0.3440   0.2856   0.2288   0.1905
    0.1597   0.1338   0.1111   0.0960
    0.2132   1.0899   2.8047   3.5339   3.4935   3.4518   3.3727   3.2503
    3.1405   3.0159   2.9085   2.8180   2.5774   2.3437   2.1203   1.9183
    1.6836   1.5144   1.3129   1.1428
```

Cu

```
 0  20
   4.2208    2.9713    3.0479    3.1365    3.2392    3.3383    3.4215    3.5259
   3.6035    3.6617    3.7049    3.7456    3.7876    3.8130    3.8265    3.8405
   3.8243    3.8457    3.8743    3.8838
   1.5903    0.7742    0.8920    0.9685    1.0033    1.0227    1.0292    1.0254
   1.0277    1.0267    1.0178    1.0086    0.9876    0.9640    0.9395    0.9159
   0.9007    0.8767    0.8532    0.8348
 5  20
   4.2083    2.9391    2.9802    3.0292    3.1108    3.1871    3.2475    3.3262
   3.3786    3.4123    3.4303    3.4440    3.4296    3.3967    3.3504    3.3019
   3.2345    3.1868    3.1451    3.0921
   1.5895    0.7760    0.8927    0.9713    1.0078    1.0288    1.0368    1.0342
   1.0378    1.0382    1.0305    1.0224    1.0032    0.9813    0.9584    0.9361
   0.9202    0.8992    0.8783    0.8615
10  20
   4.1710    2.8439    2.7831    2.7226    2.7475    2.7641    2.7666    2.7815
   2.7737    2.7513    2.7139    2.6702    2.5446    2.4083    2.2683    2.1340
   2.0125    1.8957    1.7945    1.7040
   1.5870    0.7817    0.8947    0.9805    1.0228    1.0493    1.0624    1.0642
   1.0730    1.0783    1.0756    1.0723    1.0625    1.0506    1.0381    1.0274
   1.0178    1.0164    1.0162    1.0182
15  20
   4.1093    2.6910    2.4740    2.2588    2.2094    2.1530    2.0892    2.0354
   1.9700    1.8998    1.8215    1.7408    1.5639    1.4068    1.2734    1.1673
   1.0950    1.0378    0.9994    0.9714
   1.5828    0.7914    0.8983    0.9986    1.0530    1.0923    1.1178    1.1309
   1.1530    1.1713    1.1823    1.1933    1.2148    1.2378    1.2629    1.2919
   1.3132    1.3526    1.3889    1.4174
20  20
   4.0237    2.4883    2.0797    1.6996    1.5823    1.4693    1.3636    1.2739
   1.1923    1.1197    1.0511    0.9894    0.8797    0.8121    0.7716    0.7491
   0.7371    0.7236    0.7057    0.6806
   1.5770    0.8056    0.9035    1.0317    1.1122    1.1805    1.2369    1.2796
   1.3350    1.3851    1.4293    1.4737    1.5601    1.6325    1.6858    1.7247
   1.7506    1.7748    1.8013    1.8306
25  20
   3.9150    2.2460    1.6328    1.1152    0.9623    0.8394    0.7475    0.6846
   0.6499    0.6287    0.6157    0.6087    0.5993    0.5980    0.5898    0.5728
   0.5502    0.5170    0.4854    0.4594
   1.5694    0.8249    0.9108    1.0985    1.2428    1.3896    1.5313    1.6511
   1.7719    1.8642    1.9353    1.9893    2.0536    2.0706    2.0776    2.0959
   2.1434    2.1961    2.2647    2.3359
30  20
   3.7844    1.9761    1.1683    0.5778    0.4574    0.4178    0.4278    0.4476
   0.4729    0.4857    0.4949    0.5000    0.4972    0.4812    0.4531    0.4227
   0.3994    0.3804    0.3665    0.3542
   1.5600    0.8502    0.9214    1.2738    1.6345    2.0075    2.2846    2.4294
   2.4919    2.5038    2.4918    2.4673    2.4157    2.3969    2.4398    2.5302
   2.6441    2.7369    2.7979    2.8281
35  20
   3.6329    1.6912    0.7198    0.2087    0.3004    0.3999    0.4627    0.4817
   0.4806    0.4654    0.4520    0.4376    0.4062    0.3750    0.3536    0.3392
   0.3227    0.3072    0.2946    0.2848
   1.5489    0.8831    0.9390    2.1315    2.8139    3.0046    3.0482    3.0213
   2.9513    2.8770    2.8079    2.7531    2.7143    2.7920    2.9361    3.0618
   3.1392    3.1744    3.2038    3.2470
```

Cu

```
40  20
    3.4621    1.4038    0.3162    0.3329    0.4680    0.5247    0.5349    0.5056
    0.4629    0.4210    0.3918    0.3672    0.3329    0.3176    0.3055    0.2841
    0.2624    0.2503    0.2500    0.2504
    1.5357    0.9258    0.9839    3.4743    3.4662    3.4141    3.3417    3.2526
    3.1491    3.0710    3.0246    3.0188    3.1120    3.2840    3.4111    3.4735
    3.5193    3.5784    3.6545    3.7193
45  20
    3.2733    1.1256    0.0310    0.5273    0.5854    0.5705    0.5279    0.4669
    0.4059    0.3593    0.3328    0.3156    0.2982    0.2814    0.2516    0.2271
    0.2238    0.2279    0.2267    0.2159
    1.5206    0.9815    3.2915    3.7336    3.6476    3.5527    3.4510    3.3549
    3.2744    3.2534    3.2836    3.3609    3.5457    3.6925    3.7783    3.8535
    3.9393    4.0229    4.0836    4.1205
50  20
    3.0684    0.8668    0.2812    0.6202    0.6028    0.5340    0.4625    0.3969
    0.3414    0.3083    0.2951    0.2878    0.2644    0.2277    0.2012    0.2012
    0.2094    0.2059    0.1938    0.1867
    1.5032    1.0562    3.9930    3.8096    3.7102    3.6023    3.4963    3.4282
    3.4183    3.4900    3.5911    3.7096    3.8982    4.0429    4.1765    4.2888
    4.3613    4.4205    4.4795    4.5208
55  20
    2.8491    0.6355    0.4548    0.6188    0.5439    0.4491    0.3740    0.3235
    0.2874    0.2724    0.2640    0.2520    0.2091    0.1801    0.1828    0.1935
    0.1891    0.1795    0.1761    0.1760
    1.4834    1.1599    4.0276    3.8303    3.7209    3.6082    3.5183    3.5134
    3.6008    3.7466    3.8729    4.0001    4.2324    4.4578    4.6060    4.6811
    4.7611    4.8313    4.8670    4.8800
60  20
    2.6172    0.4382    0.5474    0.5485    0.4417    0.3475    0.2854    0.2569
    0.2393    0.2309    0.2131    0.1912    0.1563    0.1621    0.1775    0.1778
    0.1707    0.1692    0.1675    0.1606
    1.4607    1.3134    4.0384    3.8151    3.6893    3.5797    3.5314    3.6144
    3.7879    3.9724    4.1213    4.2924    4.6718    4.9040    4.9814    5.0681
    5.1689    5.1957    5.1950    5.2308
65  20
    2.3748    0.2806    0.5682    0.4382    0.3260    0.2497    0.2050    0.1921
    0.1803    0.1651    0.1390    0.1241    0.1335    0.1585    0.1671    0.1651
    0.1630    0.1608    0.1539    0.1471
    1.4347    1.5645    4.0377    3.7563    3.5989    3.5045    3.5268    3.7038
    3.9485    4.1901    4.4427    4.7596    5.2142    5.2954    5.3713    5.4818
    5.5089    5.4993    5.5453    5.5879
70  20
    2.1238    0.1719    0.5312    0.3160    0.2210    0.1666    0.1323    0.1197
    0.0999    0.0798    0.0756    0.0962    0.1353    0.1560    0.1619    0.1613
    0.1558    0.1495    0.1436    0.1369
    1.4046    2.0263    4.0240    3.6183    3.3957    3.3316    3.4461    3.7240
    4.1027    4.6097    5.2457    5.5987    5.7002    5.6942    5.7969    5.8326
    5.8034    5.8427    5.8974    5.8922
75  20
    1.8663    0.1310    0.4528    0.2108    0.1498    0.1087    0.0708    0.0401
    0.0112    0.0588    0.0988    0.1224    0.1472    0.1617    0.1652    0.1573
    0.1491    0.1426    0.1342    0.1268
    1.3689    2.7861    3.9885    3.3084    2.9666    2.9180    3.0241    3.2925
    5.5847    6.3385    6.3815    6.2876    6.1104    6.1159    6.1590    6.1181
    6.1308    6.1906    6.1918    6.2049
```

Cu

```
 80   20
   1.6046    0.1509    0.3510    0.1599    0.1359    0.0994    0.0713    0.0727
   0.1031    0.1402    0.1584    0.1610    0.1677    0.1760    0.1664    0.1524
   0.1453    0.1350    0.1247    0.1205
   1.3256    3.4552    3.9099    2.7154    2.3435    2.2030    1.8605    1.3336
   0.8651    0.6577    0.5077    0.3563    0.1860    0.1852    0.1491    0.1212
   0.1751    0.1965    0.2092    0.2593
 85   20
   1.3413    0.1889    0.2458    0.1798    0.1673    0.1394    0.1396    0.1669
   0.1952    0.2124    0.2076    0.1966    0.1939    0.1867    0.1628    0.1491
   0.1380    0.1243    0.1185    0.1131
   1.2705    3.8548    3.7344    2.1388    1.9054    1.7131    1.3972    1.1580
   0.9551    0.8170    0.6881    0.5752    0.4676    0.4374    0.3860    0.4106
   0.4552    0.4738    0.5319    0.5626
 90   20
   1.0794    0.2263    0.1639    0.2268    0.2108    0.1970    0.2182    0.2528
   0.2673    0.2626    0.2430    0.2280    0.2156    0.1864    0.1573    0.1428
   0.1246    0.1150    0.1110    0.1033
   1.1958    4.1196    3.3324    1.8442    1.6910    1.5083    1.2808    1.1326
   1.0000    0.8998    0.8029    0.7353    0.6682    0.6279    0.6170    0.6725
   0.7183    0.7789    0.8404    0.8628
 95   20
   0.8238    0.2612    0.1436    0.2686    0.2522    0.2559    0.2907    0.3205
   0.3160    0.2929    0.2675    0.2531    0.2236    0.1770    0.1494    0.1288
   0.1105    0.1067    0.1006    0.0978
   1.0844   -1.9642   -3.6111   -4.5715   -4.6979   -4.8617   -5.0395   -5.1535
  -5.2544   -5.3270   -5.3941   -5.4280   -5.4727   -5.4899   -5.4451   -5.3662
  -5.2756   -5.1840   -5.1252   -5.0935
100   20
   0.5837    0.2956    0.1792    0.2964    0.2874    0.3082    0.3483    0.3655
   0.3434    0.3080    0.2825    0.2673    0.2153    0.1620    0.1349    0.1095
   0.0993    0.0972    0.0933    0.0954
   0.8933   -1.8096   -4.0682   -4.6306   -4.7531   -4.9008   -5.0532   -5.1540
  -5.2360   -5.2847   -5.3282   -5.3421   -5.3670   -5.3347   -5.2361   -5.0989
  -4.9524   -4.8424   -4.7744   -4.7633
105   20
   0.3861    0.3307    0.2226    0.3088    0.3139    0.3485    0.3854    0.3869
   0.3527    0.3115    0.2866    0.2659    0.1930    0.1418    0.1117    0.0906
   0.0899    0.0898    0.0922    0.0945
   0.5032   -1.6951   -4.2462   -4.6515   -4.7817   -4.9188   -5.0602   -5.1586
  -5.2271   -5.2549   -5.2830   -5.2873   -5.2871   -5.1887   -5.0160   -4.7700
  -4.5907   -4.4582   -4.4196   -4.4319
110   20
   0.3119    0.3669    0.2516    0.3071    0.3296    0.3723    0.3987    0.3855
   0.3464    0.3038    0.2768    0.2465    0.1600    0.1138    0.0822    0.0763
   0.0832    0.0890    0.0951    0.0965
  -0.2449   -1.6193   -4.2850   -4.6478   -4.7943   -4.9277   -5.0680   -5.1714
  -5.2297   -5.2413   -5.2629   -5.2653   -5.2331   -5.0495   -4.7229   -4.3499
  -4.1632   -4.0567   -4.0804   -4.1081
115   20
   0.4190    0.4044    0.2614    0.2929    0.3323    0.3767    0.3877    0.3633
   0.3250    0.2831    0.2506    0.2098    0.1188    0.0757    0.0547    0.0701
   0.0845    0.0952    0.1009    0.1024
  -0.9018   -1.5800   -4.2429   -4.6230   -4.7951   -4.9337   -5.0815   -5.1971
  -5.2490   -5.2523   -5.2781   -5.2881   -5.2220   -4.8918   -4.1870   -3.8164
  -3.6827   -3.6859   -3.7516   -3.8173
```

Cu

```
120  20
     0.6089    0.4433    0.2544    0.2681    0.3202    0.3604    0.3542    0.3230
     0.2886    0.2477    0.2083    0.1600    0.0708    0.0285    0.0502    0.0779
     0.0972    0.1068    0.1109    0.1110
    -1.2167   -1.5730   -4.1319   -4.5743   -4.7856   -4.9408   -5.1065   -5.2435
    -5.2959   -5.3048   -5.3533   -5.3950   -5.3331   -4.4886   -3.2359   -3.2367
    -3.2441   -3.3576   -3.4499   -3.5721
125  20
     0.8165    0.4844    0.2376    0.2349    0.2924    0.3237    0.3021    0.2688
     0.2392    0.1995    0.1561    0.1080    0.0300    0.0366    0.0793    0.1023
     0.1187    0.1235    0.1247    0.1213
    -1.3765   -1.5927   -3.9423   -4.4916   -4.7643   -4.9532   -5.1532   -5.3275
    -5.3947   -5.4374   -5.5569   -5.7165   -6.2391   -8.4638   -8.8428   -9.0589
    -9.1931   -9.3550   -9.4788   -9.6341
130  20
     1.0221    0.5287    0.2228    0.1967    0.2494    0.2689    0.2373    0.2078
     0.1842    0.1497    0.1149    0.0876    0.0685    0.0996    0.1229    0.1380
     0.1452    0.1446    0.1408    0.1335
    -1.4718   -1.6321   -3.6603   -4.3523   -4.7258   -4.9773   -5.2437   -5.4875
    -5.6023   -5.7455   -6.0562   -6.4925   -7.4921   -8.1565   -8.5250   -8.7557
    -8.9488   -9.1187   -9.2744   -9.4368
135  20
     1.2181    0.5770    0.2246    0.1592    0.1936    0.2000    0.1687    0.1527
     0.1416    0.1282    0.1266    0.1313    0.1385    0.1662    0.1720    0.1785
     0.1742    0.1681    0.1578    0.1467
    -1.5357   -1.6841   -3.3062   -4.1126   -4.6552   -5.0294   -5.4360   -5.8102
    -6.0295   -6.3436   -6.7821   -7.1488   -7.6786   -8.0489   -8.3427   -8.5568
    -8.7644   -8.9325   -9.1093   -9.2693
140  20
     1.4007    0.6294    0.2522    0.1342    0.1294    0.1230    0.1130    0.1285
     0.1432    0.1635    0.1885    0.2041    0.2145    0.2322    0.2233    0.2195
     0.2039    0.1914    0.1746    0.1601
    -1.5818   -1.7416   -2.9603   -3.7171   -4.5032   -5.1671   -5.8872   -6.3820
    -6.6444   -6.9250   -7.2008   -7.4013   -7.7165   -7.9787   -8.2191   -8.4175
    -8.6198   -8.7878   -8.9728   -9.1309
145  20
     1.5672    0.6852    0.3022    0.1370    0.0684    0.0534    0.1061    0.1541
     0.1923    0.2331    0.2664    0.2831    0.2914    0.2951    0.2743    0.2584
     0.2331    0.2130    0.1904    0.1719
    -1.6166   -1.7989   -2.6912   -3.2217   -4.0371   -5.7678   -6.6670   -6.9444
    -7.0853   -7.2234   -7.3790   -7.5026   -7.7188   -7.9243   -8.1273   -8.3134
    -8.5036   -8.6741   -8.8588   -9.0220
150  20
     1.7154    0.7424    0.3642    0.1693    0.0615    0.0700    0.1526    0.2095
     0.2622    0.3117    0.3452    0.3608    0.3654    0.3537    0.3231    0.2943
     0.2606    0.2323    0.2049    0.1823
    -1.6435   -1.8517   -2.5043   -2.8281   -2.7811   -0.9555   -0.9086   -0.9790
    -1.0275   -1.0812   -1.1788   -1.2660   -1.4267   -1.5973   -1.7733   -1.9495
    -2.1264   -2.3004   -2.4803   -2.6503
155  20
     1.8436    0.7984    0.4290    0.2169    0.1174    0.1399    0.2153    0.2731
     0.3366    0.3891    0.4197    0.4338    0.4336    0.4071    0.3677    0.3268
     0.2858    0.2496    0.2180    0.1915
    -1.6645   -1.8977   -2.3779   -2.5906   -2.2132   -1.3089   -1.1399   -1.1455
    -1.1486   -1.1554   -1.2224   -1.2898   -1.4149   -1.5626   -1.7185   -1.8866
    -2.0514   -2.2282   -2.4023   -2.5784
```

```
160  20
    1.9506    0.8501    0.4897    0.2675    0.1785    0.2062    0.2770    0.3353
    0.4088    0.4607    0.4870    0.4991    0.4931    0.4547    0.4067    0.3556
    0.3080    0.2651    0.2297    0.1996
   -1.6806   -1.9355   -2.2924   -2.4535   -2.0395   -1.4058   -1.2544   -1.2423
   -1.2224   -1.1992   -1.2483   -1.3036   -1.4036   -1.5364   -1.6773   -1.8388
   -1.9941   -2.1719   -2.3411   -2.5215
165  20
    2.0350    0.8943    0.5414    0.3135    0.2311    0.2612    0.3311    0.3909
    0.4736    0.5230    0.5447    0.5540    0.5419    0.4949    0.4388    0.3801
    0.3265    0.2787    0.2399    0.2077
   -1.6926   -1.9646   -2.2351   -2.3729   -1.9623   -1.4441   -1.3192   -1.3035
   -1.2715   -1.2274   -1.2653   -1.3125   -1.3944   -1.5183   -1.6478   -1.8043
   -1.9527   -2.1304   -2.2958   -2.4767
170  20
    2.0959    0.9281    0.5806    0.3497    0.2710    0.3020    0.3732    0.4354
    0.5258    0.5722    0.5897    0.5960    0.5781    0.5260    0.4627    0.3991
    0.3407    0.2896    0.2478    0.2141
   -1.7009   -1.9853   -2.1982   -2.3257   -1.9222   -1.4612   -1.3577   -1.3427
   -1.3041   -1.2459   -1.2767   -1.3186   -1.3880   -1.5068   -1.6284   -1.7816
   -1.9252   -2.1023   -2.2655   -2.4461
175  20
    2.1328    0.9493    0.6050    0.3729    0.2958    0.3269    0.4000    0.4643
    0.5601    0.6039    0.6184    0.6224    0.6004    0.5457    0.4774    0.4112
    0.3496    0.2968    0.2531    0.2187
   -1.7775   -1.9975   -2.1775   -2.3008   -1.9019   -1.4686   -1.3787   -1.3649
   -1.3232   -1.2568   -1.2834   -1.3223   -1.3844   -1.5006   -1.6176   -1.7688
   -1.9099   -2.0861   -2.2482   -2.4274
180  20
    2.1451    0.9566    0.6133    0.3808    0.3041    0.3353    0.4093    0.4744
    0.5721    0.6149    0.6284    0.6315    0.6080    0.5526    0.4824    0.4154
    0.3526    0.2994    0.2551    0.2209
   -1.7074   -2.0016   -2.1709   -2.2930   -1.8956   -1.4707   -1.3854   -1.3721
   -1.3295   -1.2604   -1.2857   -1.3236   -1.3833   -1.4986   -1.6141   -1.7646
   -1.9051   -2.0808   -2.2425   -2.4203
```

Pt

```
 0  20
    3.8801    2.9805    3.7492    4.6898    5.1161    5.5412    5.7043    5.8852
    5.9561    5.8845    5.9308    5.9002    5.8872    5.8307    5.7875    5.7790
    5.7823    5.7916    5.8049    5.8424
    1.2779    0.6979    1.2777    1.3768    1.3434    1.3094    1.2771    1.2468
    1.2198    1.2002    1.1624    1.1381    1.0947    1.0596    1.0282    0.9981
    0.9739    0.9532    0.9298    0.9107
 5  20
    3.8692    2.9288    3.6436    4.5263    4.9133    5.2892    5.4141    5.5514
    5.5824    5.4793    5.4728    5.3982    5.2821    5.1093    4.9592    4.8361
    4.7239    4.6228    4.5209    4.4654
    1.2773    0.6990    1.2843    1.3881    1.3561    1.3231    1.2918    1.2624
    1.2365    1.2174    1.1817    1.1581    1.1183    1.0866    1.0583    1.0323
    1.0122    0.9949    0.9778    0.9651
10  20
    3.8365    2.7771    3.3385    4.0635    4.3458    4.5936    4.6230    4.6531
    4.5913    4.4204    4.2987    4.1344    3.8221    3.4582    3.1557    2.8919
    2.6628    2.4720    2.3004    2.1847
    1.2757    0.7023    1.3055    1.4244    1.3971    1.3678    1.3398    1.3139
    1.2923    1.2751    1.2475    1.2278    1.2040    1.1906    1.1812    1.1807
    1.1886    1.2007    1.2218    1.2459
15  20
    3.7824    2.5357    2.8678    3.3801    3.5271    3.6171    3.5413    3.4573
    3.3108    3.0937    2.8872    2.6727    2.2829    1.9103    1.6447    1.4541
    1.3239    1.2398    1.1826    1.1405
    1.2729    0.7082    1.3456    1.4937    1.4764    1.4558    1.4354    1.4177
    1.4060    1.3945    1.3872    1.3811    1.4026    1.4480    1.5023    1.5761
    1.6552    1.7301    1.7992    1.8677
20  20
    3.7073    2.2205    2.2839    2.5911    2.6175    2.5792    2.4390    2.2899
    2.1185    1.9144    1.7177    1.5381    1.2593    1.0585    0.9391    0.8683
    0.8196    0.7801    0.7443    0.7160
    1.2689    0.7170    1.4149    1.6142    1.6167    1.6146    1.6095    1.6097
    1.6170    1.6166    1.6504    1.6748    1.7763    1.8996    2.0171    2.1310
    2.2370    2.3393    2.4304    2.5381
25  20
    3.6117    1.8522    1.6539    1.8317    1.7938    1.7010    1.5609    1.4156
    1.2788    1.1266    1.0058    0.8994    0.7510    0.6513    0.5839    0.5415
    0.5253    0.5244    0.5264    0.5364
    1.2636    0.7299    1.5389    1.8244    1.8634    1.8962    1.9149    1.9448
    1.9741    1.9816    2.0566    2.1036    2.2339    2.3634    2.5331    2.7305
    2.9404    3.1279    3.2612    3.3819
30  20
    3.4964    1.4541    1.0633    1.2464    1.2185    1.1451    1.0375    0.9242
    0.8176    0.6946    0.6163    0.5407    0.4251    0.3633    0.3732    0.4137
    0.4475    0.4567    0.4418    0.4336
    1.2571    0.7489    1.7897    2.1922    2.2779    2.3480    2.3751    2.4231
    2.4463    2.4409    2.5055    2.5428    2.7123    3.0031    3.3672    3.6306
    3.7843    3.8769    3.9395    4.0111
35  20
    3.3622    1.0507    0.6550    0.9532    0.9605    0.9037    0.7870    0.6708
    0.5493    0.4269    0.3446    0.2812    0.2547    0.3262    0.3943    0.4173
    0.4066    0.3904    0.3796    0.3743
    1.2492    0.7787    2.3632    2.7342    2.8136    2.8668    2.8619    2.9029
    2.9183    2.9421    3.0404    3.1815    3.7265    4.0919    4.2450    4.3314
    4.4026    4.4672    4.5267    4.5976
```

```
40   20
  3.2103    0.6658    0.6340    0.9031    0.8808    0.7782    0.6127    0.4773
  0.3522    0.2731    0.2447    0.2655    0.3692    0.4247    0.4219    0.3968
  0.3755    0.3618    0.3461    0.3250
  1.2397    0.8333    3.2043    3.2464    3.2655    3.2790    3.2641    3.3483
  3.4858    3.7375    4.0992    4.3764    4.6210    4.7061    4.7814    4.8601
  4.9283    4.9759    5.0142    5.1053
45   20
  3.0417    0.3228    0.8385    0.8983    0.7954    0.6307    0.4409    0.3355
  0.2956    0.3240    0.3712    0.4198    0.4809    0.4579    0.4161    0.3828
  0.3521    0.3160    0.2828    0.2612
  1.2286    0.9753    3.6811    3.5951    3.5858    3.6084    3.6761    3.9442
  4.3286    4.6020    4.8118    4.8936    4.9733    5.0833    5.1845    5.2497
  5.3076    5.4029    5.5292    5.7051
50   20
  2.8578    0.0819    1.0066    0.8284    0.6501    0.4645    0.3268    0.3270
  0.3853    0.4445    0.4898    0.5136    0.5029    0.4432    0.3891    0.3386
  0.2894    0.2510    0.2274    0.2129
  1.2157    2.0696    3.9052    3.8427    3.8592    3.9791    4.2691    4.6598
  4.9005    4.9784    5.0661    5.1057    5.2169    5.3615    5.4353    5.5099
  5.6643    5.8741    6.0366    6.1966
55   20
  2.6597    0.2215    1.0654    0.6858    0.4765    0.3483    0.3319    0.4020
  0.4755    0.5135    0.5273    0.5182    0.4628    0.3895    0.3165    0.2504
  0.2073    0.1824    0.1704    0.1720
  1.2007    3.5354    4.0378    4.0675    4.1889    4.5190    4.9069    5.0908
  5.1492    5.1686    5.2257    5.2682    5.4100    5.5139    5.5879    5.7844
  6.0793    6.3450    6.5703    6.8026
60   20
  2.4491    0.3691    1.0103    0.5063    0.3408    0.3354    0.3929    0.4591
  0.5063    0.5104    0.4920    0.4599    0.3812    0.2920    0.2034    0.1439
  0.1183    0.1191    0.1423    0.1587
  1.1833    3.7090    4.1351    4.3443    4.7009    5.1288    5.2929    5.3144
  5.3041    5.3271    5.3785    5.4175    5.5168    5.5646    5.7484    6.1622
  6.6231    7.0456    7.3145    7.4792
65   20
  2.2274    0.4483    0.8588    0.3452    0.3075    0.3811    0.4355    0.4667
  0.4765    0.4522    0.4115    0.3656    0.2682    0.1644    0.0745    0.0496
  0.0834    0.1243    0.1482    0.1535
  1.1629    3.7783    4.2207    4.7906    5.3676    5.5519    5.5205    5.4864
  5.4632    5.5004    5.5321    5.5300    5.5039    5.5352    6.0068    7.2389
  7.8874    8.0156    8.0552    8.1481
70   20
  1.9961    0.4607    0.6379    0.2752    0.3538    0.4223    0.4418    0.4329
  0.4089    0.3647    0.3076    0.2511    0.1443    0.0417    0.0558    0.1111
  0.1495    0.1699    0.1759    0.1763
  1.1390    3.8247    4.3119    5.5020    5.8720    5.8215    5.7116    5.6835
  5.6684    5.6949    5.6637    5.5609    5.2751    4.8272    2.6861    2.5242
  2.3798    2.3208    2.3088    2.3025
75   20
  1.7571    0.4156    0.3779    0.3167    0.4091    0.4399    0.4215    0.3817
  0.3318    0.2664    0.1933    0.1328    0.0740    0.1127    0.1695    0.2007
  0.2134    0.2101    0.2000    0.1852
  1.1102    3.8673    4.4446    6.1469    6.1785    6.0364    5.9254    5.9368
  5.9300    5.8991    5.7369    5.4000    4.1464    3.1335    2.9212    2.7793
  2.6782    2.6620    2.6278    2.6259
```

Pt

```
80  20
   1.5121    0.3262    0.1184    0.3932    0.4447    0.4385    0.3927    0.3349
   0.2614    0.1670    0.0819    0.0594    0.1445    0.2166    0.2540    0.2602
   0.2512    0.2287    0.2021    0.1765
   1.0746    3.9182    4.9327    6.5260    6.3937    6.2470    6.1769    6.2312
   6.2246    6.1047    5.6359    4.3305    3.2639    3.0816    3.0010    2.8976
   2.8716    2.8578    2.8492    2.9096
85  20
   1.2633    0.2087    0.1795    0.4570    0.4587    0.4280    0.3673    0.2953
   0.1968    0.0735    0.0352    0.1149    0.2255    0.2894    0.3003    0.2857
   0.2575    0.2177    0.1805    0.1458
   1.0287    4.0028    7.1271    6.7658    6.5772    6.4675    6.4426    6.5125
   6.5122    6.3580    3.7468    3.3275    3.1248    3.1137    3.0435    2.9954
   2.9948    2.9865    3.0377    3.0961
90  20
   1.0132    0.0829    0.4034    0.4981    0.4573    0.4145    0.3424    0.2507
   0.1317    0.0153    0.1033    0.1782    0.2788    0.3221    0.3087    0.2771
   0.2313    0.1808    0.1382    0.1025
   0.9654   -2.0003    1.1196    0.6644    0.4708    0.4062    0.4008    0.4543
   0.4954    2.1666    3.2085    3.1569    3.1311    3.1389    3.0763    3.0671
   3.0579    3.0946    3.2031    3.3605
95  20
   0.7656    0.0592    0.5883    0.5186    0.4469    0.3966    0.3078    0.1926
   0.0665    0.0693    0.1504    0.2141    0.2976    0.3143    0.2832    0.2357
   0.1781    0.1229    0.0767    0.0501
   0.8683    0.1586    1.2310    0.8204    0.6475    0.6137    0.5965    0.6134
   0.7976    3.1159    3.1904    3.1490    3.1734    3.1453    3.0968    3.0829
   3.0641    3.1550    3.3307    3.8343
100  20
   0.5284    0.1563    0.7213    0.5226    0.4306    0.3692    0.2569    0.1233
   0.0153    0.1056    0.1705    0.2212    0.2810    0.2720    0.2284    0.1691
   0.1078    0.0504    0.0049    0.0367
   0.6939    0.5761    1.3129    0.9627    0.8218    0.7970    0.7457    0.7007
   2.0500    3.3217    3.2585    3.2120    3.2138    3.1271    3.0704    2.9876
   2.9286    2.9479    3.9410    5.6571
105  20
   0.3261    0.2243    0.7988    0.5132    0.4081    0.3275    0.1908    0.0526
   0.0487    0.1217    0.1686    0.2044    0.2331    0.2050    0.1545    0.0990
   0.0540    0.0502    0.0724    0.0942
   0.2974    0.6869    1.3869    1.0948    0.9879    0.9551    0.8514    0.6040
   3.7391    3.5166    3.3940    3.3285    3.2319    3.0578    2.8982    2.5643
   1.9923    0.9188    0.3036    0.0183
110  20
   0.2548    0.2483    0.8215    0.4912    0.3768    0.2702    0.1154    0.0270
   0.0869    0.1258    0.1543    0.1716    0.1636    0.1281    0.0959    0.0933
   0.1103    0.1337    0.1486    0.1586
  -0.5855    0.7687    1.4607    1.2176    1.1416    1.0917    0.8965   -1.2690
  -2.2413   -2.5233   -2.6735   -2.7822   -3.0798   -3.4560   -4.0281   -4.7716
  -5.3172   -5.6376   -5.8687   -6.0208
115  20
   0.3852    0.2218    0.7937    0.4557    0.3341    0.1995    0.0399    0.0805
   0.1187    0.1297    0.1393    0.1328    0.0873    0.0796    0.1227    0.1631
   0.1955    0.2100    0.2136    0.2083
  -1.2842    0.8757    1.5400    1.3317    1.2816    1.2123    0.6649   -1.7228
  -2.0321   -2.2277   -2.3741   -2.5346   -3.2436   -4.2353   -4.9362   -5.2809
  -5.5000   -5.6609   -5.7971   -5.9128
```

```
120  20
    0.5887   0.1462   0.7224   0.4054   0.2781   0.1199   0.0442   0.1295
    0.1484   0.1415   0.1323   0.0986   0.0346   0.1138   0.1924   0.2372
    0.2655   0.2675   0.2584   0.2381
   -1.5662   1.1255   1.6323   1.4392   1.4097   1.3222  -1.4636  -1.7316
   -1.8655  -1.9415  -2.0403  -2.1839  -4.2332  -5.0694  -5.2690  -5.4229
   -5.5444  -5.6518  -5.7504  -5.8469
125  20
    0.8037   0.0804   0.6174   0.3397   0.2093   0.0371   0.1087   0.1699
    0.1768   0.1590   0.1315   0.0761   0.0641   0.1711   0.2501   0.2908
    0.3078   0.2978   0.2757   0.2469
   -1.7014  -3.9262  -4.5326  -4.7382  -4.7526  -4.8500  -7.9393  -7.9836
   -8.0304  -8.0195  -8.0475  -8.0359  -5.3575  -5.3140  -5.3759  -5.4678
   -5.5599  -5.6428  -5.7296  -5.8027
130  20
    1.0147   0.2195   0.4915   0.2592   0.1296   0.0432   0.1633   0.2003
    0.1999   0.1729   0.1280   0.0615   0.0971   0.2106   0.2806   0.3142
    0.3174   0.2970   0.2627   0.2230
   -1.7809  -2.8969  -4.3607  -4.6221  -4.6265  -1.6402  -1.6481  -1.6689
   -1.6790  -1.6295  -1.6026  -1.3782   0.7789   0.9187   0.8746   0.8014
    0.7107   0.6343   0.5399   0.4617
135  20
    1.2158   0.4427   0.3653   0.1665   0.0430   0.1165   0.2034   0.2177
    0.2114   0.1748   0.1144   0.0428   0.1089   0.2226   0.2801   0.3041
    0.2950   0.2661   0.2246   0.1801
   -1.8339  -2.6897  -4.0709  -4.4559  -4.3974  -1.5367  -1.6078  -1.6458
   -1.6563  -1.6129  -1.5803  -1.2097   0.8464   0.9468   0.8786   0.7954
    0.6869   0.5952   0.4751   0.3730
140  20
    1.4034   0.7016   0.2803   0.0714   0.0496   0.1797   0.2270   0.2194
    0.2062   0.1616   0.0921   0.0173   0.1033   0.2069   0.2500   0.2625
    0.2450   0.2113   0.1692   0.1267
   -1.8722  -2.6081  -3.5496  -3.9778  -1.5536  -1.4390  -1.5536  -1.6328
   -1.6787  -1.6939  -1.7709  -1.8666   1.0906   1.0400   0.9122   0.7843
    0.6321   0.4840   0.2920   0.0862
145  20
    1.5749   0.9795   0.2948   0.0733   0.1398   0.2318   0.2349   0.2052
    0.1842   0.1392   0.0804   0.0476   0.0983   0.1702   0.1948   0.1949
    0.1751   0.1472   0.1198   0.1017
   -1.9011  -2.5637  -2.8898  -2.1730  -1.3632  -1.3414  -1.4880  -1.6322
   -1.7600  -1.9198  -2.3211  -3.3048  -4.6990  -5.0470  -5.2949  -5.5321
   -5.7908  -6.0932  -6.4801  -6.9752
150  20
    1.7281   1.2612   0.3954   0.1754   0.2265   0.2737   0.2300   0.1784
    0.1526   0.1268   0.1102   0.1108   0.1222   0.1304   0.1222   0.1092
    0.1032   0.1126   0.1289   0.1487
   -1.9236  -2.5355  -2.4493  -1.7167  -1.2721  -1.2453  -1.4109  -1.6512
   -1.9400  -2.3451  -2.9186  -3.4255  -4.1040  -4.6318  -5.1051  -5.6505
   -6.2316  -6.8171  -7.2768  -7.6300
155  20
    1.8611   1.5314   0.5236   0.2805   0.3061   0.3069   0.2173   0.1456
    0.1271   0.1469   0.1713   0.1847   0.1789   0.1254   0.0564   0.0309
    0.0952   0.1578   0.2023   0.2364
   -1.9411  -2.5163  -2.2179  -1.5810  -1.2046  -1.1549  -1.3242  -1.7055
   -2.2732  -2.8174  -3.1935  -3.4269  -3.7087  -3.9389  -4.3086  -6.7497
   -7.3259  -7.5487  -7.7412  -7.9114
```

Pt

```
160  20
    1.9723     1.7757     0.6478     0.3768     0.3758     0.3332     0.2020     0.1148
    0.1253     0.1943     0.2418     0.2615     0.2513     0.1735     0.0942     0.1011
    0.1730     0.2445     0.2956     0.3324
   -1.9546    -2.5028    -2.0897    -1.5106    -1.1523    -1.0752    -1.2345    -1.8201
   -2.6978    -3.1087    -3.2972    -3.4109    -3.4884    -3.4015    -2.9834    -1.9144
   -1.6096    -1.5754    -1.6467    -1.7449
165  20
    2.0602     1.9809     0.7539     0.4580     0.4330     0.3535     0.1881     0.0932
    0.1460     0.2485     0.3090     0.3330     0.3244     0.2425     0.1777     0.1942
    0.2644     0.3348     0.3861     0.4165
   -1.9647    -2.4933    -2.0138    -1.4671    -1.1132    -1.0107    -1.1524    -2.0077
   -3.0082    -3.2447    -3.3351    -3.3938    -3.3582    -3.1431    -2.6744    -2.0605
   -1.7870    -1.7080    -1.7341    -1.7987
170  20
    2.1238     2.1360     0.8341     0.5196     0.4755     0.3683     0.1777     0.0834
    0.1720     0.2952     0.3641     0.3914     0.3866     0.3059     0.2518     0.2717
    0.3428     0.4108     0.4623     0.4957
   -1.9717    -2.4869    -1.9686    -1.4398    -1.0861    -0.9641    -1.0886    -2.2221
   -3.1623    -3.3023    -3.3470    -3.3801    -3.2788    -3.0202    -2.5739    -2.1048
   -1.8588    -1.7724    -1.7793    -1.8324
175  20
    2.1624     2.2325     0.8840     0.5580     0.5017     0.3773     0.1714     0.0816
    0.1913     0.3262     0.4001     0.4297     0.4284     0.3495     0.3017     0.3229
    0.3953     0.4616     0.5121     0.5414
   -1.9758    -2.4833    -1.9443    -1.4245    -1.0702    -0.9364    -1.0495    -2.3758
   -3.2247    -3.3240    -3.3495    -3.3714    -3.2350    -2.9629    -2.5325    -2.1224
   -1.8902    -1.8034    -1.8014    -1.8469
180  20
    2.1753     2.2652     0.9008     0.5711     0.5106     0.3803     0.1694     0.0819
    0.1982     0.3370     0.4126     0.4430     0.4431     0.3649     0.3193     0.3409
    0.4138     0.4798     0.5289     0.5489
   -1.9771    -2.4821    -1.9366    -1.4196    -1.0650    -0.9272    -1.0365    -2.4281
   -3.2409    -3.3293    -3.3496    -3.3685    -3.2209    -2.9459    -2.5207    -2.1272
   -1.8992    -1.8129    -1.8079    -1.8485
```

APPENDIX VII

Graphs of

Scattering Amplitude, $F(\beta,k)$ in Å,

and Phase, $\theta(\beta,k)$ in Radians,

vs. Photoelectron Wavevector, k in Å^{-1},

for Some Elements

at Different Scattering Angles, β in Degrees

The scattering angles are: $\beta = 0°$ ($\square$), $15°$ ($\bigcirc$), $30°$ ($\triangle$), $45°$ (+), $60°$ (×), $75°$ (<>), and $90°$ (∇). The corresponding bond angles are $\alpha = 180° - \beta$.

AMPLITUDE

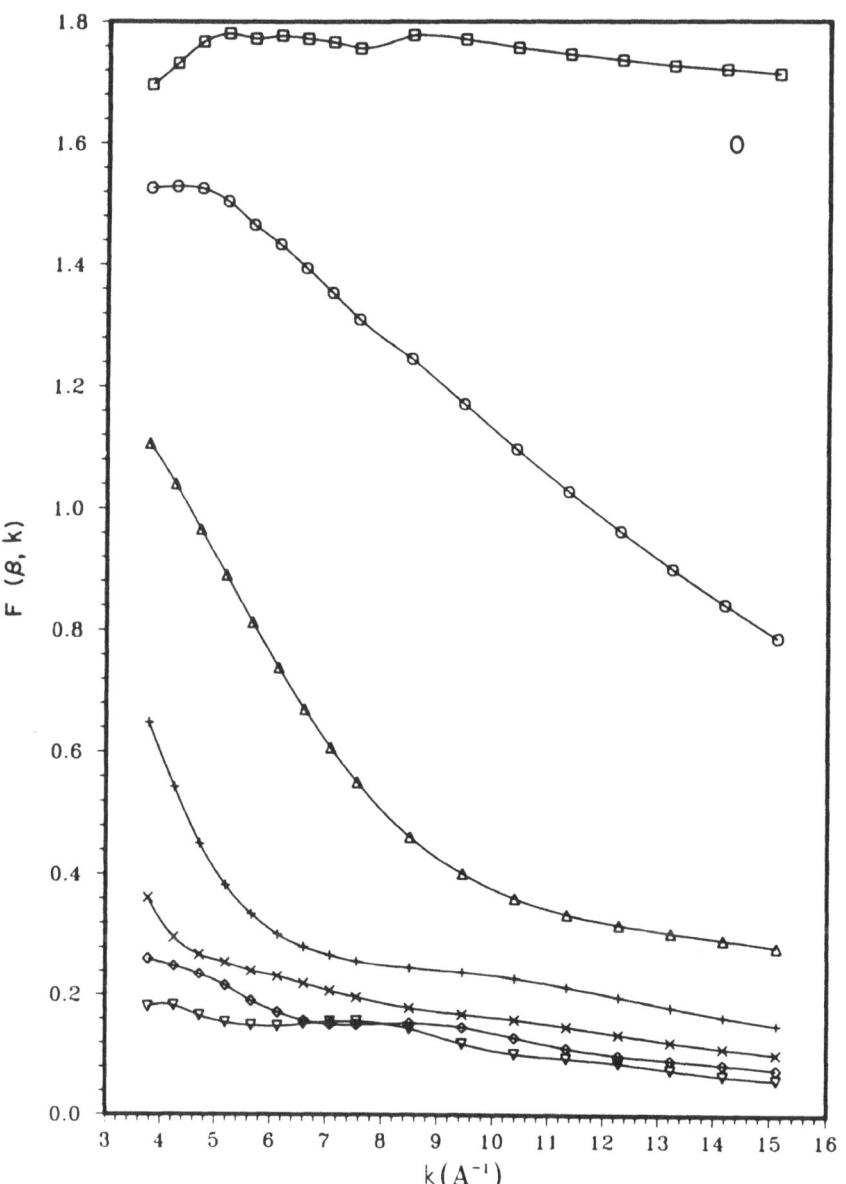

SCATTERER PHASE

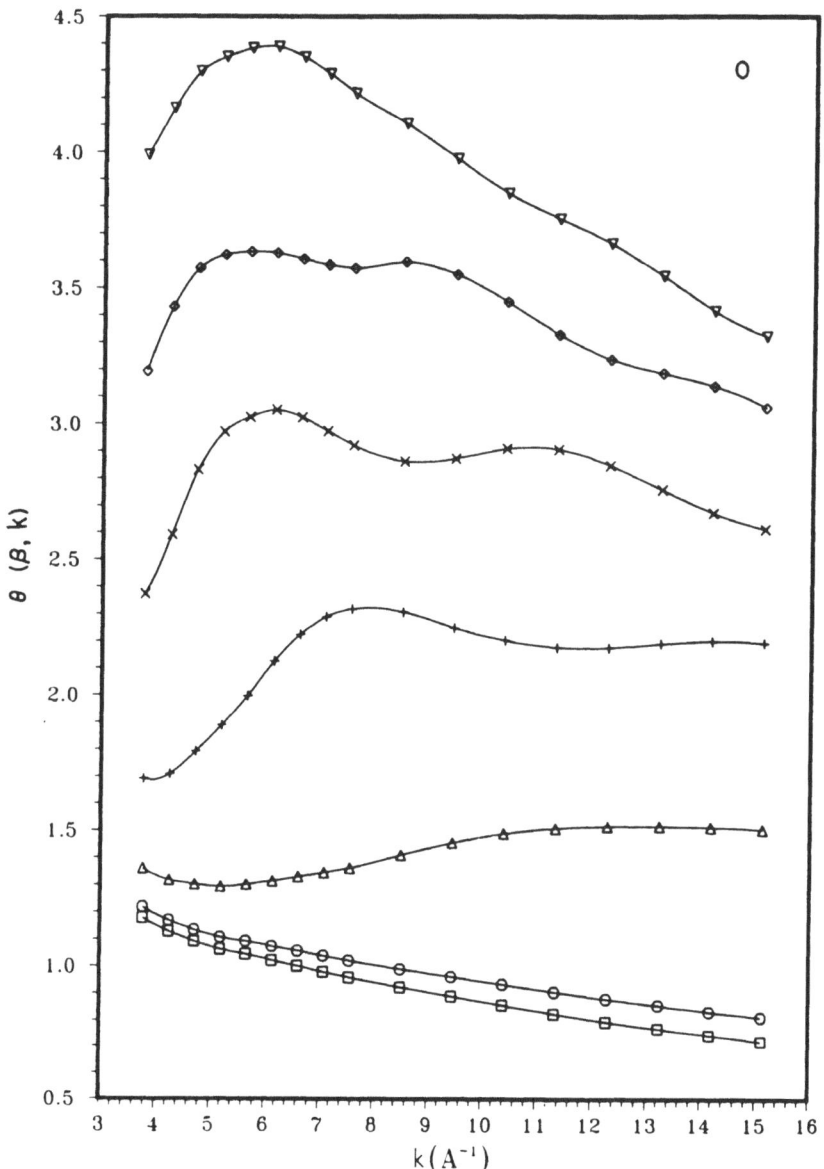

AMPLITUDE

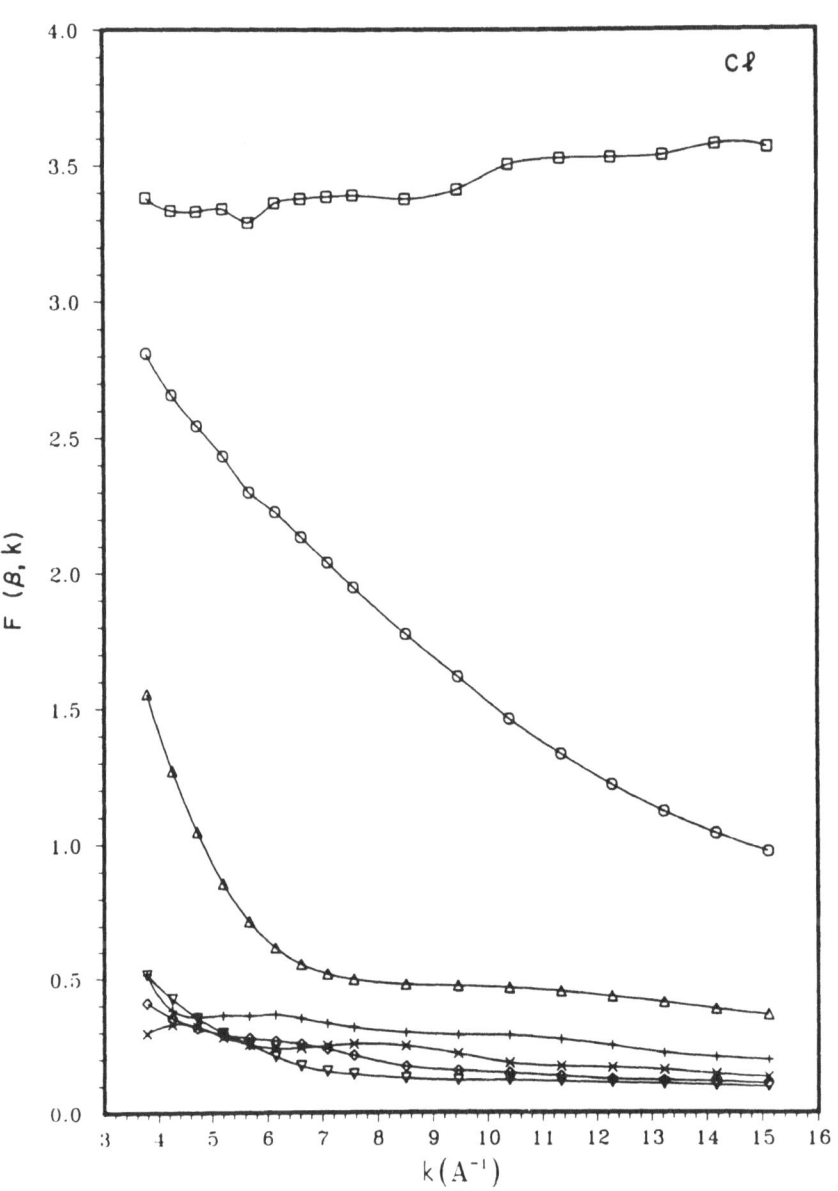

SCATTERER PHASE

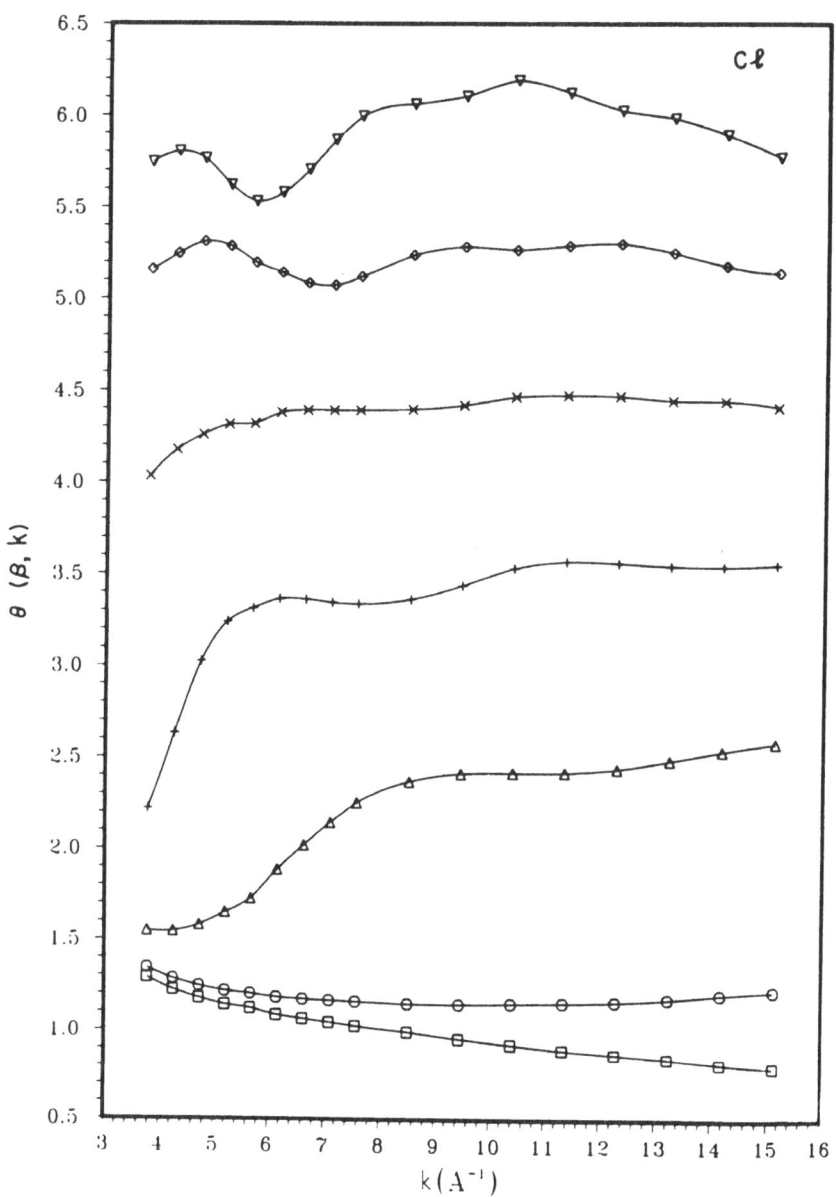

AMPLITUDE

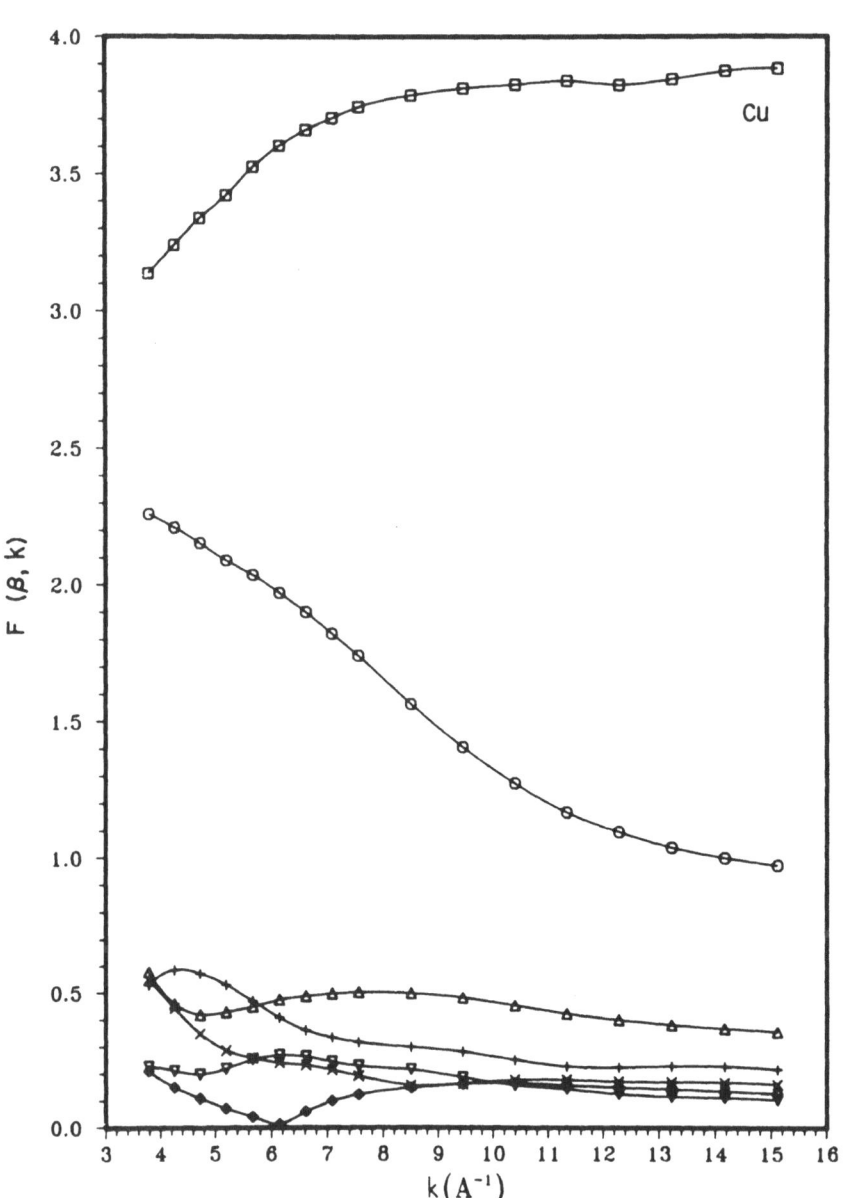

SCATTERER PHASE

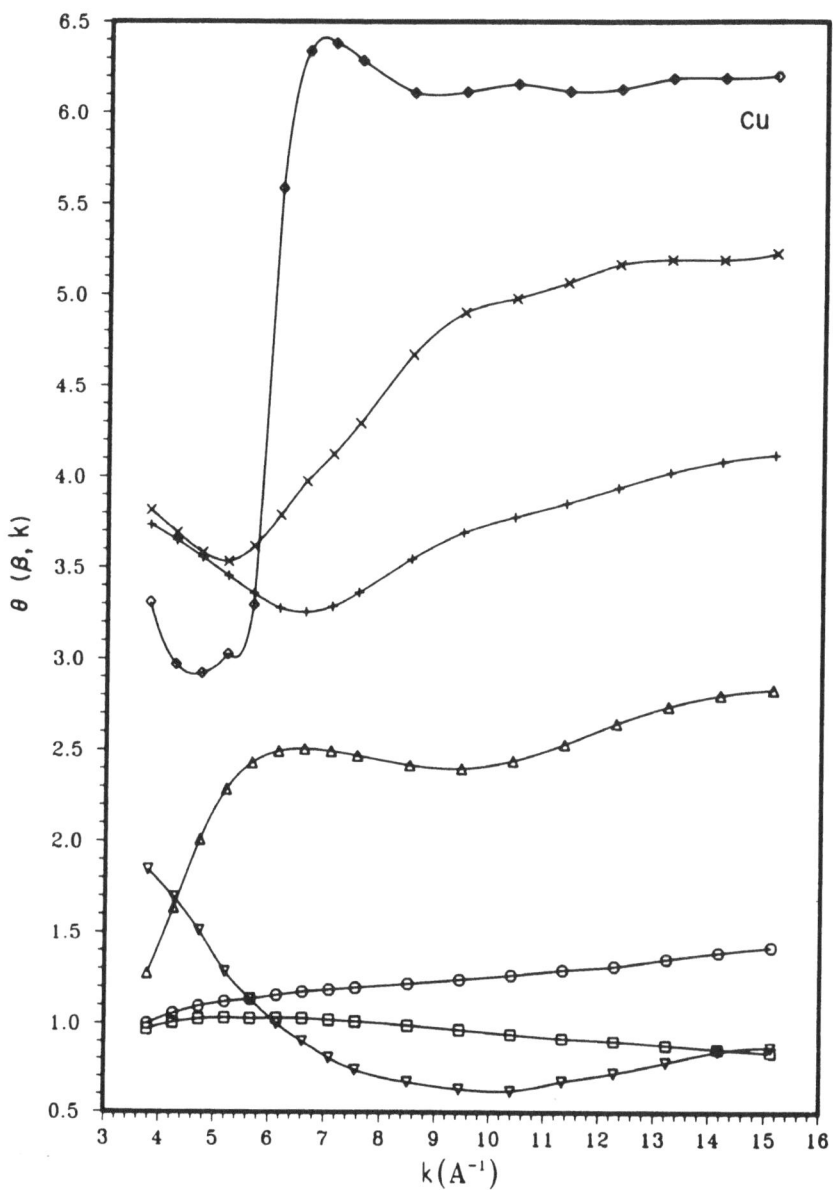

AMPLITUDE

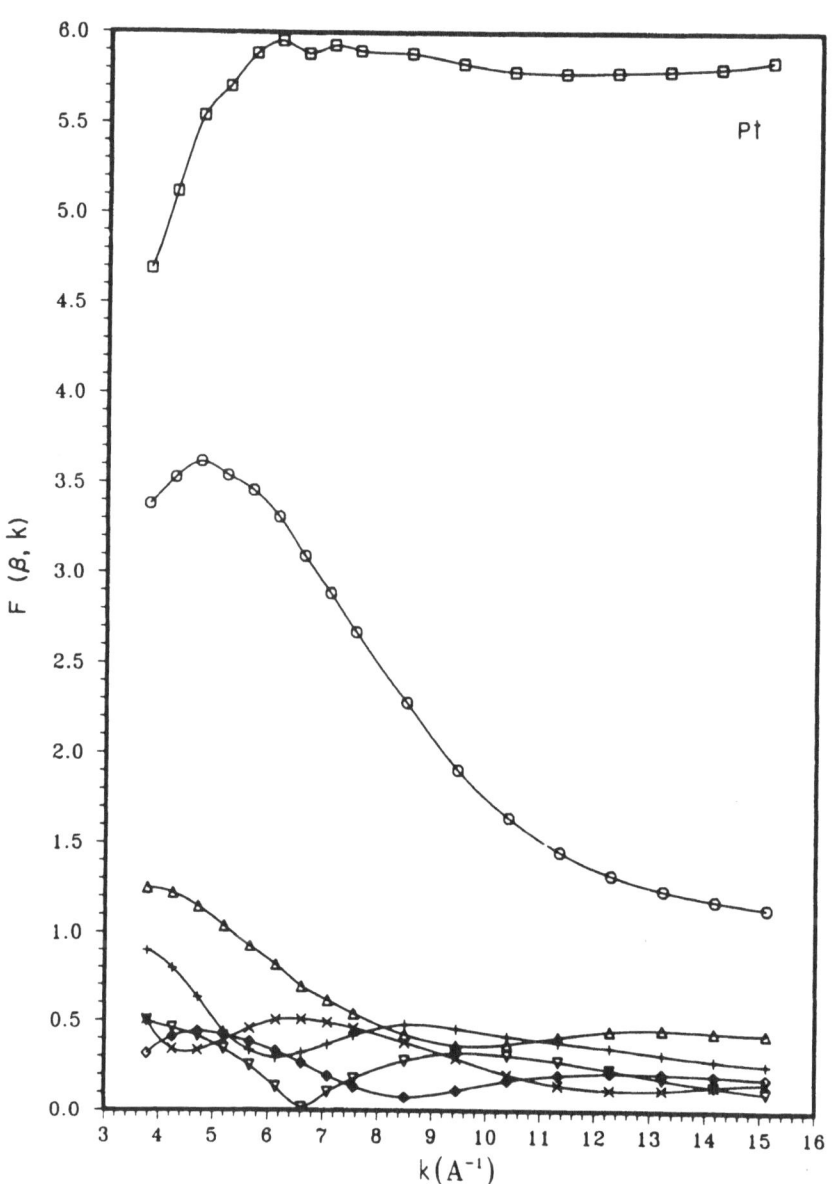

Pt

SCATTERER PHASE

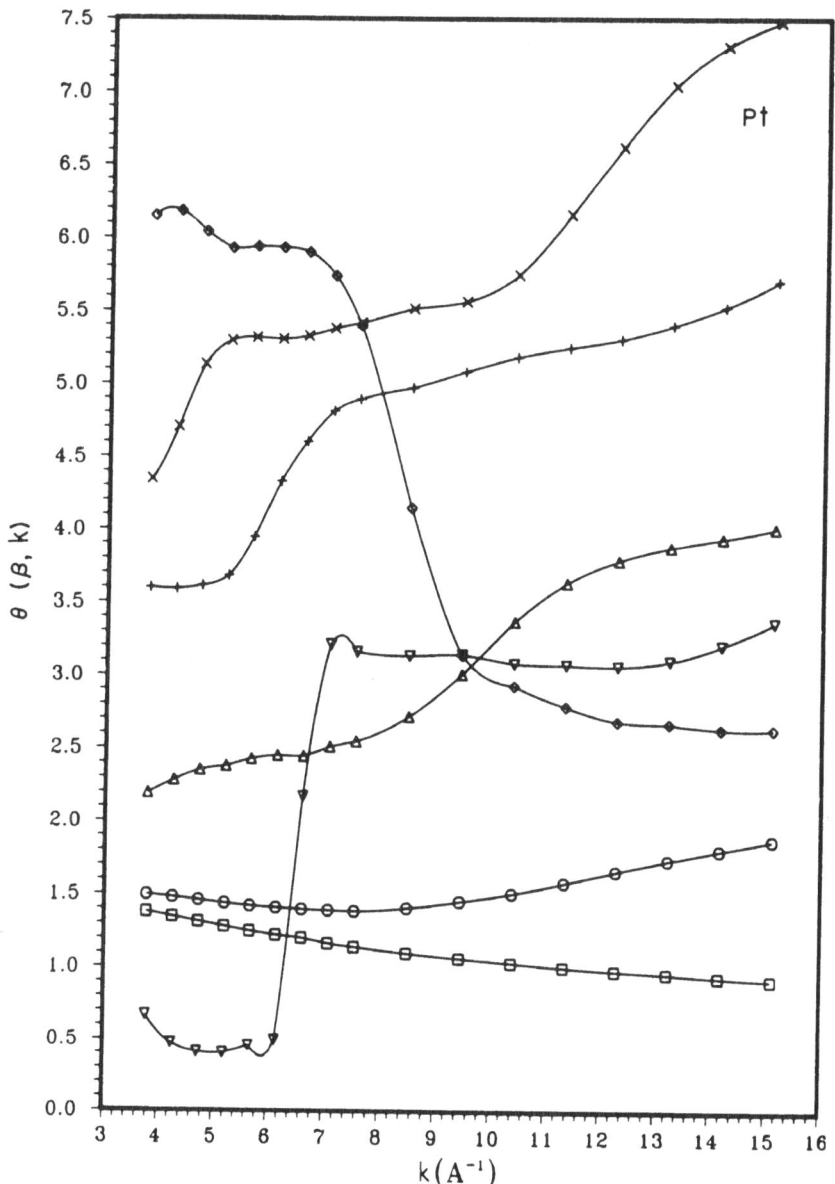

INDEX

Amplitude modification factor   197

Amplitude function   128, 163

Amplitude transferability   84

Amplitude reduction factor   26, 54, 63

Atomic orbitals   6

Absorption coefficient

— atomic   11

— linear   10

— mass   11

— true   12

Auger electrons   14

Auger transition   16

Atomic absorption coefficient   11

Aufbau principle   5

Backscattering amplitude   26, 38, 54

Beat-node method   151

Bent crystal optics   3

Best fit based on theory   133

Born approximation   129

Bragg diffraction   14, 18

Bremsstrahlung spectrum   2

Characteristic lines   1, 3

Clementi-Roetti wave functions   163

Coherent scattering   14, 18

Compton scattering   14, 18

Core hole

— lifetime   75

— relaxation   77

Coster-Kronig transition   17

Continuous spectrum   1

Critical potentials   2

Curve fitting   128

Curved-wave approach   71

Debye-Waller factor   26, 38, 54, 91

Difference technique   139

Dipole approximation   55

Dipole transition   8

Duane-Hunt law   2

Electric-dipole transition   8

Electronic shells (or levels)   5

Final-state interference effect   55

Fine adjustment based on model   133

Fluorescence yield   15

Fluorescence X-rays   14

Focusing effect   183

Fourier filtering   127

Hard X-rays   1, 4

Hartree-Fock method   160

Hartree-Fock-Slater method   160

High-energy approximation   71
    see "small-atom approximation"
Herman-Skillman wave
    functions   163

Incoherent scattering   14, 18
Inelastic electron mean free path   89
Inelastic scattering   14, 18, 54, 63,
    84, 89

Klein-Nishina coefficient   13, 19

Laser-produced plasma   4
Lee and Beni method   153
Lifetime of core hole   75
Linear absorption coefficient   10
Low energy electron diffraction
    (LEED)   20, 23

Mass absorption coefficient   11
Min-max method   142
Moseley's law   10
Multiple scattering   113, 192

Nonradiative process   14

Pair distribution function   91
Pauli's exclusion principle   6
Plane-wave approximation   56
    see "Small-atom approximation"
Phase function   128, 163
Phase Linearization method   157
Phase modification factor   197
Phase shift   26, 38, 54, 167
Phase transferability   80
Photoionization   14

r space method   156
Radial distribution function   106,
    112
Radiative process   14
Rayleigh scattering   14, 18
Regularization algorithm   157
Relaxation of core hole   77
Resonance transition   8
Rotating anode   3

Scattering coefficient   12
Secondary electrons   14
Shake-up/off processes   26, 54,
    63, 85
Single-electron
    single-scattering theory   24, 53
Siegbahn   8, 9
Single-shell model   142
Slater's local X-α
    approximation   160
Small-atom approximation   56, 71
Soft X-rays   1, 4
Sommerfeld   9
Spherical-wave approach   71
Spline-fitting method   118
Surface EXAFS (SEXAFS)   23
Synchrotron radiation   4

Target anode   2
Threshold energy   10, 38, 80
Transition probability   8
True absorption coefficient   12

White radiation   2
Window function   125

X-ray absorption edge 10

X-ray Absorption Near Edge
   Structure (XANES) 23

X-ray absorption spectrum 10

X-ray photoelectron spectroscopy
   (XPS) 1